International Trends in Manufacturing Technology

PROGRAMMABLE ASSEMBLY

Edited by Professor Wilfred B. Heginbotham

IFS (Publications) Ltd., UK
Springer-Verlag
Berlin Heidelberg New York Tokyo
1984

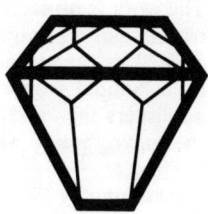

British Library Cataloguing Publication Data

Programmable Assembly. — (International trends in manufacturing technology)
   1.   Assembling machines — Automatic control
   I.   Heginbotham, Wilfred   II. Series
670.42'7   TJ1317

   ISBN 0-903608-65-0 IFS (Publications) Ltd.
   ISBN 3-540-13479-4 Springer-Verlag Berlin Heidelberg New York Tokyo
   ISBN 0-387-13479-4 New York Heidelberg Berlin Tokyo

© 1984 **IFS (Publications) Ltd.**, 35–39 High Street, Kempston, Bedford MK42 7BT, UK
   and **Springer-Verlag** Berlin Heidelberg New York Tokyo

The work is protected by copyright. The rights covered by this are reserved, in particular those of translating, reprinting, radio broadcasting, reproduction by photo-mechanical or similar means as well as the storage and evaluation in data processing installations even if only extracts are used. Should individual copies for commercial purposes be made with written consent of the publishers then a remittance shall be given to the publishers in accordance with §54, Para 2, of the copyright law. The publishers will provide information on the amount of this remittance.

Typesetting by Fleetlines Typesetters, Southend-on-Sea, UK
Printed and bound by Butler and Tanner Ltd., Frome, UK

International Trends in Manufacturing Technology

The advent of microprocessor controls and robotics is rapidly changing the face of manufacturing throughout the world. Large and small companies alike are adopting these new methods to improve the efficiency of their operations. Researchers are constantly probing to provide even more advanced technologies suitable for application to manufacturing. In response to these advances IFS (Publications) Ltd, is to publish a series of books on topics that highlight the developments taking place in manufacturing technology. The series aims to be informative and educational.

Subjects to be covered in the series include:
- Robot vision
- Programmable assembly
- Robotic assembly
- Electronics assembly
- Flexible manufacturing systems
- Lasers in manufacturing
- Robotic welding

The series is intended for manufacturing managers, production engineers and those working on research into advanced manufacturing methods. Each book will be published in hard cover and will be edited by a specialist in the particular field.

This, the second in the series – Programmable Assembly – is under the editorship of Professor Wilfred Heginbotham. The series editors are: Jack Hollingum, John Mortimer, Brian Rooks and Michael Innes.

Finally, I express my gratitude to the authors whose works appear in this publication.

John Mortimer,
Managing Director,
IFS (Publications) Ltd.

Acknowledgements

IFS (Publications) Ltd. wishes to express its acknowledgement and appreciation to the following publishers/organisations for supplying some of the articles reprinted within this book.

British Robot Association
28–30 High Street
Kempston
Bedford MK42 7BT
England

CIRP
International Institution for
 Production Engineering Research
19, rue Blanche
75009 Paris
France

Industrial Business Machines
 Corporation
1701 North Street
Endicott NY 13760
USA

Japan Society of Precision Engineering
Ceramics Building
2-22-17 Hyakunin-cho
Shinjuku-ku
Tokyo 160
Japan

Society of Manufacturing Engineers
One SME Drive
P.O. Box 930
Dearborn MI 48128
USA

GEC Research Laboratories
Marconi Research Centre
West Hanningfield Road
Chelmsford
Essex CM2 8HN
England

Institution of Production Engineers
Rochester House
66 Little Ealing Lane
London W5 4XX
England

Fraunhofer Institut für Produktions-
 technik und Automatisierung (IPA)
Postfach 800 469
Nobelstrasse 12
D-7000 Stuttgart 80
West Germany

PREFACE

When one thinks of the assembly problem there is a natural tendency to place undue emphasis on the assembly machine, its mechanisms and its immediate surroundings. A better appreciation of the potential effectiveness of automatic assembly is obtained if one considers the elements of the industrial handling process as a whole. This comprises many functions which are peripheral to, and just as important as, the functions performed by assembly machinery. In other words, the overall marshalling and control of component parts in industry is vital at all levels and is a vital prerequisite to success. The consideration of assemblies as a separate individual activity can lead to wrong conclusions. In order to appreciate all the interactions, the problem can be divided up as follows:

- *Economics*
  Choice of the correct system and appropriate architecture. This implies the optimal selection of the level of sophistication which is required to obtain sensible financial payback. The level of sophistication (and, therefore, cost) required will depend mainly on the cost of labour, interest rates and the versatility required to cater for a particular batch production situation.

- *Parts control – coarse and fine*
  Coarse – basic design, scheduling, inventory, ordering, stores control, goods and parts received, general production control and 'in-process' handling.
  Fine – parts feeding, bowl feeders, magazines, pallets and conveyors of all types, and placement.

- *Inspection – explicit and implicit*
  Explicit – is the component the correct one? Are all the features present? (holes, screw threads, etc.)
  Implicit – will the parts go together as expected in view of the tolerances which are attainable and the general implications of component quality.

- *Mechanisms*
  Motions to bring the component parts together using placement devices and the need to use jigs and fixtures for high quantities or 'soft' tooling involving microprocessor control and sensory interaction for lower batches.

- *Communication – direct and global*
  Direct – machines requiring frequent changes to deal with smaller batches require easy reprogramming and instructional facilities. A 'black box' approach with built-in microprocessor interfacing to enable shopfloor workers

to change functions by simple procedural means based on properly designed software. Diagnostic capabilities are also necessary to encourage fast responses to malfunctions.

Global – the days of the 'free-standing' isolated control system are numbered and interfacing facilities between individual control centres must be available to enable a move to be made towards complete factory integration.

The evolution of assembly machine technology follows from the development of 'hard' dedicated machinery, up to the more versatile systems which are currently available. Typical machines as developed in the mid-1960s were in-line or rotary transfer machines and such technologies in the right context are just as effective today. Such equipment is mechanically actuated and built around modular elements comprising a main chassis, a set of placement devices and feeders. However, even though these elements of modularity exist it is not practicable on economic grounds to alter such hard automation in order to respond to product changes; i.e. one can consider the problems by a numerical yardstick as follows:

Let P = cost of the standard modular elements of the system, and Ps = cost of adapting the standard system to do a particular job, then a 'versatility index' can be expressed quite simply as $I = P/Ps$.

Thus a large value for I implies good versatility, and vice versa. For the type of automation previously referred to the ratio I is in the range $0.1 - 0.2$ (i.e. 80 – 90% of the installation is special purpose), hence there is no effective inter-product versatility (i.e. the ability to change the line from one product to another).

There are very many systems available today which use a higher level of modularisation usually in the form of 'free-standing' standard mechanical motions backed up by standard control system packages, be they hard wired sequential, simple memory (i.e. plugboard) or computational. None of these machines, however, can be considered to be truly 'programmable', they have a reasonable element of 'reclaimability' but could not be said to penetrate deeply into the small batch variable product production requirement because the amount of special adaption needed to adapt such devices to a particular need is still too high. However, they exhibit extensive 'inter-process versatility', i.e. they are capable of adaptation over a wide range of relatively dedicated activities such as press feeding, 'in-process' handling, machine serving and assembly. It is possible in certain application areas to achieve an 'I' of from 0.5 to 1.0.

Programmable assembly machines have, as yet, made little penetration into the everyday scene in industry; this is perhaps because the philosophy of their application is so little understood. There are generally three main approaches to programmable assembly automation:

(1) Machines architecturally designed to be capable of adaptation based on fixed structural form such as the Olivetti SIGMA, the IBM 7565 and the Bosch system. These machines have significant sensor capabilities with standard functions like 'part not placed', 'part not present', etc.

It is clear that the situation will ultimately resolve itself into 'horses for courses' and that such machines will find their niche in certain activities.

The principle by which such machines can be successfully employed is that a machine should be capable of being fooled into thinking it is handling the same things when it is not! In other words, the need for special workholder design and gripper redesign or autochanging is reduced to a minimum thus creating a large value for the factor I. Therefore this machinery is good for dealing with 'families' of assemblies.

(2) Design the machine architecture to suit the particular job so as to create a 'versatile' dedicated machine.

The machines in (1) and (2) are both generally successful as 'programmable dedicated machines' and there is considerable scope for the economic exploitation of technology where problems of this type can be identified. The 'programmable dedicated' concept, however, becomes even more powerful when such units are integrated together to form a complete line. By this means, the technology of mass production can be applicable to a batch production situation. Because 'station can talk to station', rapid reprogramming is possible as far as functioning is concerned. Thus, the restriction on total reprogramming of the system depends on the ease of mechanical adaptation that can be achieved which depends once more on the degree to which 'family' relationships can be established for the product.

(3) Robot arms for assembly with extensive software back-up and control.

This technology carries with it a number of question marks at this time. Not the least of these questions is that concerned with the real necessity to have a multi-axis complex arm to carry out the majority of assembly processes in industry. Of course, with the computing power of the microprocessor as it is currently available, it is possible to carry enough computing potential to control such an arm. But in the majority of cases, assembly insertion is a straight-line process and to generate an accurate straight line via a six-axis robot would seem to be an unnecessary complication. By appropriately designed software, it is relatively easy to endow the machine with a number of 'instinctive' routines which can be selected to deal with faults which are expected or anticipated.

However, having said this, if we have a continual train of potential malfunctions and the machine spends most of its time using its fault correcting functions, it is not spending its time on production. Therefore, the need for machines to be able to *interpret what is wrong* after having been sent round an interrupt loop and initiate diagnostic correction is very much a problem for the future.

This brings us to the very important subject of 'sensory' interaction in relation to assembly devices. There are around 40 'machine vision' systems worldwide and it is important to appreciate just what a vision system really achieves. On a close examination, what has really been created is a 'universal escapement' but most of the old problems of total versatility remain, i.e. problems of mechanical adaptation of grippers and other peripherals if really small batches of very different components are required to be handled.

There have also been significant developments in robot 'feel' or tactile sense over the last decade; Hitachi were the first with a commercially available system in the Hi-T-Hand Expert 2 System, which is a fully reactive feedback system for 'plug-in-the-hole' type insertion. The Draper Laboratories in the USA have a remote centre compliance (RCC) device which achieves a similar effect for relative positional errors in insertion.

A machine which adopts a particular architecture, i.e. a robot arm design which is not based on the fully articulated human arm, is the IBM 7530/40 series. This is based on the SCARA robot (selective compliance assembly robot arm) originally conceived by Professor H. Makino of Yamanashi University.

This system has an effective insertion 'compliance' characteristic by design and not as an 'add-on' feature. The jointed arms are constrained to move in a horizontal plane and to have a high stiffness in a direction normal to the plane of movement, coupled with a very great resistance to angular rotation. Control during insertion is achieved in a 'free' state (i.e. without actual drives energised) in a direction parallel to the plane of rotation. Thus the small forces involved (low horizontal stiffness) and no significant angular deflection enables the system to 'comply' in a horizontal direction and achieve the same effect as an RCC device. This is a good example of how intelligent mechanical design can utilise the natural characteristics of a particular piece of robot architecture.

Programmable assembly machines will become economic inside a *total systems concept* for a factory whereby instructions will be able to be accepted by the machine in response to the input from a central source concerned with product design and controlling production. A CAD/CAM terminal is the likely mode for the input of design and functional information. However, the whole integrated factory concept implies a greater discipline from engineering as a whole and it is true to say that, unless this discipline in terms of product design and manufacture to facilitate assembly and handling takes place, then the future for flexible assembly systems remains uncertain.

A rule which is generally true in terms of applying automation is that, if the machine produces more parts in a given time than its human counterpart, then it will succeed economically. Now this does not mean that a direct comparison should be made between the rate of manual working and the *rate of machine working* in terms of the cycle time to produce an assembly. The overall output over a longer period is the figure to look for. This is because that although the machine may be working on a longer cycle time than its human counterpart, but so long as it only takes a fraction of a human being to supervise it and can be left for long periods to look after itself, then its *daily production* could be significantly in excess of that which could accrue from a human workforce.

Thus the beneficial effect of increasing automaticity can be missed by one's preoccupation with comparing man and machine directly on the wrong basis. It is also true to say that machines and systems presented here, once having been 'taught' the job, give their full production rate straight away, whereas a human workforce has a learning curve and a performance which varies throughout the working period. Clearly, to achieve the advantages that programmable assembly can offer requires a whole saga of re-thinking, re-assessment and evolutionary interaction. Success will be achieved by evolution not revolution. It is hoped that this book will assist in stimulating this process.

May 1984
W. B. Heginbotham

CONTENTS

Chapter 1 – Flexible assembly systems

The use of modular flexible assembly systems as a half-way path between special design and robots 3
    F. J. Riley, The Bodine Corporation, USA

Meeting a variety of future needs with flexible assembly 11
    M. Schröder, Robert Bosch GmbH, West Germany

Conception and realisation of flexible assembly systems 21
    R. Goebel, Burkhardt and Weber GmbH, West Germany

Design of a flexible assembly system for a small company 37
    S. J. Mallin, Orbit Controls Ltd, UK and
    P. J. Sackett, University of Bath, UK

Flexible assembly system for unmanned factory 45
    H-J. Warnecke, R. D. Schraft, E. Abele and J. Spingler,
    Fraunhofer Institut für Produktionstechnik und Automatisierung
    (IPA), West Germany

Chapter 2 – Assembly applications

The development of an automatic assembly line for VTR mechanisms 53
    T. Ohashi, S. Miyakawa, Y. Arai, S. Inoshita and A. Yamada,
    Hitachi Ltd, Japan

Motor fan flexible automatic manufacturing system   63
    I. Del Gaudio and S. Del Sarto, Olivetti Controllo Numerico –
    Divisione Robotica, Italy

Automated assembly in the electrical industry 75
    B. Lotter, EGO Elektrogerätebau GmbH, West Germany
Versatile assembly by dedicated programmability 93
    W. B. Heginbotham, D. Gatehouse, D. Law and G. Wakefield, PERA, UK

Chapter 3 – Design for assembly
Design of data processing equipment for automated assembly 105
    F. L. Bracken and G. E. Insolia, IBM Corporation, USA
Product design for automatic assembly 127
    T. Lund and S. Kähler, Danish Technology Ltd, Denmark
Assembly oriented design 143
    W. Eversheim and W. Müller, Technical University Aachen, West Germany
Design for automation: The competitive edge 155
    J. A. Behuniak, Digital Equipment Corporation, USA

Chapter 4 – Sensors in assembly
Adaptive robot and sensory system for the attachment of electrical connectors to solar arrays 161
    T. Brooks and R. Cunningham, Jet Propulsion Laboratory, USA
A research programme in sensor guided assembly ... 183
    A. Pugh, P. M. Taylor, J. J. Hill, G. E. Taylor, D. G. Whitehead, A. M. Ali, I. Mitchell, D. C. Burgess, K. K. W. Selke, D. R. Kemp and C. Stubbings, University of Hull, UK
A modular programmable assembly station 197
    R. C. Smith and D. Nitzan, SRI International, USA
Precise assembly by low precision machines 217
    A. Romiti, G. Belforte and N. D'Alfio, Politecnico di Torino, Italy

Chapter 5 – Parts handling and feeding
Computer-controlled magazining system 233
    M. Schweizer and I. Schmidt, Fraunhofer Institut für Produktionstechnik und Automatisierung (IPA), West Germany
Programmable feeder for non-rotational parts 247
    D. Pherson, G. Boothroyd and P. Dewhurst, University of Massachusetts, USA

A computer controlled reconfigurable gripper 257
  A. Butcher and P. Fehrenbach, GEC Research Laboratories/
  Marconi Research Centre, UK

The flexible parts feeder which helps a robot assemble automatically 267
  T. Suzuki and M. Kohno, Hitachi Ltd, Japan

Development of a flexible parts feeding system 281
  S. Hara, K. Azuma and K. Hironaka, Mitsubishi Electric Corporation, Japan

Chapter 6 – Economics

Equipment justification for an assembly system 291
  C-S. Ho, Unimation Inc, USA

Automatic or manual assembly? Boundaries of economy at middle or low batch production 305
  D. Elbracht and H. Schacher, University of Duisburg, West Germany

Automated assembly *can* equate with short payback periods 315
  A. E. Owen, Tony Owen MBA Ltd, UK

Chapter 7 – Social aspects

People performance in automated assembling 327
  J. Wood, Rank Xerox Ltd, UK

Perspectives and social implications of assembly automation in West Germany 333
  D. Seitz and V. Volkholz, Gesellschaft für Arbeitsschutz und Humanisierungsforschung (GfAH), West Germany

Changes in assembly work environments 343
  R. G. Davison, University of Salford, UK

Chapter 1
FLEXIBLE ASSEMBLY SYSTEMS

Units in isolation may be dedicated but when linked together can create a flexible system. This chapter examines the systems aspect. The five papers illustrate that flexibility can be achieved with a variety of basic methods – fixed indexing, free-flow pallet and linked cells.

THE USE OF MODULAR FLEXIBLE ASSEMBLY SYSTEMS AS A HALF-WAY PATH BETWEEN SPECIAL DESIGN AND ROBOTS

F. J. Riley, The Bodine Corporation, USA

First presented at the 3rd International Conference on Assembly Automation,
25–27 May 1982, Boeblingen, Stuttgart, West Germany

Abstract

Much of the media coverage of robotics contrasts their flexibility with the so called 'hard' automation of dedicated one-up specially designed assembly machines. These articles tend to ignore the existence of modular universal assembly systems capable of meeting a broad spectrum of assembly requirements. This paper proposes a viewpoint that broad advances in the use of automated rather than manual assembly will come not from robots, but from standard modular assembly systems. The choice will be an economic one and even a political one; not what is technically feasible, but what is economically practical. Standard assembly machines designed for specific products, sizes and volumes rather than for specific industries are based on broad experience in special design and pay full attention to the problems of debugging, operation and maintenance. The use of such standard systems allows the tool designer to concentrate on the real problems of product and product component design and quality.

Introduction

For the last few years those of us involved in the construction of automatic assembly systems seem to have been bombarded by press reports of the advent of the age of robots. Certainly the sales of robots have been incredible. In some areas, particularly in those of spot welding and painting, they offer definite advantages over manual techniques, not only by cost reduction but in improved quality.

These reports seem to infer that robots are also increasingly active in discrete parts assembly. One article is quoted by the next and it becomes increasingly difficult to determine the extent of the specific applications of robotic assembly.

Even more so, is it to find out if there are any economically justifiable installations of discrete parts assembly through the use of robots.

As children we were entertained by reading fairy tales and other folklore. I remember specifically the tale of the emperor who walked into his court stark naked and asked his courtiers how they liked his new clothes. Each member of the court outdid the others in praising the beauty and the fit of the clothing until a small child spoke out and said the emperor was naked.

It seems we have a close similarity here to the present top management enthusiasm for the use of robots in assembly, and the ensuing staff level lip service to the future role of robots in assembly. This paper is intended to be one small voice questioning the substance of these claims. To be successful in this attempt, the topic of automatic assembly must be made clear.

What is automatic assembly?

Automatic assembly may be interpreted as the mechanised placement of individual discrete components into a specific spatial relationship to form an end item with some specified function. This automatic progressive placement of one component after another may include the orientation of components from bulk storage, the retrieval of oriented components from storage, the fabrication of components on the assembly line, the monitoring of the physical placement of each component, the joining of the components to form the assembly and inspection for functionality of the assembled product. An increasingly important requirement for product documentation such as serialisation or date coding may be included among the operations of an assembly system.

While we stand in awe at some of the larger car assembly lines having a high degree of mechanisation, the vast bulk of all assembly work is done on products or subassemblies of relatively small size. One does not assemble a car from discrete components, but rather from a series of major subassemblies which in themselves are made up of smaller assemblies. The vast bulk of assembly labour lies in the preparation of relatively small subassemblies.

Many widely used products are in themselves quite small, for example switches, circuit breakers, writing instruments and toys.

If mechanised assembly is to aid in the increase of productivity necessary to maintain a good standard of living expected in the industrial world and desired by the Third World, it must assemble products on a competitive basis to that of hand assembly. Harsh as it may sound to some academic ears, mechanised assembly has a socially useful role only if it is economically justifiable.

The great fallacy of proposed robotic assembly is not its failure to face up to the problems of debugging and start-up; not its failure to recognise that most assembly problems are parts feeding rather than parts transfer; not in its proponent's failure to include quality audits as part of assembly; but basically because *robotic assembly probably is not cost efficient.*

Let us put this into a simple context, contrasting so called 'hard' automation at its present state with proposed robotic assembly.

A comparison of technologies

The Bodine Corporation builds many complex assembly systems. These machines produce assemblies at a rate of 30 to 65 assemblies per minute. These assemblies generally have from six to 15 components. Machine prices range from $300,000 to $650,000 (US) for a fully debugged system with full documentation installed on a customer's floor. At the present time, in addition to many smaller rotary machines, the American plant ships approximately 35 machines of this size yearly.

For purposes of comparison take a most typical example of a 12 part assembly produced at a rate of 55 assemblies per minute with a typical machine price of $500,000 (US). Over half of the machines to be produced this year would fall within ±10% of these figures. The licensees in England and in Japan would normally be somewhat lower in price for comparable machines to those made in the USA.

Looking now to the world leaders in the use of robots. We are told that there are approximately 14,500 true robots in use in Japan (December 1981) of which 30% are involved in assembly. If this figure is accurate it would mean 4350 assembly robots in operation. Assume that each of these robots is capable of assembling a 12 part assembly in one minute while joining it and functionally testing it as part of the assembly process. Also assume that these robotic systems were placed on the production floor in operation with all necessary tooling, feeders and so on, for $35,000 (US) average price. This is a generous assumption.

Comparing this assumed total Japanese use of assembly robots with the production of assembly machines by the Bodine Corporation alone in the last 27 months, we have:

Robots:

$$\begin{array}{rl} 4350 & \text{robots} \\ \times 12 & \text{pieces/min/robot} \\ \hline 52{,}200 & \text{components assembled/min (gross)} \end{array}$$

Automatic assembly machines:

$$\begin{array}{rl} 12 & \text{pieces/assembly} \\ \times 55 & \text{cycles/min} \\ \hline 660 & \text{pieces/min/machine (gross)} \end{array}$$

then,

$$\begin{array}{rl} 52{,}200 & \\ \div 660 & \\ \hline \simeq 79 & \text{assembly machines required to assemble same} \\ & \text{number of parts as 4350 robots in one minute} \end{array}$$

Cost comparison:

$$\begin{array}{lll} 4350 \text{ robots} & @\ \$35{,}000 = & \$152{,}250{,}000 \\ 79 \text{ assembly machines} & @\ \$500{,}000 = & \$39{,}500{,}000 \end{array}$$

Therefore the potential capital reduction through use of so called 'hard' automation is $112,750,000 (US).

Adjusting the assumptions in favour of robots to any reasonable level and downgrade the performance of so called 'hard' assembly to any reasonable level, will not significantly change the results. *Robots will not be as cost effective as mechanised assembly for most types of production suitable for automatic assembly.*

Robot supporters may call 'foul'. They will say that the flexibility of the robot and the rapid changes in the market place have been ignored and that a 'hard' dedicated machine useful only for high volume production of a specific product has been compared with the flexibility of the programmable robot.

So at last we come to the point of this paper. Manufacturers it would seem are faced with the choice of 'hard' automation or flexible robots as replacements for the high cost and poor quality of human assembly.

Modular assembly systems

There is, however, a most viable third choice: standard modular mechanical assembly systems. These machines are designed to meet the broad spectrum of assembly operations with mechanical simplicity and be fully capable of producing a family of similar parts; capable of accepting running product changes and capable of being retooled for new products.

From the beginning of the industrial revolution man has sought to eliminate tedious repetitive operations using machinery designed for specific industries such as weaving, bottle filling, can making, and other areas not related to any country or company. Early prototypes of such machinery were refined time and time again till today we have extraordinarily efficient machines for specific industries.

Metal cutting, metal forming and moulding machinery became universal in an opposite sense. Machines were designed not for any given industry but for specific types of operations.

Discrete parts assembly, if mechanised, remained for the most part, however, in the realm of the special machine. A machine was designed to produce a specific product for a specific company.

Some industries such as lamp bulb manufacturing, spark plug manufacturing and tyre valve production, reached such high internal levels of production that large companies developed machinery for their own requirements which could be adopted readily for different models of similar products. These machines have reached high levels of productivity and their designs are often closely guarded secrets. Hence major international competitors often use totally different systems to produce very similar products.

The market place, however, has changed with increasing rapidity since World War II, as consumers demand a greater diversity of products, while at the same time revolutionary changes in material and product design are constantly obsoleting older product designs.

It is no longer economically sound to develop dedicated unique assembly machinery for the majority of products now coming into the market place.

In March 1982 at the IPE's PEP II Conference in London, I stated my belief that continued progress in productive, economically justifiable automatic assembly would come through the use of standard modular assembly machinery not designed for any specific industry, but rather designed to reproduce the functions of the human assembler, whatever the product.

In the USA, there was early adaptation of lamp bulb manufacturing machinery to other types of assembly. Sylvania and Swanson-Erie were representative of such manufacturers. Intermittent motion devices used in the basic design of high volume automatic machines such as matchmaking machinery were adapted for the market place (for example, Ferguson indexers).

For the most part, however, these early machines were only intermittent motion devices with some simple tool mounting surfaces, power shafts and reciprocating tool platforms. They remain a major element of the present assembly machine market.

Russell Gilman in the USA must be credited for the first real successful attempt to produce a standard assembly system, providing not only intermittent motion of fixtures but also standardised tooling stations. His machinery, together with the GM assembly machine chassis used so successfully by Delco Remy, was extremely important to early attempts at significant assembly mechanisation.

In the late 1950s Bodine built several highly successful assembly machine chassis on rotary machine bases providing good facilities for tool mounting and tool actuation but without any standardisation of the tooling stations. These machines were very successful from an operations standpoint but the very success in the market place brought increasing difficulty in maintaining profitability. The engineering content became staggering as unique tooling stations for each specific project were designed.

In the period from 1960 to 1965 we started a design effort to build a totally standard assembly system requiring a minimum of adaptation to the specific product being assembled.

This mechanical system had one basic problem. We could not come up with a standardised electrical system to match the mechanical standardisation.

The advent of the microprocessor based programmable controller allowed us to produce in the late 1970s a totally standard control panel. This effort was significantly assisted by incorporating magnetic shaft encoders to report the mechanical system condition to the programmer for control action.

This equipment has met with wide market acceptance and is representative of what will be the true basis of growth in assembly productivity. A broad range of such standardised assembly systems is likely to become available. Each of these systems will be oriented toward a certain size of parts, specific operations and specific production volumes rather than to any specific type of product.

The Model-64 system

The Bodine Model-64 system was built around certain assumptions. It was oriented towards products produced in excess of one million or more units per year which would weigh less than 5lb (22.5kg) and fit within an 8in. (20.3cm) cube. These were not arbitrary limitations. They were based on market research indicating that a broad spectrum of parts could be assembled on such a machine, including subassembly work representing over three quarters of all the assembly labour in an automobile.

By limiting the size and weight to the limits mentioned we could build a totally synchronous system without any requirement to locate each fixture through the use of shot pins or other locating devices. This in turn meant reduced inertia, higher utilisation of dwell times and permitted intricate operations at high speed through coordinated cam actuation without sequential electrical control. This opened up a vast number of application opportunities for complex assembly not practical on smaller relay-controlled machines. It also meant higher operating speeds than were previously practical for linear machinery.

It also meant that on relatively simple work the machine would be at a price disadvantage when compared to rotary machines.

Each modular system will have to address itself to a basic decision of what size and what volume ranges it is intended to produce.

In evaluating the worth of any proposed standard modular system the customer must consider several key areas:

- ☐ Has modularity been achieved at the price of durability?
- ☐ Does the modularity in the proposed machine have the capability of accepting running product changes without extensive disruption to normal production schedules?
- ☐ When major product changes occur, to what degree will the system be reusable?
- ☐ Is the nature of the modular construction such that the maintenance and operation can be done successfully if it is necessary to move the machine from a skilled labour market to one with lesser skill levels?

These operating considerations focus on one aspect of automatic assembly construction that is routinely ignored. *The difficulty of adapting to the routine variations in product components in the parts feeder, tracks, escapements, transfer unit fingers and in the matching of the part being fed to those parts already fed, is enormous.*

This difficulty with parts variations calls for the highest tool design skills. For success it requires a happy blend of experience and creative design effort. It will not work if the underlying machine system is in itself a source of problems.

In my own experience most special purpose machines which are not successful fail not because tool design was poor but because the underlying machine system was inadequate to meet production requirements. Systems failures may include lack of durability, lack of access for maintenance and adjustment, and lack of ability to coordinate all of the operating stations into a cohesive system.

It is likely that special purpose machine builders will become increasingly unable to supply operational machinery rapidly enough to match the evolutionary nature of many products. Shorter lead times, running product changes and rapid product obsolescence will not tolerate prototype machine development, and short pay-back periods will require high operating rates for quick capital recovery.

The use of standard proven assembly systems leaves the tool designer free to concentrate on the problems of the product.

Future trends

The single biggest jump in productivity in assembly will come from product design improvements not only in the light of product function and fabrication but in an increasing awareness of the need for more efficient assembly. This may mean more efficient joining, reduced number of components or even the elimination of assembly through the use of new materials (such as silicon chip devices). Those with batch order manufacturing will standardise components and sub-assemblies to facilitate mechanised assembly.

Secondly, attempts to automate the assembly of complex products made in low volumes through the use of robots will in most instances not stand up to economic audits. Over-capitalisation and extensive overhead costs are a very real danger for those considering robot assembly.

Products made in quantities of 500,000 to 1,000,000 or so units a year will gravitate to producers willing to accept lower returns on investment than are typical in many Western countries with the significant exception of Japan. Freedom from anti-trust considerations and a simple entrepreneurial approach to capital recovery coupled with a desire to attain market share will favour Japanese producers of complex products in this situation. This is not a technical situation, but rather a managerial one. The assembly machinery used for these products will probably be fairly conventional non-synchronous transfer machines often internally tooled by the producer, and will make modest sensible use of robots essentially as material handling devices at specific workstations.

In high production manufacturing where volumes exceed 1,000,000 or more units a year, increasing political instability in Third World countries with some major exceptions such as India, Argentina and Brazil will lead to wider utilisation of standard assembly machinery in the industrial nations and less dependence on transferring assembly operations to low cost labour markets.

When production levels fall below 500,000 units a year, programmable conveyor systems coupled with ergonomically designed workstations will probably continue to be the most cost efficient way to assemble small products. There remains a real need for better industrial engineering.

Even where products are expected to have a very long market life and designs are exceedingly stable, standard modular machines will find broad acceptance because of their durability and ease of maintenance. The heavy educational emphasis on computers, CAD-CAM and electronics has produced a real shortage of competent mechanical designers. This in turn has led to a reduced ability to properly design one-of-a-kind machinery. The initial higher costs of standard modular systems will be returned.

Finally, the wider use of standard systems will lead to and require a greater use of in-house tooling efforts than now exists.

MEETING A VARIETY OF FUTURE NEEDS WITH FLEXIBLE ASSEMBLY

M. Schröder, Robert Bosch GmbH, West Germany

First published in *wt. – Z. ind. Fertig.* 71 (1981) in German, and subsequently in *Assembly Automation* (February 1982)

Abstract

A flexible assembly system which is constructed on the module principle, with which the most varied requirements for future-orientated production can be met, is described. For the use of a flexible assembly system with its various possible layouts the following criteria are the most important: good possibilities of arranging the manual workstations ergonomically; many possible combinations of manual workstations and automatic stations within an assembly line; stage-by-stage automation and hence investment; rapid setting-up of an assembly line by using standard modules; economic assembly of mass-produced products.

Introduction

When planning a component assembly line or rationalising an existing one, one of the most important decisions to be made is the choice of the correct assembly system. While in a fully automatic system technical considerations are in the forefront; when the assembly is only partly automatic good work conditions for the operators carrying out manual operations as part of the assembly process must also be provided. Significantly the demands being made on systems are constantly becoming more stringent.

Two other important criteria in evaluating a system are those concerning the equipment and the personnel, respectively.

Equipment-dependent criteria can be used to evaluate the technical and the organisational possibilities. These are :

☐ Flexibility regarding variation of quantity and the number of operators employed, variety of types of new or modified products, mechanisation and automation.
☐ Manufacturing reliability with regard to faults and scrap.
☐ A high utilisation ratio resulting from a low effect of breakdowns.

12 PROGRAMMABLE ASSEMBLY

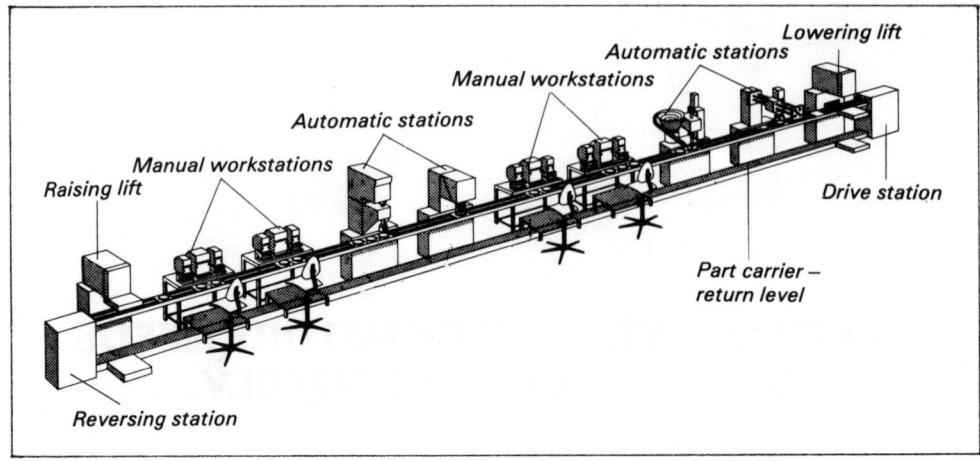

Fig. 1   Double-belt assembly conveyor arranged in-line

By using personnel-dependent criteria it is possible to judge the individual development of the productivity of the operators. Among these are:

☐ Possibility of individual development by independence from the machine cycle and individual choice of rest periods.
☐ Possibility of communication between the operators.
☐ Ergonomic design of the workplace.

Design and function

One basis for a flexible assembly system which can be adapted to meet a particular assembly is a double-belt arrangement. The design of such a system on modular lines makes it possible to use standard units for component transfer devices arranged in-line or in a rectangle, manual workstations, and automatic stations.

In the case of the double-belt assembly line shown in Fig. 1, the parts to be assembled are transferred from one workstation to the next on part carriers or pallets which rest loosely on two constantly moving belts. The carriers consist of plates each of which contains one or more component jigs.

At individual manual or automatic workstations an escapement device stops the pallets while the belts continue moving. On completion of the operation at the station and after the escapement has been released, the carrier together with the component is moved to the storage position at the following workstation by its transfer device.

At the manual stations the escapement device is operated by a foot release while at automatic stations it is operated by the sequence control of the station.

In the case of in-line assembly belts when the pallets reach the end of the belt they are conveyed back to the starting level of the assembly belt by means of various lowering and raising lifts.

On a double-belt assembly conveyor built in a rectangular manner the pallets are transferred by two double-belt conveyors identified as conveyor tracks 1 and 2 in Fig. 2 and arranged at the same level. Both conveyor tracks are connected at the beginning and end by means of pneumatically or electrically driven cross feeders.

MEETING FUTURE NEEDS FOR FLEXIBLE ASSEMBLY 13

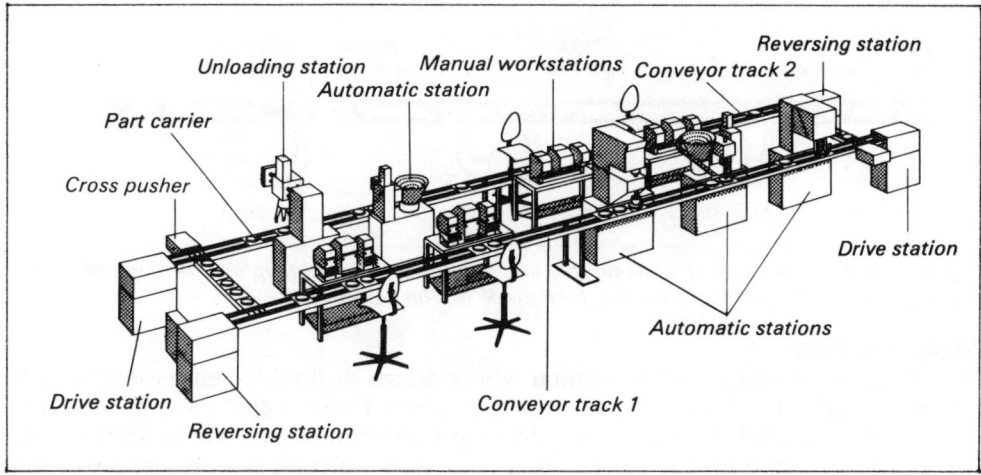

Fig. 2   Double-belt assembly conveyor arranged in a rectangle

Since it is possible with this arrangement to use the belt track of the carrier return section as workstations the total layout is therefore much shorter than for the in-line arrangement.

Fig. 3   Manual workstation in a double-belt assembly line conveyor used for the assembly of pumps for motor vehicles

14 PROGRAMMABLE ASSEMBLY

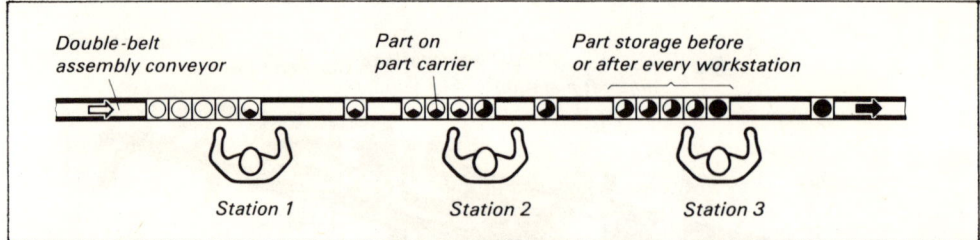

Fig. 4   Work carried out at an assembly line arranged on the principle of 'division of work'. The circles can be used to show progress made on the component

Manual stations

For the correct fitting out of manual workstations at double-belt assembly conveyors, standardised units have to be provided. These include work seat, foot pedal, foot-operated release, work table and parts containers. These units should be designed on ergonomic principles. Fig. 3 shows a typical workstation of this type.

When the principle of working at an assembly conveyor is 'work division' each worker carries out the next operation in the workcycle of the assembly of the product (Fig. 4). This gives the worker a degree of freedom of action with the loose inter-linking between the workstations on an assembly conveyor resulting from the cycle time and the number of pallets to be stored before and after the workstations.

If a greater freedom of action is required for the operators, including the disengagement of the workstations from each other, then so-called cycle-independent workstations can be set up. At these stations the work is arranged according to the principle of 'total work' in which every operator completes the same work cycle. The pallets are fed from the main conveyor through the cross-feed sections (Fig. 5) to an auxiliary conveyor at which similar manual workstations are situated.

The control signals for initiating the transfer to a workstation is operated manually from each workstation; that is the pallets are allocated according to the readiness of the operators. According to this principle the operators can work at rates which are independent from each other (Fig. 6).

With the use of cycle-independent workstations it is possible to satisfy requirements for widening and enriching the work for manual operators and, to a large extent, the disengagement of the workstations from each other.

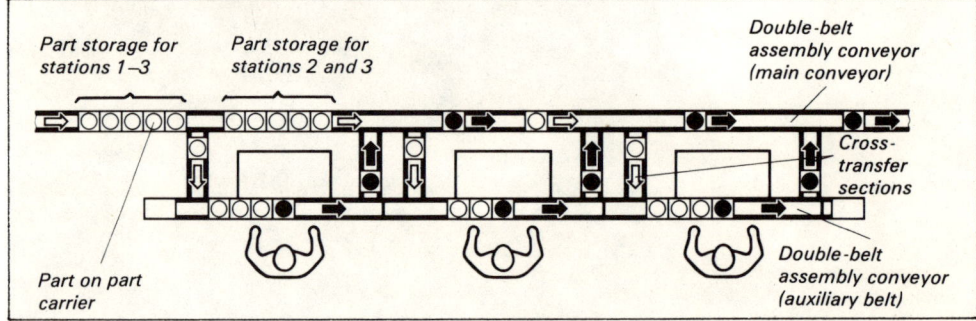

Fig. 5   Work carried out on an assembly line arranged on the principle of 'total work'. Again the circles can show work progress

MEETING FUTURE NEEDS FOR FLEXIBLE ASSEMBLY 15

Fig. 6   Three 'cycle-independent' workstations for a double-belt assembly conveyor arranged in a rectangle for the assembly of water pumps for motor vehicles

Cycle-independent workstations increase flexibility of an assembly line with respect to:
☐ changes in numbers assembled,
☐ variety of types,
☐ number of operators employed,
☐ introduction of new manufacturing processes.

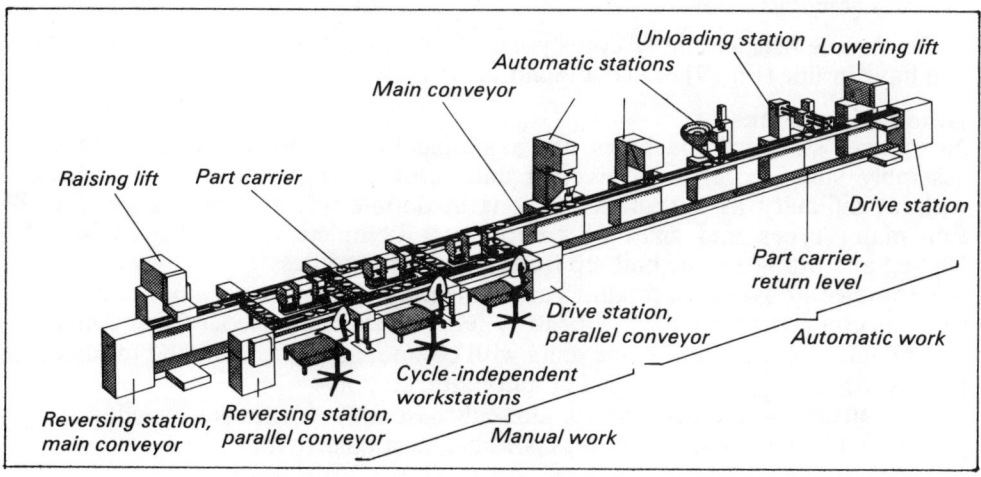

Fig. 7   Double-belt assembly line arranged in-line with three 'cycle-independent' manual workstations and four automatic stations

Fig. 8 Automatic station with vertical and horizontal multiple screwdriving units for the assembly of motors

On double-belt assembly conveyors, cycle-independent manual workstations can have in-line (Fig. 7) or rectangular arrangements.

Automatic stations

Nowadays standard machines are available for assembly processes in series assembly such as pressing, riveting and screwdriving, which can be included without difficulty as automatic stations in double-belt assembly lines (Fig. 8). For many types and sizes of components complete automatic stations for feeding and fitting can be built up from standard modules.

Examples of available modules are: assembly frames; separators and escapement mechanisms for pallets; vibratory feeders, elevating feeders, belt conveyors and magazines; pick-and-place units with components grippers; and modules for pneumatic and electrical control of automatic stations.

Even in automatic stations for difficult assembly porocesses existing design principles make the use of modules possible, particularly for:

☐ Assembly after blanking from strip (such as sealing, insulating and spacing washers).

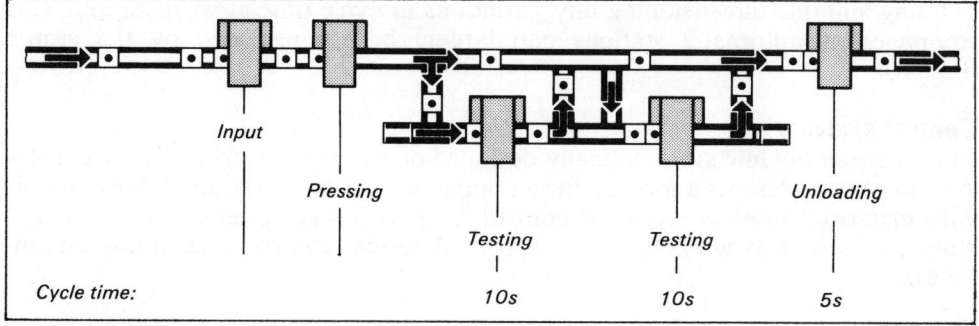

Fig. 9 Arrangement of two cycle-independent testing stations with long cycle times where the remaining stations have short cycle times, arranged on the double-belt assembly conveyor

☐ Fitting with assembly sleeves (for example for O-rings, sealing rings and slotted retaining rings).
☐ Assembly with expanding pliers (as with spring clips and sealing bellows).

The growing demand for automatic stations, especially for parts feeding and assembly has led to the development of NC-controlled manipulating modules. These dc motor-driven modules are an addition to the modules which up till now have been pneumatically operated. They have an advantage when several positions have to be approached with each axis. For example, for orientated loading of pallets and magazines; and when assembling different types of parts having different heights, lengths of axis and angular positions.

Generally, automatic stations are placed in the production line at the main conveyor. If, however, owing to different cycle times it becomes necessary to install several similar stations, these can be placed at an auxiliary conveyor independent of the workcycle of the main conveyor (Fig. 9).

Component stores

In an assembly system with component stores the components can be stored through pallets and double-belt conveyors. This method of storage retains the workpiece position and allows gentle handling, as well as making disengagement possible between individual manual workstations or complete manual workstation groups and automatic stations; or between individual manual workstations and complete manual workstation groups.

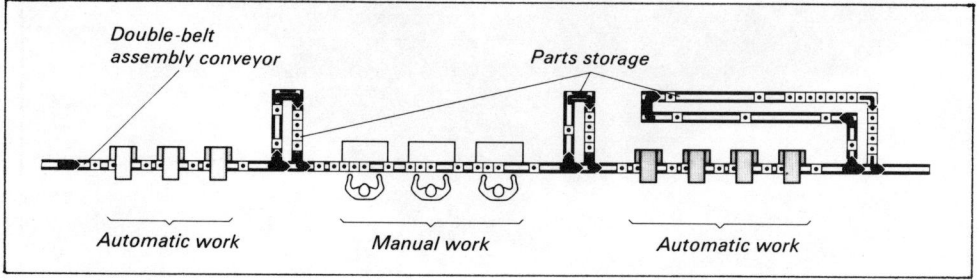

Fig. 10 Parts storage between automatic and manual workstations on a double-belt assembly conveyor

18 PROGRAMMABLE ASSEMBLY

Using suitable dimensioning any variations in cycle time at workstations and stoppages at automatic stations can largely be compensated by the stores (Fig. 10).

Control systems
For an assembly line systematically designed on the module principle the control system also represents a module for a complete solution. Programmable controls with memories used as sequence control for part-feeder systems and automatic stations, as well as adaptive controls for numerical control, offer many advantages.

Planning

When planning assembly lines with a high degree of automation it often turns out that manual stations can be placed between automatic stations for operations which cannot be automated, or are difficult to automate economically. In this case an arrangement arises in which manual workstations (for difficult assembly and test operations) and automatic stations (for screwdriving, riveting and press operations) are in a mixed sequence (Fig. 1).

When the planning of manufacturing operations is combined with product design, then a better work structure can be obtained. By grouping all the manual

Fig. 11 Pallet for the assembly of electric components with three different assembly fixtures

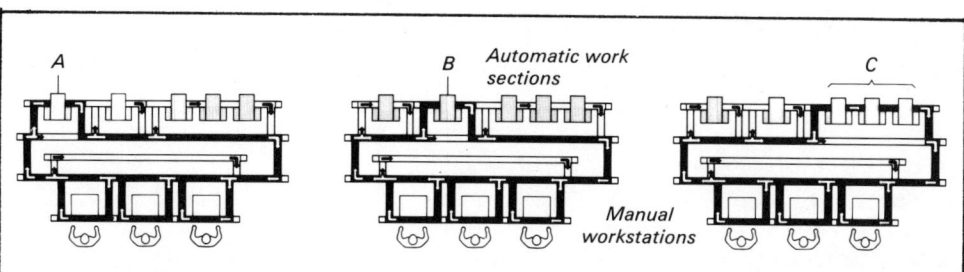

Fig. 12 Multiple circulation of a pallet with parts at three manual workstations each with the automatic sections A, B and C

operations on one side and the automatic ones on the other, manufacturing groups for manual operations and those for fully automated ones are created (Fig. 7). The formation of such groups avoids isolated manual workstations and improves communication between operators.

On an assembly line planned on the division of work principle, in which individual manual workstations and automatic stations are arranged in a varying sequence, an even work distribution between the manual workstations is often not possible. The output of such an assembly line and the wage contents of the product being assembled are governed by the workstation with the greatest work content. This causes lost time or waiting times at individual manual workstations.

Compared with the principle of division of work with its unequal work content at all workstations, the principle of combining the work at workcycle-independent manual workstations has, in addition to the advantages already mentioned, the same work content at all workstations and therefore a lower wages content.

The most important factors during planning which decide the operational sequence with regard to the formation of manual and automatic manufacturing blocks are: size of product, cycle time, number of workstations, and the type of work process.

The layout of assembly lines with manual and automatic work groupings can be arrived at in several ways.

For small products a solution can be found in which the pallets can carry several parts. With these all the manual production processes at the different production stages can be completed without interruption using automatic production processes. The number of parts for each pallet must be equal to the number of manual production stages to be combined (Fig. 11). Every time a pallet completes a cycle a fully assembled product is produced.

With large products however this solution, using multi-position pallets, generally is not feasible. To implement the separation of man and machine in this situation, the principle of multiple circulation of pallets is chosen. Here a pallet passes backwards and forwards between manual workstations and automatic stations until completion of the product as required by the number of automatic production stages interspersed between the manual operations (Fig. 12).

The movements of the pallets are controlled by appropriate coding on each unit.

CONCEPTION AND REALISATION OF FLEXIBLE ASSEMBLY SYSTEMS

R. *Goebel, Burkhardt and Weber GmbH, West Germany*

First presented in German at MHI Congress, April 1983, Hannover

Abstract

An attempt is made to define the requirements for a modern assembly system. The realisation of such systems is described with the aid of two examples.

The future trends in development in this field will be characterised in particular by the increasing influence of electronics. On the one hand more and more industrial robots will be employed, especially for medium cycle times, which can carry out assembly and handling operations with a certain 'intelligence' depending on the product variant to be assembled.

On the other hand systems are being conceived, especially for very short cycle times, which make 'chaotic' assembly with a simultaneous multiplicity of variants possible. For this, specially extensive data processing with high information interchange between the routing computer and the individual assembly stations, will be required. This presupposes both powerful hardware and especially problem-orientated and extensive software. This type of system will increasingly take into account the desire for higher productivity together with high flexibility.

Introduction

With the background of constantly hardening competition, especially from low-wage countries, the necessity of decreasing production costs is growing constantly. As manufacturing processes have generally been rationalised more thoroughly the interest in solving corresponding problems in assembly increases. An analysis of the production costs for different products indicates that these still contain considerable reserves for rationalisation, particularly that of assembly.

Why so few automated assembly plants, capable of reducing the cost of assembly, have been used until now is due to:

☐ too few product designs orientated towards assembly,
☐ too many variants of the product, and hence
☐ insufficient commercial efficiency of the assembly plants on offer.

Requirements for modern assembly systems

The requirements which enable a rationalisation of the assembly operation to be made are:

○ high assembly quality (i.e. high assembly security),
○ high productivity,
○ flexibility adapted to the requirements of production,
○ ergonomics of the remaining manual workplaces, and
○ economy of the total system.

The most important aspect is the economic operation which is nowadays 'loaded', especially by the requirements for maximum flexibility. A contribution to the suitable calculation of the economic operation is supplied by constant product quality which eliminates considerable reworking, recalling and correction costs.

Suitably adapted flexibility can exert a noticeable influence on intermediate storage, on tying-up of capital, and hence on the interest cost.

The ergonomic layout of the manual workplaces eventually reduces workplace stress costs especially for work requiring high physical effort. Also, the high productivity of an assembly system eventually reduces the unit cost of the product.

In general a step taken towards assembly automation, even when commercial efficiency is not 100% guaranteed, must be regarded as a 'down payment' for securing future risks, as certain cost factors (e.g. personnel costs) are bound to rise. It is, however, doubtful whether the qualifications of the assembly personnel will rise to the same extent.

Requirements for the products to be assembled

To enable an assembly system to be designed sensibly, i.e. with the minimum investment, certain preconditions should also be satisfied by the product to be assembled:

□ the design of the product should be assembly orientated (direction of assembly, modular construction, efficient transportation, etc.),
□ the variants should be designed with assembly in mind, and
□ all component parts should be suitable for assembly with respect to handling and quality.

It must be said in this connection that there should be a constant interchange of information between the production and development departments of the user and with the project engineers of the supplier of the assembly system. Only in this way is it possible to ensure that the design of new parts and subassemblies is influenced in the early stages, i.e. already in the development stage and before the startup of new products, hence considerably reducing the investment costs of a new assembly system and thus the unit cost. This is a constant learning process involving 'give and take' between manufacturer and user.

Layout of an assembly system

As the products to be assembled generally are not designed in accordance with the basic principles described above, the layout of an assembly system must be adapted to the given conditions. Experience has shown that even tiny modifica-

CONCEPTION AND REALISATION OF FAS 23

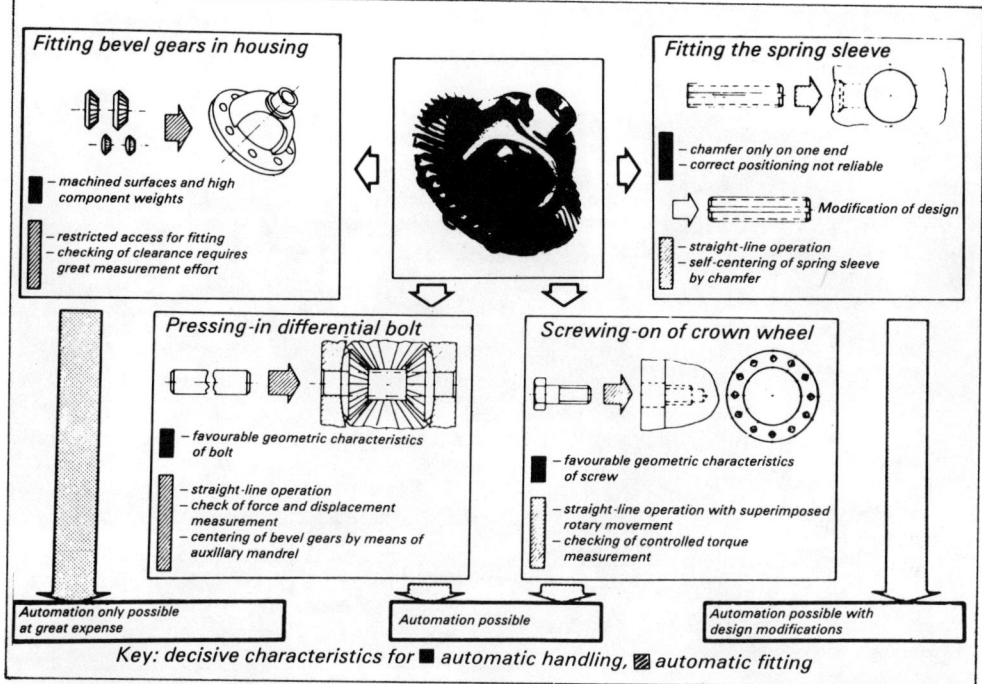

Fig. 1 Assembly possibilities for a differential gear – investigation into suitability for automation

tions of existing parts, e.g. in the automobile industry, are hardly possible during series production.

During the conceptualisation of an assembly system all the criteria on whether or with what consequences the automation of an individual assembly process is possible at all, must be defined:

○ assembly processes which can be automated without problems (these usually include all screwing processes),
○ assembly processes which can be automated after certain design changes, and
○ assembly processes which can only be automated with maximum effort, i.e. which cannot be justified economically or which are incapable of being automated (e.g. parts which handle badly, very flexible parts).

The 'assemblability' of a differential gear in a gearbox is shown in Fig. 1. For the fitting or insertion of the pair of bevel gears or the selection of suitable shims, highly skilled assembly personnel are required who cannot be replaced by an assembly unit. Even an assembly robot would be unable to carry out this process, i.e. it is incapable of being automated. On the other hand the screwing-on of the crown wheel after fitting can be automated without difficulty.

After a design change, namely the incorporation of chamfers on both sides, the insertion of the spring dowel pins for securing the bevel gear shaft could be done automatically. Previously the insertion of this pin was more like a broaching operation rather than one of fitting.

In many cases the retention of manual workplaces as a technical necessity is entirely welcome, as complete automation is often connected with certain social

24 PROGRAMMABLE ASSEMBLY

Fig. 2   Ergonomically optimised manual assembly workplace

problems. In these cases, such remaining manual workplaces are designed on the most modern ergonomic principles.

The requirements for such workplaces are:

- ☐ greater work contents and extent,
- ☐ separation from the workcycle,
- ☐ adaptation to the human work tempo, and
- ☐ independence of assembly workers from each other.

This type of manual workplace is shown in Fig. 2, which includes mechanical aids and the tying-in of its function into the automatic flow of components.

Using ergonomically optimal workplaces arranged in parallel, the workers can determine their own working tempo and any rest periods required.

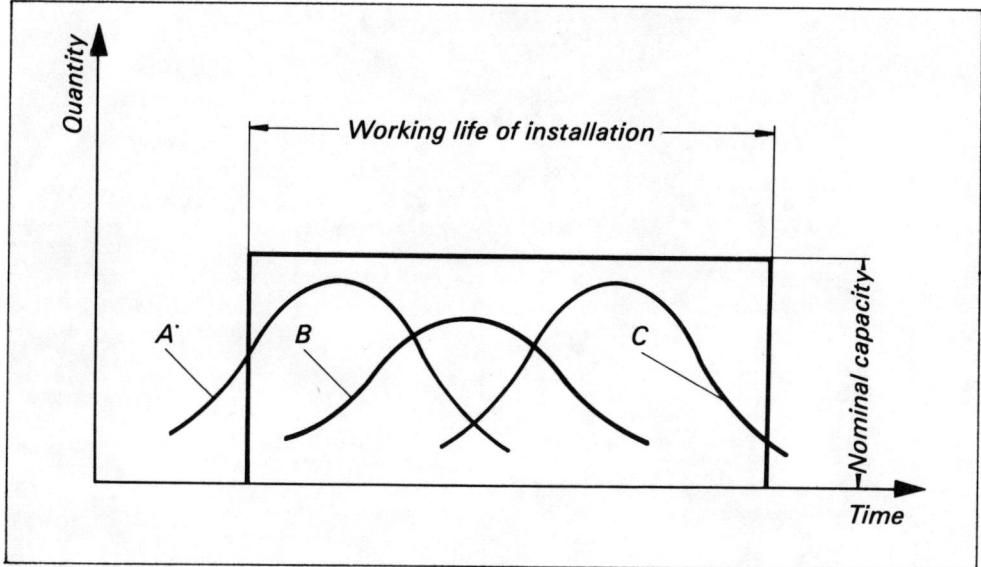

Fig. 3   Momentary and time-dependent flexibility of an assembly system

Flexibility

The conception of an assembly plant must not only encompass the solution of the immediate assembly tasks but should also be able to tackle problems of a similar type and with similar products arising in the future.

Fig. 3 shows a family of curves for the lifespan of some products. A vertical line at any point in time gives the respective product mix.

The greater the flexibility of individual stations and assembly units the easier is their adaptation to any changes in the product or the market. It cannot be denied in this context that increasing flexibility requires a corresponding higher investment.

Implemented examples

The following examples show clearly how the flexibility in an assembly system adapted to a particular problem can be achieved in accordance with the requirements of the user.

Assembly system for different types of vane pumps

This is an installation of high productivity and relatively high flexibility (Fig. 4). The main demand is the ability to reset to a new product in the shortest time possible. Because of cost considerations the operation should not be 'chaotic' but in batches.

In this design the automatic, manual assembly and repair sections were strictly segregated (Fig. 5). For this type of assembly system this solution has been extremely successful.

An essential component is the workpiece pallet. By accommodating three workpieces in three different positions, three different assembly operations can be carried out at each station. Pallets are loaded with preassembled housings in the manual assembly areas (in parallel arrangement), and circulate continuously.

26 PROGRAMMABLE ASSEMBLY

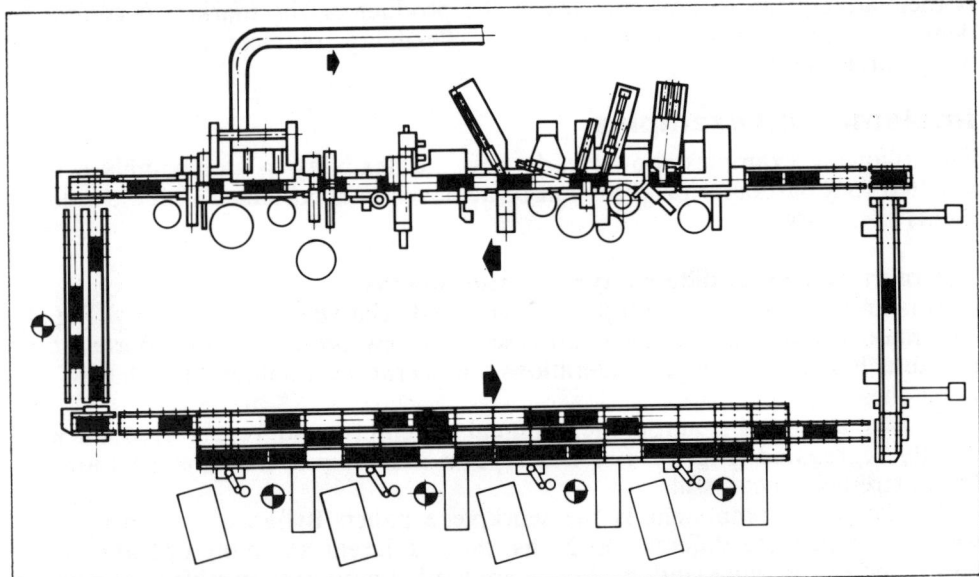

Fig. 4 Assembly system for different vane pumps

Fig. 5 Layout of the vane-pump assembly system

Fig. 6   Component parts of a vane pump

When an empty pallet is required by the operator it is channelled-in from the waiting area to the manual assembly area by pressing a button, and simultaneously an assembled pallet is released. All the parts at the assembly position are arranged ergonomically so that the assembly operation is optimised.

The component parts for one type of vane pump are shown in Fig. 6.

In the adjoining automatic section the following assembly operations are carried out:

Fig. 7   Automatic pressing-in of needle bearings

28 PROGRAMMABLE ASSEMBLY

Fig. 8   Simultaneous unloading or relocating of vane pumps in the unloading station

○ pressing-in of needle bearings (Fig. 7),
○ pressing-in of restrictors,
○ fitting of sealing cones and slide bushing,
○ insertion of previously greased radial sealing rings,
○ insertion of circlips, and
○ screwing-in of grub screws, studs, etc.

All movements and functions are controlled and when a fault occurs a repair code is set which interrupts the assembly process. At the end of the automatic section the faulty part is detected and channelled to the repair section where the correction takes place. At the end of the line the finished pumps are removed after the third circuit and placed on a conveyor belt. Simutaneously the other two workpieces are moved to the new assembly positions (Fig. 8).

The requirements for the shortest resetting time were dealt with on this installation as follows:

Different assembly programs are stored in the machine control and are available for any variant. At the start of the assembly the respective program is switched-on through a selection switch. The automatic stations are partly laid out as so-called alternative stations which are approached

Fig. 9   View of assembly system for different truck engines

for the workpieces in the line. Magazining and feeding elements for the various components can be replaced complete as they are designed for a particular part.

Assembly system for the complete assembly of six-cylinder engines
This installation is a larger recently installed assembly system of medium productivity and high flexibility. The requirements include the assembly of widely differing variations of petrol and diesel engines with a cycle time of about 1 min/unit (Fig. 9). The installation was designed to enable it to operate in the so-called 'chaotic mode', i.e. different engine variants can be assembled as desired. The particularly important assembly operations for assembly quality were to be replaced by automatic stations without difficulty at a later date.

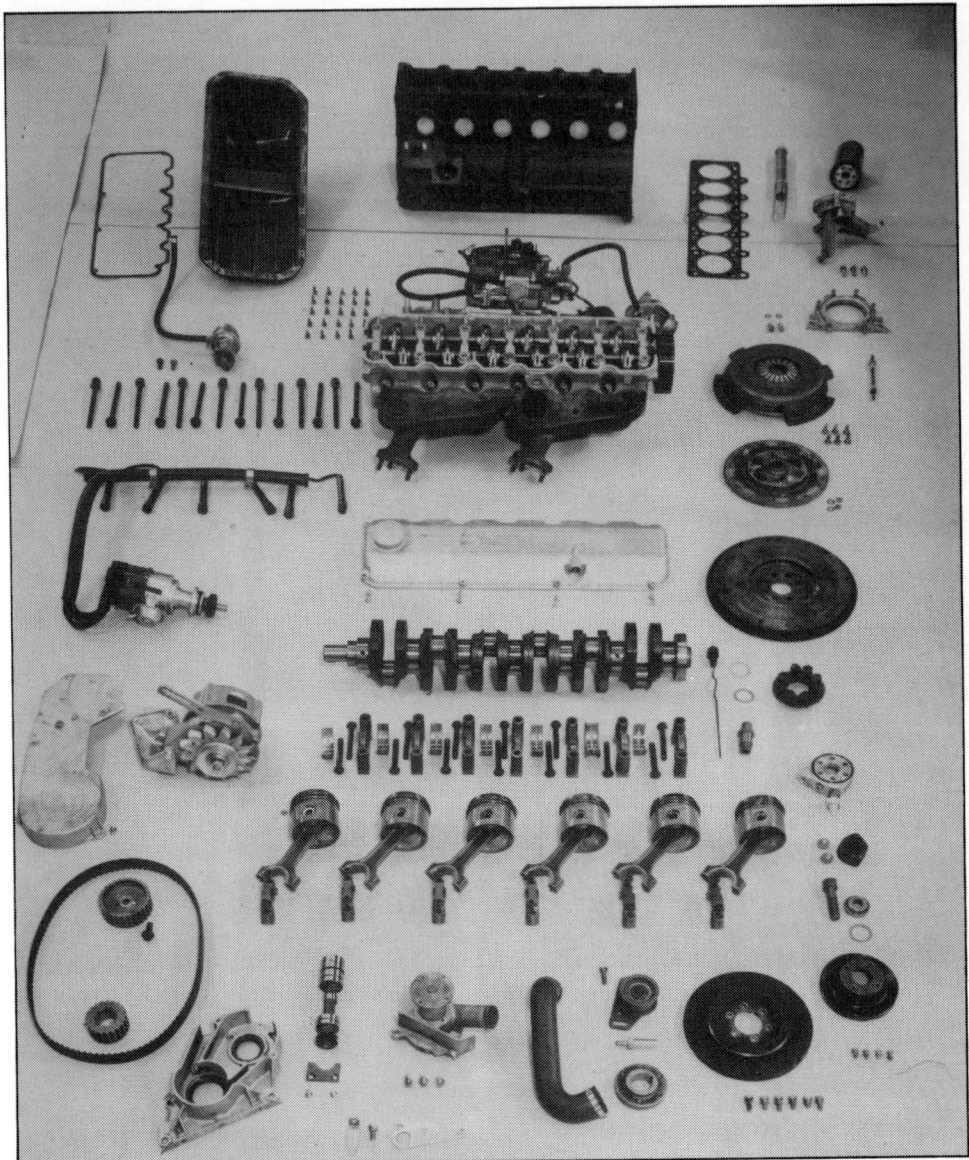

Fig. 10 Component parts of a six-cylinder truck engine

The component parts of the assembly operation are shown in Fig. 10.

As can be seen from the layout the system is constructed in rectangular form so that the workpiece pallets can follow a closed circuit (Fig. 11). Chain-driven friction roller conveyors transport the pallets and provide the buffers between the assembly stations. There are two overhead sections in the centre of the installation, facilitating access and which also serve as storage. An additional pallet buffer section is integrated in the pallet circuit which can also be used as a parking track. This also incorporates a pallet washing station.

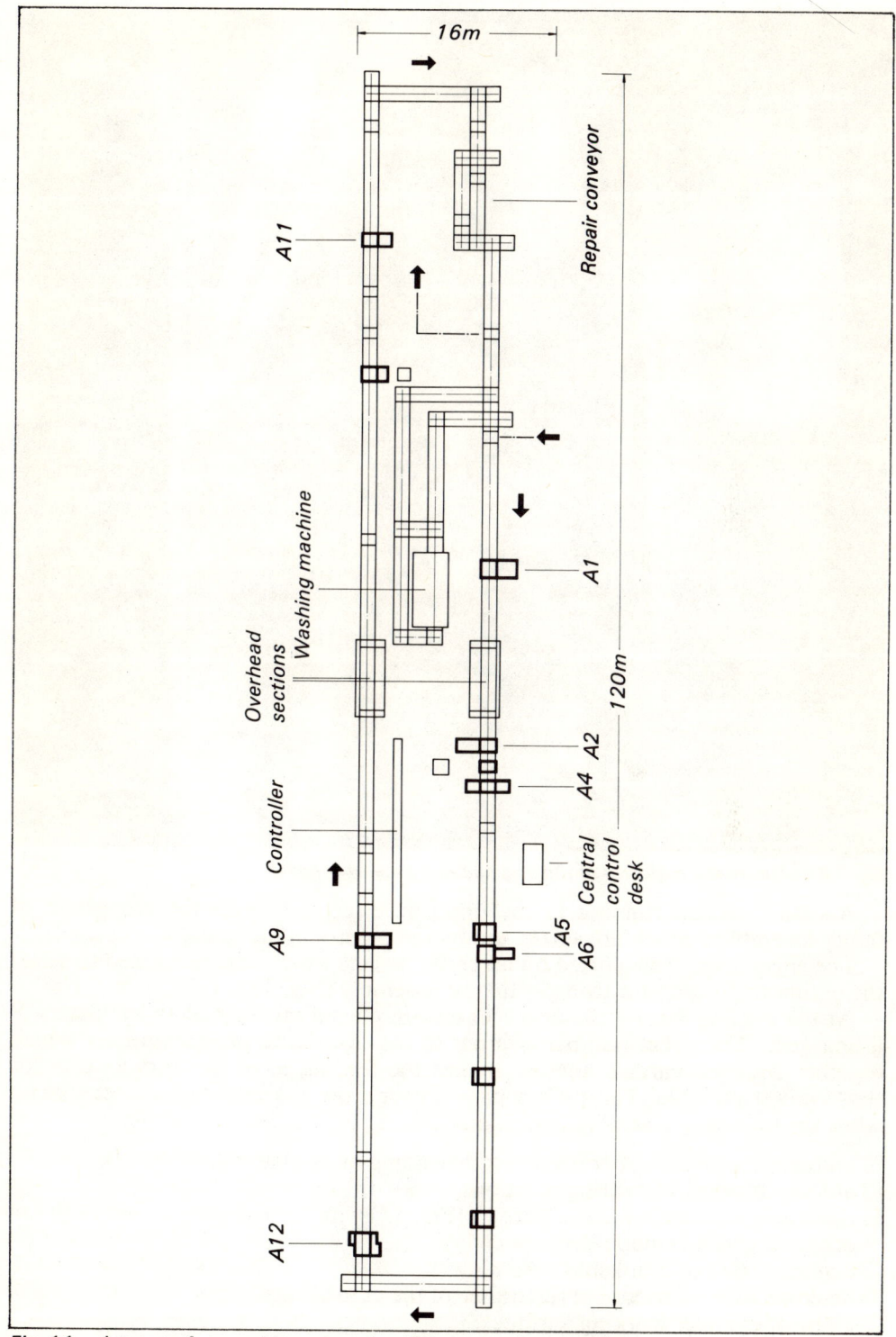

Fig. 11   Layout of assembly system for different truck engines

Fig. 12   Automatic handling of the engines and assembly pallet

A repair section running in the opposite direction serves the correction of faulty assemblies which are sorted out by the control in the repair conveyor.

The engines are transported on assembly pallets which can be rotated to select the optimum position for transporting or assembly (Fig. 12).

At the loading station the crank cases are placed on the pallets by means of lifting gear. The pallet number is input to the overriding plant computer which together with the variant number makes the feeding in of the components for that variant possible. The pallets pass through the assembly system clockwise while the following assembly operations, among others, are carried out:

☐ automatic insertion of crankshafts depending on engine type (Fig. 13),
☐ automatic oiling of bearing positions,
☐ tightening the bearing cover screws (Fig. 14) with a close torque tolerance (an accurate check is made electronically),
☐ determination of crankshaft end play,
☐ determination of frictional resistance of the crank mechanism,
☐ oiling of cylinder working surfaces,
☐ insertion of needle bearings in distributor,

CONCEPTION AND REALISATION OF FAS   33

Fig. 13   Automatic insertion of crankshafts depending on engine type

☐ automatic measuring of piston projection with selection of suitable cylinder-head gasket (which is then placed on the joint face),
☐ tightening of flywheel screws within close tolerances,
☐ filling with engine oil depending on engine type, taking quantity and oil type into account (Fig. 15).

34 PROGRAMMABLE ASSEMBLY

Fig. 14   Tightening of bearing cover screws to close tolerances

There are also a number of manual assembly operations which are carried out in addition to these major operations.

The system is designed so that present manual stations can be replaced relatively simply by automatic stations at a later date. The power supplies for this are already prepared. All handling operations in this installation are automatic and in addition the pallets can be swivelled to take up optimum positions for improving the ergonomics (Fig. 16).

The control of the installation is embedded in three levels. The engine data are automatically transmitted by the overriding computer to the routing control of the installation – the so called master PC. If this plant computer is defective

CONCEPTION AND REALISATION OF FAS 35

Fig. 15   Filling with engine oil depending on engine type

these data can also be input manually by means of a keyboard. The master PC transmits control commands to subordinated mini PCs at each assembly station.

In addition to the fixed pallet coding, a variable reject code is provided which is interrogated for an 'OK' at every station. Then the engine characteristic data which are stored in the master PC and which are allocated to the pallet number are recalled and and the corresponding assembly operation carried out.

At a central control desk a comprehensive fault diagnosis can be carried out on a display unit. Here the different machine conditions, e.g. the preconditions for automatic operation, are displayed. In addition, comprehensive fault indications are possible, especially for the automatic stations, making rapid fault detection possible.

36 PROGRAMMABLE ASSEMBLY

Fig. 16   Swivelling station for facilitating manual assembly

DESIGN OF A FLEXIBLE ASSEMBLY SYSTEM FOR A SMALL COMPANY

S. J. Mallin, Orbit Controls Ltd., UK and P. J. Sackett, University of Bath, UK

First presented at the 5th International Conference on Assembly Automation, 22–24 May 1984, Paris, France

Abstract

This paper describes how a company with minimal experience of flexible manufacture has embarked on a programme leading to a radical departure from their existing manufacturing policy. An assembly system has been designed offering many characteristics of classical flexible manufacturing together with large capacity increase, and reduced unit cost. This will be achieved on an existing site and with existing labour. In fact, the design relies heavily on capable operators. Through a phased implementation programme and the use of standard modularised equipment, including the company's own controllers, initial capital investment is modest and risk minimised. Despite this the facility offers scope for the application of both hardware and control strategy enhancements as they become available. In the meantime, a high degree of realisable flexibility is offered.

Technology is now available to radically improve the performance of the assembly function in small companies who may not have considered automated methods. It is the will to understand, and the confidence to apply this technology, which is lacking.

Introduction

In small manufacturing organisations the assembly function remains a labour intensive area with traditional automated approaches being unsuited to fluctuating product demand. The modular design of a computer controlled assembly system for the production of a range of transducers in a small company is described. The resultant extendible flexible assembly system accommodates the product diversity of the previous wholly manual set-up whilst offering an increased volume capability. Many features of classical FMS have been realised in this assembly system. Competitiveness in both existing and potential markets is aided by the response characteristics plus quality and reliability of product supply of the flexible assembly system.

The existing system

Transducer assembly in this company has been a totally manual process requiring a high level of dexterity. Typically six small components are involved, assembled units being 5–7cm in length. Production is in batches of 25 to 100 with individual operators being responsible for most tasks in a unit process. Build instructions are issued every time a batch is processed as more than 100 variations exist, this situation has arisen because there is a market requirement for customised units in small quantities. The present system is considered to be a maximum capacity for the existing product designs and manufacturing methods; the current annual output of 17,000 units is achieved with six operators. However with lead times for new orders running at up to ten weeks the response to customer demand is poor. The lack of sophisticated production equipment means that system performance is wholly labour dependent. Quality levels are more of a problem with subcontracted components than completed assemblies which have a scrap rate of only 4–7%. There is an attempt to impress quality awareness on operators but in practice this results in a duplication of inspection and test procedures. This 100% inspection policy is costly and by no means infallible.

At an early stage in this project it was established that changes in product design and manufacturing methods could provide a ten-fold increase in capacity without the need to raise the number of operators. However this would amplify the following problems associated with the present methods of production organisation:

☐ Long lead times, due in part to the time taken in preparing schedules, asssembly kits and organising a batch for production. To overome this, large stocks of finished items are maintained.
☐ High throughput times and work-in-progress levels; it takes an average of three weeks to process a batch, whereas the unit building time is only 30 min.
☐ Batch tracking and fault traceability are difficult to the extent that the sources of reject work can rarely be determined.
☐ Manufacturing data are inconsistent and collection non-real-time.
☐ High subcontracted components and assembly processes content; results in a loss of control over key areas of the manufacturing operation.

The working environment can be considered pleasant and benefits from a flexitime attendance basis. The system can accommodate this arrangement due to the independent operation of assembly workstations. It is likely that any attempt to change this to some rigid flowline-type set-up would be met with opposition and in any case would be undesirable. The standard of work is semiskilled, though interesting, because of the variety of tasks each operator has to perform.

New system objectives

Traditional automated assembly techniques are not suited to the expected volume demand and product variety at the present time; investigations have shown that less than 10% of the product cost is direct labour and 50% is overheads. These overheads have been identified as the major area for cost savings.

The present system suffers from long lead times, high work-in-progress and poor monitoring of production performance. These problems have all been addressed by flexible manufacturing systems in the batch manufacture of machined items. The problem is how to apply these principles when designing an assembly system for a small manufacturing organisation incorporating the following systemal features:

○ Adaptable for both expansion and the further automation of individual processes.
○ Multi-product capable in high and low volumes.
○ The operators working environment must be maintained.
○ No trivialisation of the operators tasks.

The primary requirement is a greatly reduced throughput time and any tooling or process changes must be fast and straightforward. This will in turn reduce work-in-progress and its associated problems. Similarly lead times must be reduced in order to improve response to customer demand and minimise stocks. Flexibility is to be derived through the good utilisation of the human skills available from the existing operators, though in order to establish the necessary degree of control, a hierarchical computer facility must be incorporated. This aim for complete control over the manufacturing/assembly process is the key to reducing overheads.

Uncertainty about the future production requirements of the system, and the necessity to learn by experience and gain confidence, demands a modular approach utilising standard equipment to keep capital investment and risk to a minimum. This ensures that expansion can proceed rapidly when necessary. Real-time data and processed information must be available to management; system status, batch status and production performance all need to be closely monitored. Existing well publicised FMS projects have shown that the application of these techniques benefits from a step-by-step approach, modifying the tactics at each stage yet still working within a pre-determined framework. It is important to establish the overall architecture of the system at an early stage in order that future expansion is compatible with other developments in the company.

Realisable flexibility, particularly in the case of small companies without specialist maintenance facilities, demands that system reliability is very high. This requires the use of standard equipment. An additional important face of realisable flexibility in this application is capable operation, in a degraded capacity, with less than the full quota of operators or with isolated overtime working.

System design

The emphasis has been on designing a system capable of economic operation and flexibility to batch size, product type, working procedures and future demand. The use of human operators greatly aids this flexibility and it is not economically feasible to substitute these assets with automation at present.

The resultant design consists of a number of workstations linked by a sophisticated work transport capability, and controlled by a hierarchical computer system as in a clasical FMS, but there are important differences (Fig. 1). Initially these workstations will be manually operated with the facility to accommodate conversion to a high degree of automation. This requires some component redesign which is being undertaken in parallel with the manufacturing programme. Work will have to be processed in small batches, consequently procedures necessary when changing from the assembly of one batch to another

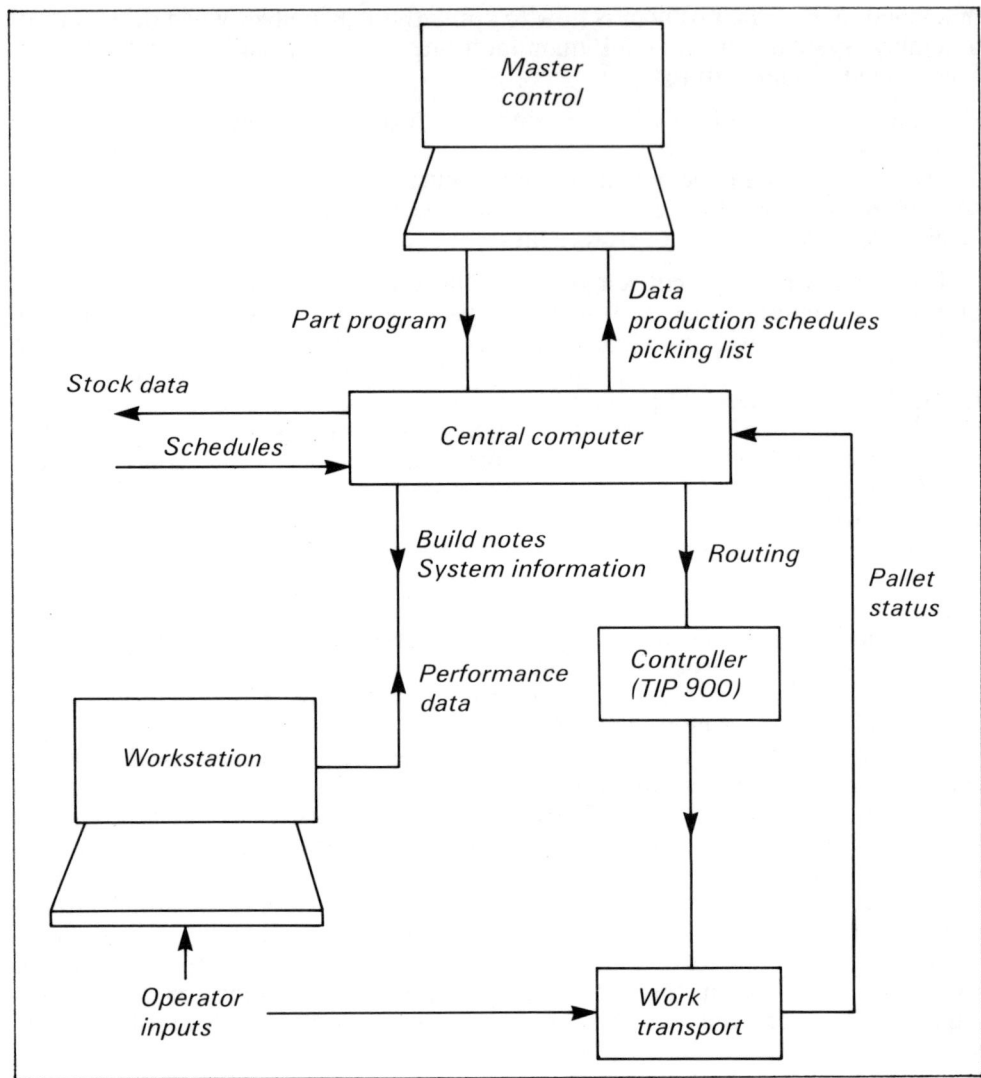

Fig. 1   Manufacturing control

must be quick and accurate. This will require a facility for communication between the computer and individual workstations – a VDU is the obvious choice as it provides a two-way link. The computer will monitor the production process and offer access to real-time information. Physical conveyance of items from one station to another demands an intelligent routine capability as the combination of products to be made require different processes.

There is the added problem of a high variety of products being made at any one time, consequently not all subassemblies in the system will be taking the same route. This requires coded pallets which can be identified at critical stages and distributed to the appropriate workstation.

The company has a minicomputer for handling various administrative tasks including stock control and production scheduling functions. It is intended that the proposed system will eventually be integrated with these. Immediate requirements will be for product routing and performance monitoring capability. A lower level of control for repetitive transportation functions will be provided by programmable controllers from the company's own TIP 900 controller range. The transportation system is based on a range of flexible modular assembly units. Typically this equipment is used in stand-alone mode but here the computer provides a much higher level of control. With flexible workstations the intention is to apply true FMS principles to this small product assembly area at a low capital cost.

Physically the system consists of a number of workstations as shown in Fig. 2, the number and layout of these stations being variable for future requirements. A buffer storage facility at each station provides some worktime flexibility and avoids the rigidly enforced discipline of a constantly moving conveyor. Most stations will be flexible in the work that can be done, they simply act as a workplace with the operator performing the necessary assembly function. Variation in the tasks required will maintain the present level of job satisfaction but where the use of certain production equipment is necessary then a station will become dedicated. During operation without a full complement of operators, the buffer stores allow those operators present to work at isolated stations.

In operation, coded pallets containing one or more subassemblies are carried on the conveyor system; code readers monitor their progress and the contents of each pallet are known by the computer (Fig. 1). Manufacturing routing details, held in the computer memory, divert pallets to the appropriate workplace queue until called for by the operator. Should space not be available in the buffer then the pallet will continue to move around the central rectangle. Once the operator has accepted the pallet, information about assembly/manufacturing details will be displayed on their workplace monitor. On job completion the operator enters information requested by the computer. This will include the number of units completed, how many are faulty and reasons for failure. No more pallets can be obtained until the computer is satisfied with the information that it has received.

Monitoring of workstation performance will provide information on processing times and advise on excess deviations. Similarly monitoring of workstation activity will isolate excessive downtime or bottlenecks. Operators will have full access to system status to provide them with the maximum flexibility of work organisation consistent with system performance objectives. When a batch is completed, instructions about the next product to be assembled, machine setting changes and components required will be displayed for the operator. This ensures continuity of production as one batch does not have to pass through the system before another can enter. The production scheduling system will already have issued a component picking list and these will have been accessed by the operator. The test and packing stations are common to all products and will have to cope with alternating requirements. By the use of automatic test equipment the test parameters corresponding to the contents of the pallet will be set; this will once again be backed up with displayed instructions. Details of the packed contents will be printed when a pallet arrives; a label to be sent with the work will indicate product type, customer, and batch number.

Throughout the system the aim is not to trivialise the operators tasks by introducing this degree of computer control, but to ensure that system performance is maintained. Indeed it is expected that operators will provide a much higher level of optimisation than current mechanistic algorithmic based approaches.

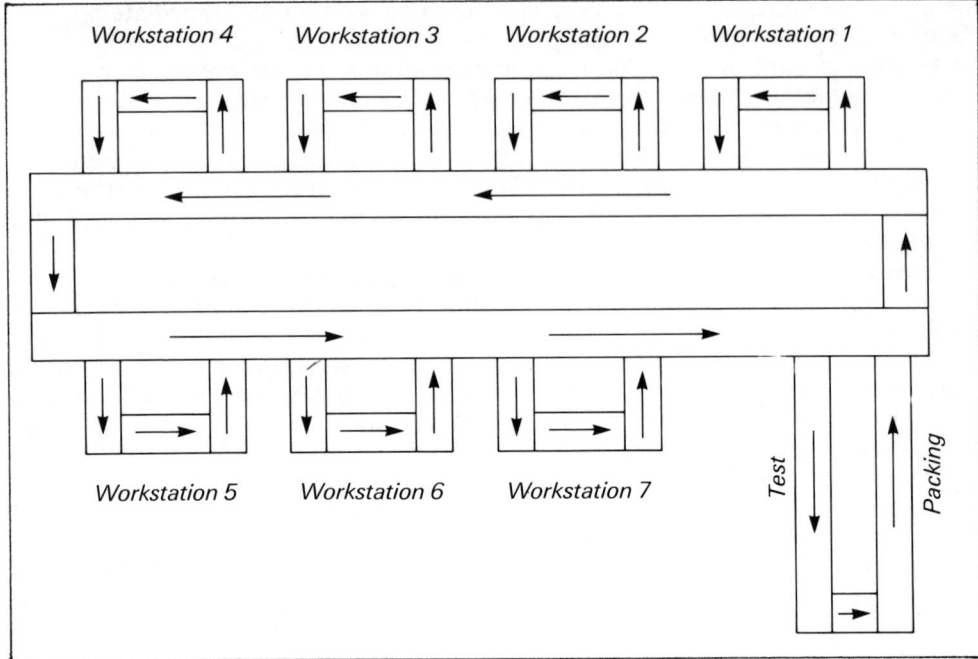

Fig. 2   Work transport

Phased implementation

Based on current marketing predictions, the company require a fully operational flexible assembly system by early 1986, but the majority of features described here will be available during 1984 in a system capable of rapid capacity expansion. This interim system will not incorporate full integration with the company's existing computer based facilities. It has been established that a staged approach is needed in order to modify the project plan in the light of market requirements and on-going experience. The plan for implementation can be seen in three parts but some concurrent running is taking place:

☐ The installation of a pilot work transport capability, to establish computer interfaces and to develop the communication routines. This provides the opportunity to become acquainted with the system hardware and to make an early start on testing of the computer software which will be the subject of considerable on-going development. It is possible to incorporate the work handling facility into the present manufacturing system from the start, operating as a simple conveyor in the initial stages.

☐ Having established the working pilot system, production engineer and automate selected workstation activities. This will include the introduction of processes which are currently subcontracted, such as plastic injection moulding. It will provide a steady increase in volume capability and spread the normal proving and commissioning procedures as well as the capital requirement.

☐ The integration of other computer based systems, including stock control and production scheduling, once the manufacturing system is on-line. These measures are really refinements and become increasingly valuable as volumes rise. In addition, this requires liaison and input by other areas of the organisation.

The use of human inputs underpins this implementation programme and the final set up will still require the existing number of operators although the output will rise substantially. Throughout the development stages it is necessary to prepare the existing manufacturing/assembly system, including the people, for the changeover and to maintain output. This is particularly critical during the initial stage when spare capacity is not available and means gradually changing production methods so that they are organised in a way similar to the new system but still on a manual basis. The changeover will then be a relatively simple process with operators only needing to become acquainted with the new equipment, not a wholly new way of working.

FLEXIBLE ASSEMBLY SYSTEM FOR UNMANNED FACTORY

H-J. Warnecke, R-D. Schraft, E. Abele and J. Spingler,
Fraunhofer Institut für Produktionstechnik und Automatisierung (IPA),
West Germany

First presented at the 4th International Conference on Assembly Automation,
11–13 October 1983, Tokyo, Japan

Abstract

The concept and features of a flexible automatic assembly system for different products are described. The assembly system consists of programmable assembly cells with industrial robots, special stations, a storage for parts, and automated guided vehicles for flexible interlinking.

Introduction

Constantly rising wage costs have for some time been forcing the manufacturing industry into automation measures. Whereas up to now it has been primarily parts manufacturing which has been automated, the automation measures are now being increasingly concentrated on the assembly sector. Assembly is, however, a branch of manufacturing where automation necessarily involves a high degree of technical complexity and high costs. It is therefore only possible to apply automated assembly systems economically if these capital-intensive means of production are used to their full capacity over a long period of time. This is made more difficult, however, by more and more frequent product changes, and by an increasing variety of products and types together with falling quantities. Moreover it is increasingly a requirement to be able to react flexibly to customer wishes[1].

Conventional automatic assembly systems are mainly designed for assembling large quantities of one product. Product changes or supplementary versions, which often result in a new assembly sequence, therefore usually necessitate complex conversion measures with regard to the existing system, insofar as it is even possible to adapt the assembly system to the new assembly sequence. Successor products can only rarely be assembled automatically with the existing system. However, automated systems which can assemble the different variants and types of a product or even of a group of products, are particularly essential in the case of small and medium lines in order to achieve sufficient capacity.

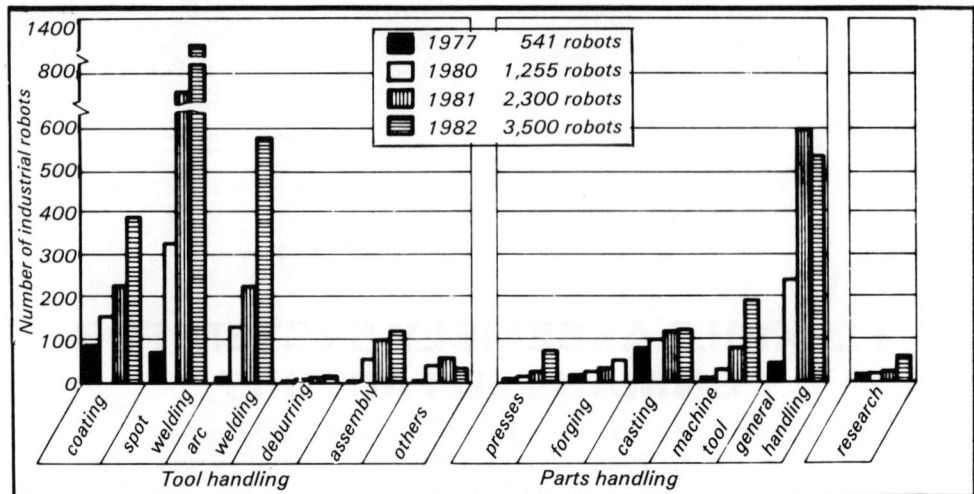

Fig. 1 Change of fields of application for industrial robots in Germany

This complex of problems has resulted in the demand for greater flexibility of automated assembly systems. One certain method of achieving more flexible automation is the industrial robot. Compared with other manufacturing sectors the number of industrial robots used in the assembly sector is still very low (Fig. 1)[2]. Nevertheless the mere use of industrial robots in assembly systems is not sufficient to maintain a flexible overall system. The assembly sector in particular requires a wide range of so-called peripheral devices (e.g. assembly devices, tools, parts supply, etc.) in addition to the industrial robot. Both this periphery and the linking system must have a flexible design in order to be able to take full advantage of the flexibility of the industrial robot.

Automatic assembly system with flexible linking structure

A concept for a flexible, automated assembly system has been developed at the Fraunhofer Institute for Manufacturing Engineering and Automation (IPA) in Stuttgart. The overall concept of this assembly system was based on the demand for a system structure independent of specific assembly sequences. This concept is described below using the example of an assembly system designed for use with various types of automobile assembly.

The assembly system shown in Fig. 2 consists of three different subsystems:

☐ programmable assembly cells,
☐ system storage, and
☐ a flexible conveying system.

In addition to these largely product-independent subsystems it is possible to integrate function-specific special stations in the assembly system (e.g. test stations, cleaning stations, presses, etc.). The central feature of this assembly system are the programmable assembly cells, where the actual assembly activities are performed with industrial robots. One of these programmable assembly cells has already been implemented at the Institute, and is currently at the testing and optimisation stage. The design and the characteristics of this programmable assembly cell are described elsewhere[3].

FAS FOR UNMANNED FACTORY 47

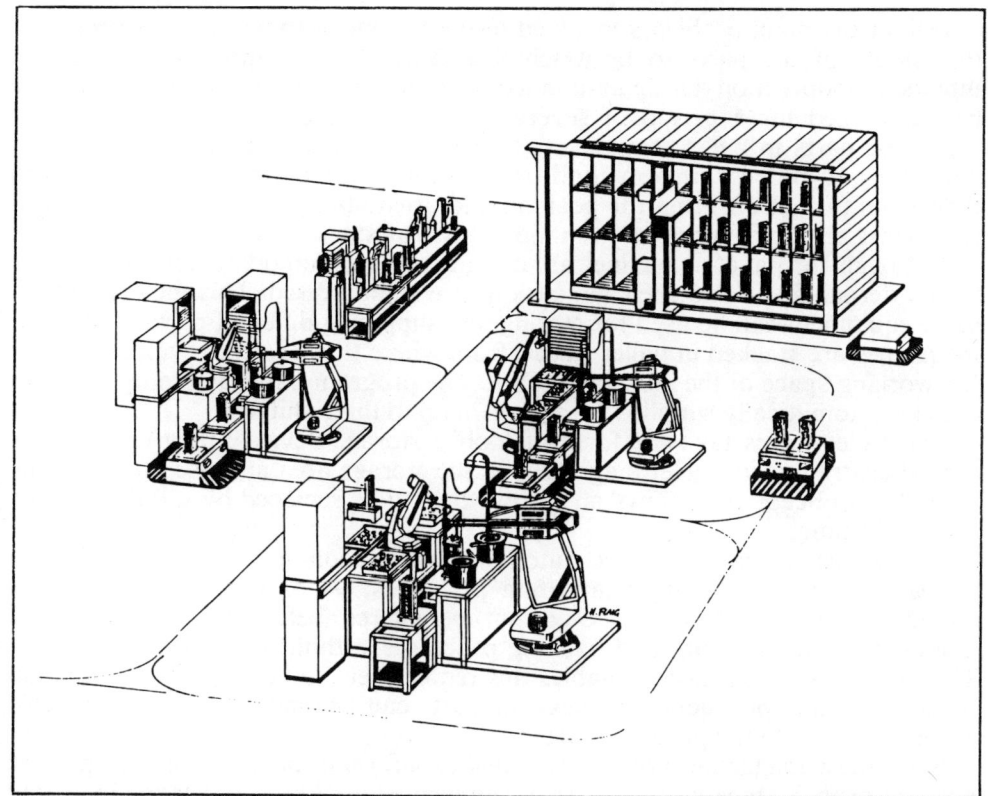

Fig. 2 Flexible automatic assembly system

The individual programmable assembly cells, the special stations and the system storage are linked together by means of automated guided vehicles equipped with a roller conveyor for automatic acceptance and transfer of the pallets with the assemblies. In addition to their linking functions within the assembly system the automated guided vehicles are also intended to link the assembly system with the preceding and subsequent production sectors. The assemblies are conveyed to the operating rooms of the industrial robots in the assembly cells and to the special stations along friction conveyors. In conjunction with indexing equipment these permit precise positioning of the assemblies at the assembly location, and also serve as buffers for the particular programmable assembly cells and special stations in case of temporary faults[4].

It should be possible for the assembly system to work automatically in three shifts independently of the adjacent production sectors. However if these production sectors are manual or semi-automated systems, they will normally be operated in a maximum of two shifts. This is the reason for the storage, which ensures that aggregates are supplied to or retrieved from the assembly system on completion of assembly, even if no production takes place in one or two shifts in the adjacent sectors. A further function of the storage is to act as a buffer for the assembly system in case of faults, the storage consists of a shelf with programmable shelf control unit, which permits random access to the individual storage locations. The number of storage locations in the shelf depends on the interval which must be bridged.

One of the main problems involved in the flexible automation of assembly is the supply of the parts to be assembled. Since the assembly system is also intended for operation during unmanned shifts, interrupt-free supply of parts must be guaranteed for a period of several hours. This problem can be solved by supplying parts in pallet magazines. These are manufactured from plastic in a deep-drawing process, this method being extremely economical. In addition to their main task of ensuring the position and orientation of the parts, these pallet magazines also fulfil subsidiary functions, such as stacking and transporting.

Although the pallet magazines are designed to accommodate particular parts, their external shape and their dimensions are standardised. This means that the same equipment can be used for storing and supplying different parts. The pallet magazines are stacked in pallet stores, from where they are conveyed directly to the working space of the industrial robot. The programmable assembly cells can thus be automatically supplied with parts in up to three shifts.

The pallet stores take the form of mobile storage devices; they can thus be loaded centrally with pallet magazines in the storage area and used for material flow. This concept also allows an empty store to be replaced by a full store in a very short time.

An important basic rule of automatic production is intended to prevent unnecessary repetition of organisation processes: 'A part position which has already been defined once should not be altered before termination of all manufacturing, assembly and packing processes without a compelling reason'. The use of pallet magazines enables this requirement to be met. On termination of the preceding production process the parts can be loaded either manually or automatically into the pallet magazine.

Since these magazines are ideally suited as a form of packing for transport on account of their stacking ability, they can also be used by suppliers or in distant production sectors.

Concept features

The assembly system can be operated automatically in several shifts, including unmanned shifts; operation is independent of the adjacent production sectors. The use of industrial robots in conjunction with sensors enables an important requirement for unmanned operation to be fulfilled; namely, automatic monitoring of the assembly process[5].

The most important features of the assembly system are:

○ Minimum proportion of product-specific devices.
○ Rapid change-over possible.
○ Structure independent of assembly sequences, and thus easy adaptation to new assembly sequences.
○ Good expansion capacity.
○ Automatic material flow.
○ Low susceptibility to faults.

This concept of a flexible, automatic assembly system is largely independent of specific products. It can therefore also be adapted to assembly systems which must be designed for another product range. The individual elements of the assembly system will then differ in their detail from those described in accordance with the requirements of the product range to be assembled; however, it will be possible to take over the overall concept with all its advantages.

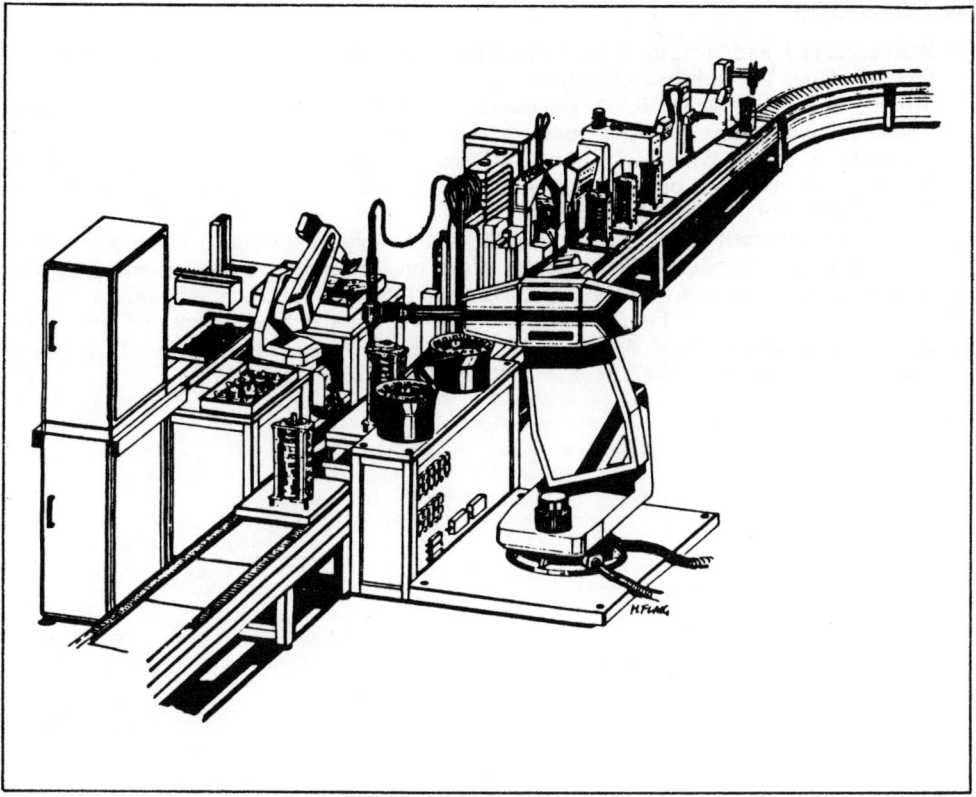

Fig. 3 Section of an automatic assembly line with programmable assembly cell

Applications

In the short term it will not always be possible to apply the flexible, automatic assembly system for a number of reasons. The most important of these concern the often restricted amount of space available in existing production shops and, for financial reasons, the need to continue using existing production facilities. However, in the long term it will be possible to apply automatic assembly systems based on the concept described, especially when new production shops are built or existing ones extended.

It is, however, the aim of many enterprises to increase productivity in the short term through a greater degree of automation in existing assembly lines. In such cases it might be possible to integrate particular subsystems from the assembly system described in the existing assembly lines, thus increasing the degree of automation and the flexibility of these lines. In particular it is possible to use the programmable assembly cells in conjunction with parts supply in pallet magazines as a substitute for manual workplaces or inflexible automatic stations in semi-automatic assembly lines[4].

Fig. 3 shows a section of an assembly line for automobile assemblies, where a programmable assembly cell has been installed between existing facilities. However the specified linking sequence in these assembly lines means that the high degree of flexibility of the flexible linking structure cannot be achieved.

References

[1] Warnecke, H-J. and Spingler, J. Industrieroboter – Ein Mittel zur flexiblen Automatisierung. Proc. Int. Conf. Die Industrieroboter, Linz, September 1982.
[2] Schraft, R-D. and Weiss, K. Angebotslücken und Entwicklungstendenzen in der Handhabungstechnik, Proc. Int. MHI-Congress, Hannover, April 1983.
[3] Warnecke, H-J. and Walther, J. Assembly of car engines by hand-in-hand working robots. In, Proc. 4th Int. Conf. on Assembly Automation, Tokyo, October 1983, pp. 15–23. IFS (Publications) Ltd., 1983.
[4] Bässler, R., Schunter, J., Spingler, J. and Walther, J. Aplication of industrial robots for the assembly of car-devices. In, Proc. 3rd Int. Conf. on Assembly Automation and 14th IPA Arbeitstagung, Böblingen, May 1982, pp. 495–508. IFS (Publications) Ltd., 1982.
[5] Abele, E. and Ahrens, U. Influence of microelectronics on the development and application of assembly robots. In, Proc. 3rd Int. Conf. on Assembly Automation and 14th IPA Arbeitstagung, Böblingen, May 1982, pp. 577–588. IFS (Publications) Ltd., 1982.

Chapter 2
ASSEMBLY APPLICATIONS

Flexible methods can be applied to many products and in most industries. The papers in this chapter illustrate this spectrum with applications to electrical, electromechanical and automotive products.

THE DEVELOPMENT OF AN AUTOMATIC ASSEMBLY LINE FOR VTR MECHANISMS

T. Ohashi, S. Miyakawa, Y. Arai, S. Inoshita and A. Yamada,
Hitachi Ltd, Japan

First presented at the 15th CIRP International Seminar on Manufacturing Systems,
20–22 June 1983, Amherst, USA

Abstract

An automatic assembly line for VTR mechanisms was developed. The introduction of assembly robots provides flexibility to the mass-production automatic assembly line. The assembly line, 88% stations of which are automated, consists of 11 robot stations, 50 pick-and-place units and specific assembly machines, and 2 variable stroke pick-and-place units, and produces 75,000 VTR mechanisms per month. The paper outlines the developed assembly line, and the systematic approach for the rationalisation activities from the product design improvement to the facility development, especially the assembly line performance simulation and its result.

Introduction

In accordance with the increasing variety of customer choice and the shortening of a product's life cycle, mass-production automatic assembly lines are required to be more flexible, less expensive to build, and quickly lined up. Recent improvements in robot technology provide a hope that robots can be the key to the realisation of flexible automatic assembly, and various technological attempts to utilise robots for automatic assembly have been made. Though it seems that robots should be ideal for flexible automatic assembly systems, not many robot installations have been used so far in mass-production automated assembly lines for the following reasons:

☐ Assembly speed of robots is considerably slower than that of dedicated assembly machines.
☐ Robots are still expensive.

In order to develop a flexible automatic assembly line to which robots are effectively installed, a systematic approach to product design to facilitate

54 PROGRAMMABLE ASSEMBLY

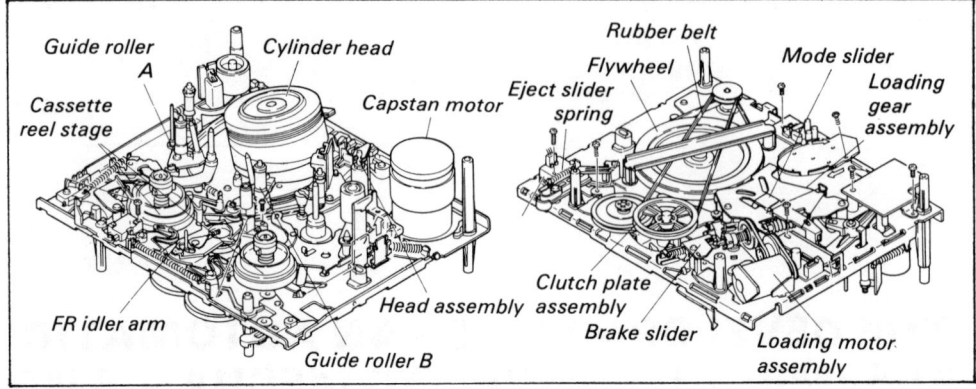

Fig. 1 VTR mechanism

development is necessary. The authors have been involved with a project to develop an automatic assembly line for VTR mechanisms which have a complicated construction, use many parts and offer a life cycle which is relatively short.

Product design improvement

Product design plays a very important role in the automation of assembly. If the structure of a product is not easy to assemble, assembling facilities must be more sophisticated and dexterous, and therefore expensive; the failure rate of assembly stations becomes high and the performance of the whole line will be low. Therefore, firstly the product design was thoroughly reviewed and a mechanism suitable for automatic assembly was developed.

Fig. 1 shows the developed VTR mechanism. Most of the parts such as pressed parts, moulded parts, rubber belts, and coil springs are assembled from the top with an easy attaching movement so that they can be attached by pick-and-place units or cheaper robots with three degrees of freedom.

For the product design review, an analytical design improvement procedure developed by Hitachi called the Assemblability Evaluation Method or AEM was used (Fig. 2). The AEM analyses assembly structure using 17 symbols and gives designers and production engineers an idea of how easily products can be

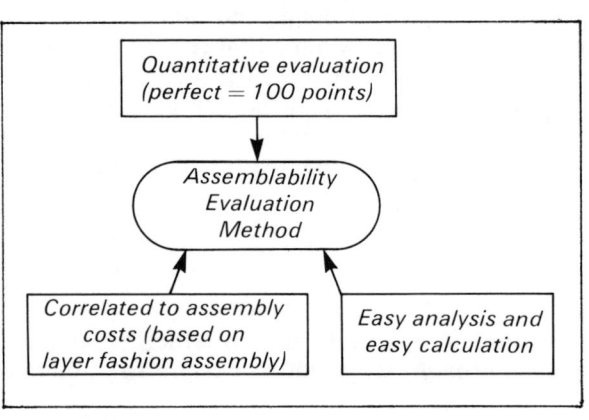

Fig. 2 General idea of the Hitachi Assemblability Evaluation Method

Fig. 3 Accuracy of cost estimation by 'A E M'

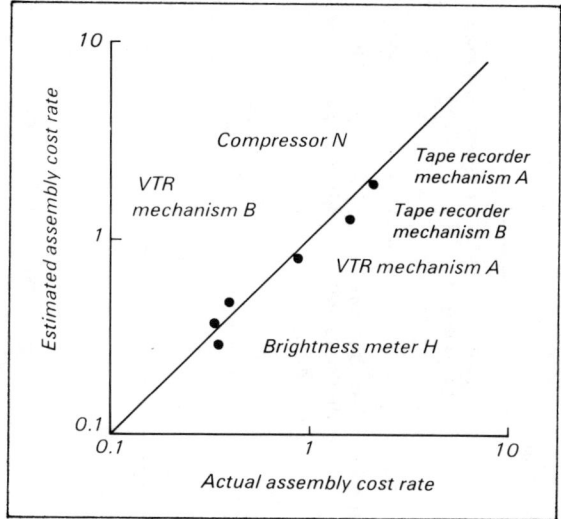

assembled. It points out weak areas of design from the assemblability viewpoint. The basic ideas of the AEM are:

☐ Quantification of the difficulty of assembly by means of a 100 point system of evaluation indices; one can easily infer from the indices the difficulty of assembly operations.
☐ Easy analysis and easy calculation making it possible for designers to evaluate the assemblability of the product in the early stages of design.
☐ Assemblability evaluation indices are correlated to assembly cost. A reasonable cost estimation can be made using Fig. 3.

By means of the evaluation and improvement iterations as shown in Fig. 4, designers can improve the product design very effectively in the early stages of

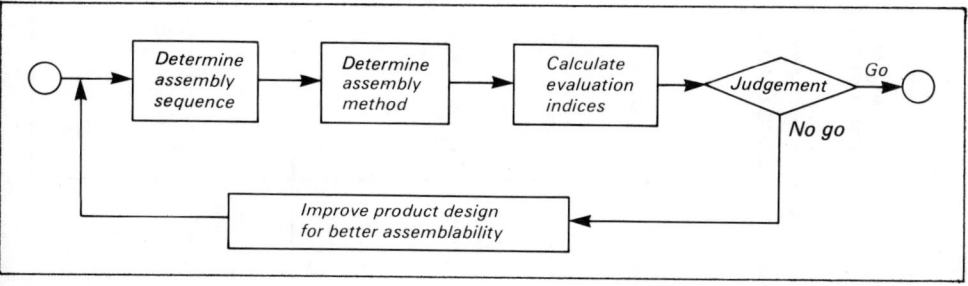

Fig. 4 Evaluation procedure

Table 1 Examples of design improvement in the VTR mechanism

Purpose	Reduction of number of parts	Stabilise parts positioning	Easy handling	Easy insertion	Prevent entanglement	Prevent jamming
Means	Unitise parts	Prepare positioning guide portion	Prepare parallel portion for gripping	Prepare chamfering	Use connector to eliminate lead wires	Increase parts thickness
Part name	Tape guide	Loading motor assembly	Capstan motor assembly	Screw	Loading motor assembly	Washer
Before design improvement						
After design improvement						

the development. Table 1 shows the examples of design improvement in the VTR mechanism, and the result of the improvement is summarised in Table 2.

Table 2   Results of assemblability improvement

	New mechanism	Previous mechanism
Number of parts	379	460
Assemblability evaluation rating	73	63

Line performance simulation

For effective development of an automatic assembly line, the following evaluations and the feedback of the results to the system design is necessary:

○ Will the assembly line be able to achieve the intended rate and volume of production?
○ Which station is the bottleneck of the assembly line?
○ Are the buffer sizes between stations pertinent?
○ What is the lead time of the assembly line?
○ The influence of the problems of individual stations on the line output.

In order to evaluate these items, an assembly line simulator was developed using the GPSS (General Purpose Simulation System) – a simulation language based on queuing theory. The principal features are:

□ Information necessary for system design such as buffer occupancy rate, blocking, and so on, as well as production volumes and production rate, are provided as the output.
□ Machine problems and the time required for part supply are taken into account.
□ Assembly operations of individual stations are expressed as a program module so that the assembly model can easily be modified.

Fig. 5 shows the flow diagram of the simulation program. Assembly operations are carried out at each station when the chassis arrives and the assembly unit is ready to assemble. If the buffer is full, the chassis cannot be passed to the next station and blocking occurs. When an empty magazine is replaced by a new one, the assembly unit must wait until the magazine replacement is completed and the part to be attached is supplied. Assembly operation data for individual stations, such as time required for assembly and buffer size, are specified in the program.

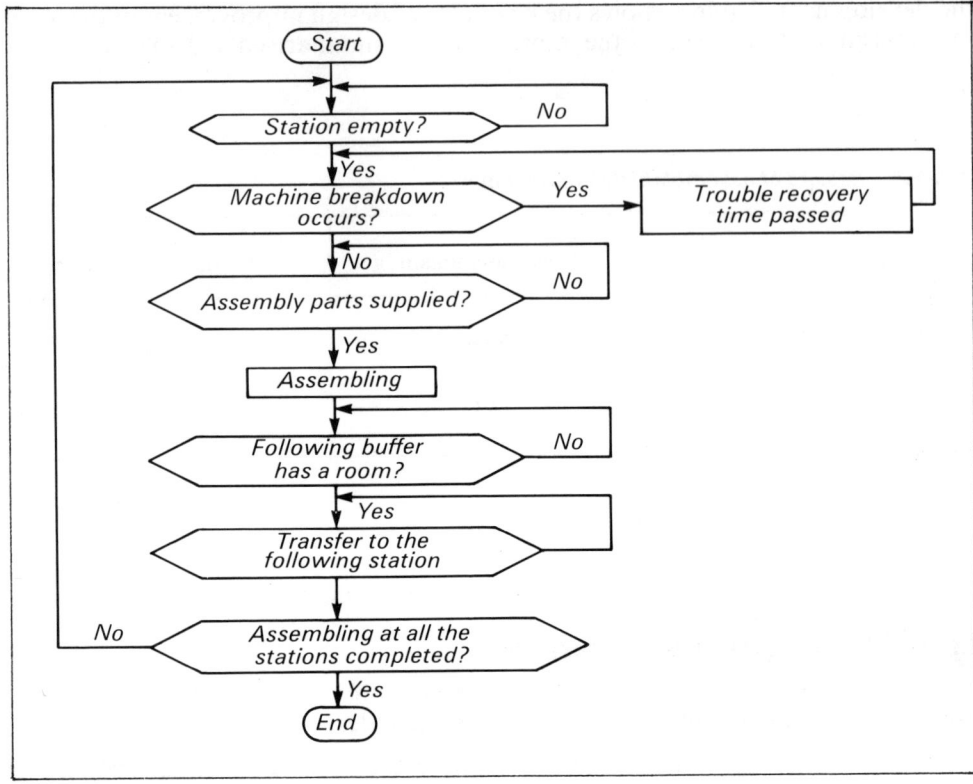

Fig. 5 Flow diagram of the simulation program

Fig. 6 shows the input and output data of the simulator. Input data were presumed based on the actual result of automatic assembly line for tape-recorder mechanisms[1,2] which have a similar line structure. Machine troubles are classified according to their time required for recovery:

Minor breakdown: breakdown caused by misassembling, parts jamming, etc., which can be recovered easily. Machine is supposed to stop for 30 sec on average.

Serious breakdown: breakdown which requires repairing such as component replacement and adjustment. After a machine stop period of 5 min an operator is meant to take over the assembly operations for the machine.

As a serious breakdown is expected to happen rarely during the simulation period, the influence of the serious breakdown is approximated as follows:

$$T = T_0 \times (MTBF + 5\ min) / MTBF$$

where T is the approximated line tact time considering the influence of serious breakdowns, T_0 is the simulated line tact time considering only minor breakdowns, and MTBF is the expected mean time between failure of the whole assembly line. (Note, tact time means cycle time for one complete assembly.)

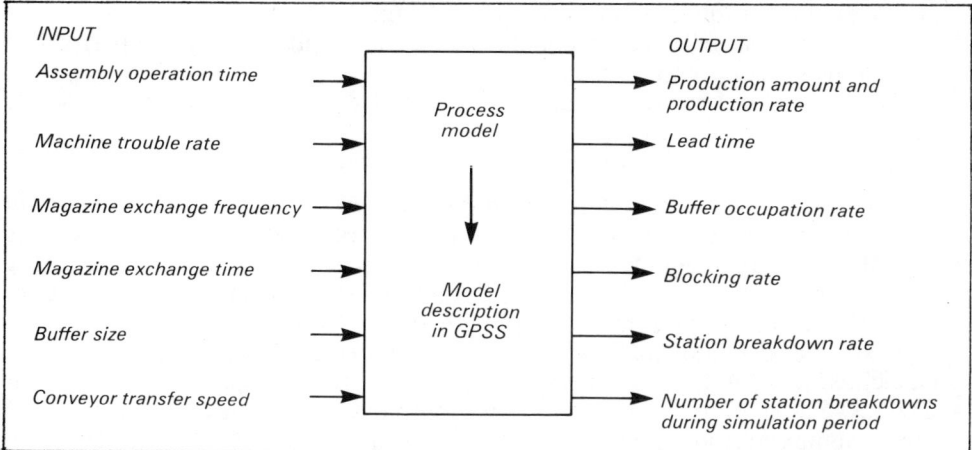

Fig. 6   Input and output data of the simulation program

The simulation is started from the initial condition which assumes that the entire assembly line is empty. Once the line operation reaches a steady condition, statistical data are calculated for 2 hours of operation. The steady condition is judged from the number of works in process. According to the simulation, the assembly line reaches steady conditions some 2 hours after the start of operations. In order to verify the accuracy of the simulation, tape-recorder mechanisms were examined by this simulator. Simulation results provided a line tact time of 8.5 sec, which is close to the actual tact time of 9.0 sec and the validity of this simulator was verified.

After several attempts at trial simulation for the automatic assembly line for VTR mechanisms, buffer size distribution was decided. The final simulation result is shown in Fig. 7. Expected line tact time is 7.5 sec when both minor and serious breakdowns are considered. By taking account of the influence of the difference between the calculated line tact time and the actual tact time of the tape-recorder mechanism assembly line, the tact time of the VTR mechanism assembly line was expected to be 7.9 sec. Thus, the planned assembly line was expected to achieve the aimed performance. After the line was developed, the simulation accuracy was again examined. The calculated line tact time based on the actual data is very close to the actual one.

Based on this simulation result, assembly machine specifications as well as the assembly line structure was determined. The calculation time of this simulation is approximately 80 min by Hitachi M160-II computer.

Automatic assembly line for VTR mechanisms

A mass production assembly line often produces a relatively small number of standardised models with minor variations introduced for a certain period, such as one or two years, and then it is modified to accept new models. Therefore, the flexibility required for mass-production assembly lines should be adaptable to these model changes, layout changes, and partial parts changes. Accordingly, quick, easy change of parts supply and assembly operations are required. In addition, the facilities are required to be re-used with minor modification and be lined up quickly. Considering this, automatic mass production assembly lines should be designed to satisfy the following:

60 PROGRAMMABLE ASSEMBLY

○ Individual assembly machines should be simple.
○ Reprogramming of each station should be easy, and it should be performed for all the stations simultaneously.
○ Maintenance and repair should be performed without influencing the whole line operation.
○ The parts supply system should be easily changed.

The layout of the automatic assembly line for VTR mechanisms is shown in Fig. 8. The line consists of 86 modularised assembly stations, 52 of which are dedicated assembly machines such as pick-and-place units, coil spring fitting machine, rubber belt fitting machine, screwdriving machines, and so on. In addition there are 11 assembly robots with three degrees of freedom for complicated shaped, delicate and precise subassemblies or parts which require a fairly sophisticated assembly movement, such parts being supplied by using flat magazines. Subassemblies which require complicated movements are attached manually at nine stations. Chassis, assembly parts and subassemblies are contained in flat magazines or bins, and supplied by automated guided vehicles from automated warehouses to each station. Production control is carried out by Hitachi minicomputer L-320 backed up by the process H-08 control computer. The mean tact time of the line achieved less than the aimed 8 sec.

	Simulation (planning)	Actual production - result*	Simulation (based on the actual data)*
Average machine cycle time	4.25sec	5.3sec	5.2sec
Average machine breakdown rate	2.2%	0.3%	0.3%
Line tact time	7.5sec	8.0sec	7.9sec

* Data 6 months after the line operation began

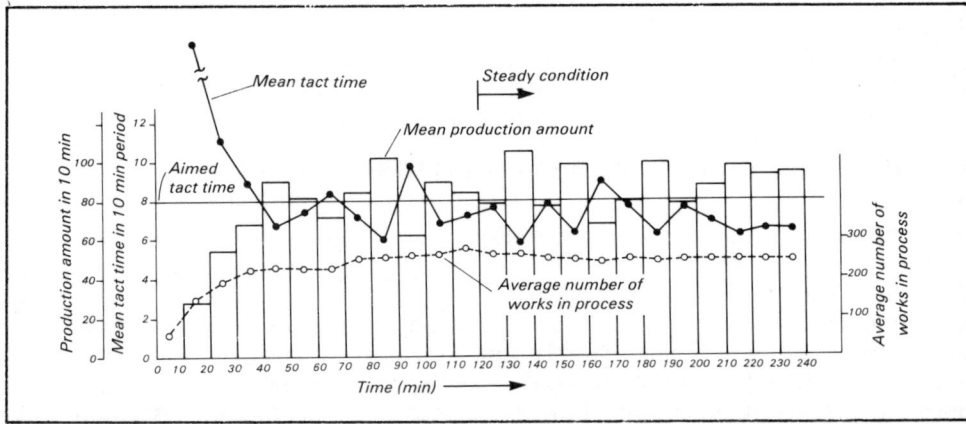

Fig. 7 Simulation result

AUTOMATIC ASSEMBLY LINE FOR VTR MECHANISMS   61

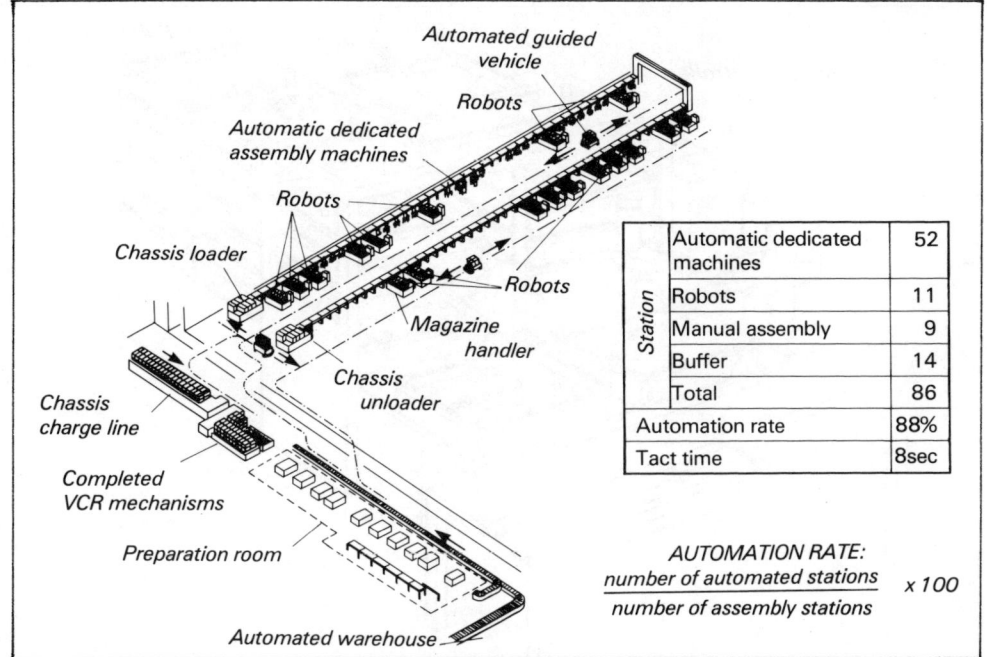

Fig. 8   Automatic assembly line for VTR mechanisms

Fig. 9 shows the construction of the assembly stations. Standardisation of the facilities are intended by the block-build type construction. These standardised units can be used even if major model changes occur. Robots and basic machines are especially adaptable for model changes following program changes and machine adjustment.

Fig. 10 shows the robot assembly station. As the assembly speed of the robot is not as fast as that of the dedicated assembly machines, the robot is placed so that the assembly path is the shortest. As a result, the robot station cycle time is shortened to 4.5–5 sec. Room is available for emergency manual operations,

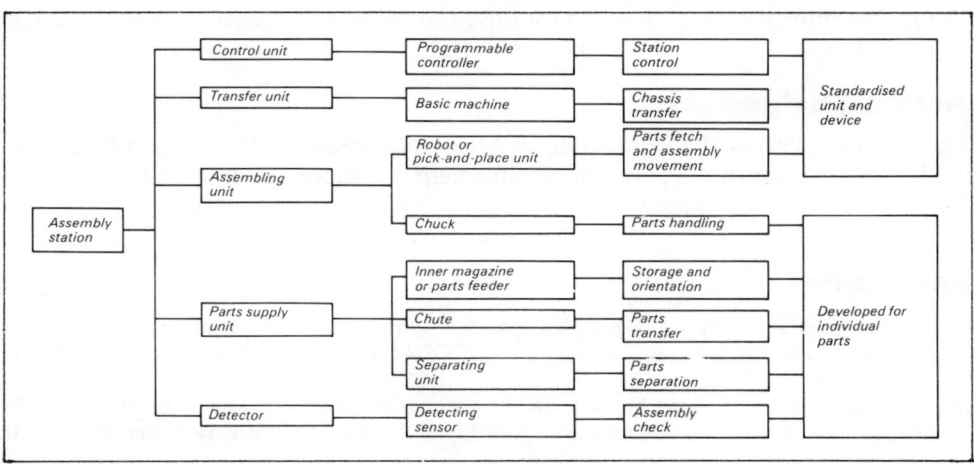

Fig. 9   Composition of the block-build assembly station

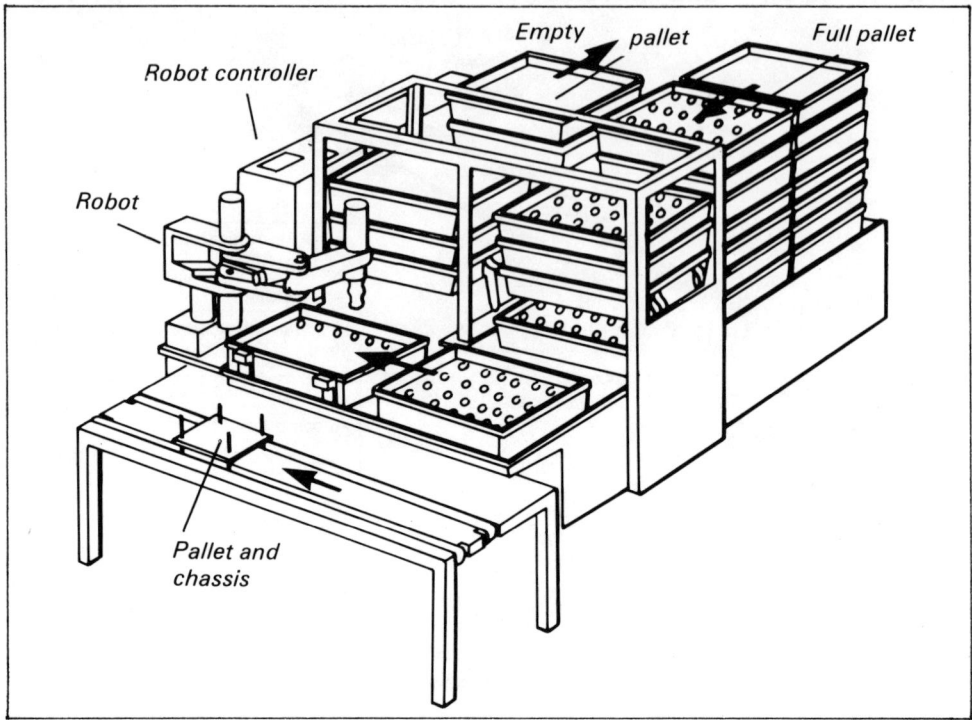

Fig. 10 Robot assembly station

such as a serious breakdown. At some stations, a palletising system is used to simplify the teaching operations. Protection cover and safety switches are equipped for operator protection purposes.

Thus a fairly flexible automatic assembly line was developed for VTR mechanisms. The results of this development are summarised as follows:

☐ Assembly labour was reduced by 83%, that is reduced by 150 workers.
☐ Area of assembly shop was reduced to one-third.
☐ Quality improvement; the need for repair of faulty final assemblies was reduced by 80%.

Acknowledgement

The authors express their gratitude to Mr. J. Nakazato of Production Engineering Research Laboratory, Hitachi, for his help in the computer simulation.

References

[1] H. Awane et al. Computer aided planning of a fully automated assembly system for tape-recorder mechanisms. In, Proceedings of AUTOFACT III, 9–12 November 1981, Detroit, USA, pp. 487–497.

[2] S. Hashizume et al. Development of an automatic assembly system for tape-recorder mechanisms, Research and Development in Japan Award the Okochi Memorial Prize 1980, Okochi Memorial Foundation, pp. 52–56.

MOTOR FAN FLEXIBLE AUTOMATIC MANUFACTURING SYSTEM

I. Del Gaudio and S. Del Sarto,
Olivetti Controllo Numerico-Divisione Robotica, Italy

First presented at the 4th International Conference on Assembly Automation,
11–13 October 1983, Tokyo, Japan

Abstract

Only a few flexible automatic manufacturing systems have been realised in practice because of the high requirement of resources in terms of people and finance. This paper describes a very innovative application of a standard SIGMA multi-arm robot performing the assembly (10 operations) and testing (10 operations) of over 15 different types of motor fans, at a rate of one assembly every nine seconds, by means of six cooperating arms (18 asynchronous, contemporary, coordinated axes), a power-free conveyor, standard grippers, etc. This application opens a new face in automatic assembly because of the high modularity and flexibility of the basic SIGMA system.

Introduction

One of the most important experiments on flexible assembly systems was carried out by James Nevins and Daniel Whitney in performing the assembly, by means of a four degrees of freedom cartesian arm (Fig. 1), of 17 parts of an ac generator for Ford cars[1,2,3]. The assembly was performed in 2 minutes and 42 seconds, but the opinion of the authors was that by improving the assembly tools a time of 1 minute and 5 seconds was attainable.

Another consideration was that from the standpoint of the pay back and profitability, the above mentioned solution was acceptable for low volumes (up to 50,000 pieces per year). For higher production rates, in order to improve these figures a multi-arm configuration system would be more suitable. In this case, each arm should be tooled to assemble a proportion of the total parts distributed to the facility in several positions. This system was able to show the possibility of performing the assembly of different types of products using a general purpose arm, equipped with specific tooling designed in such a way to allow slightly different parts to be accepted.

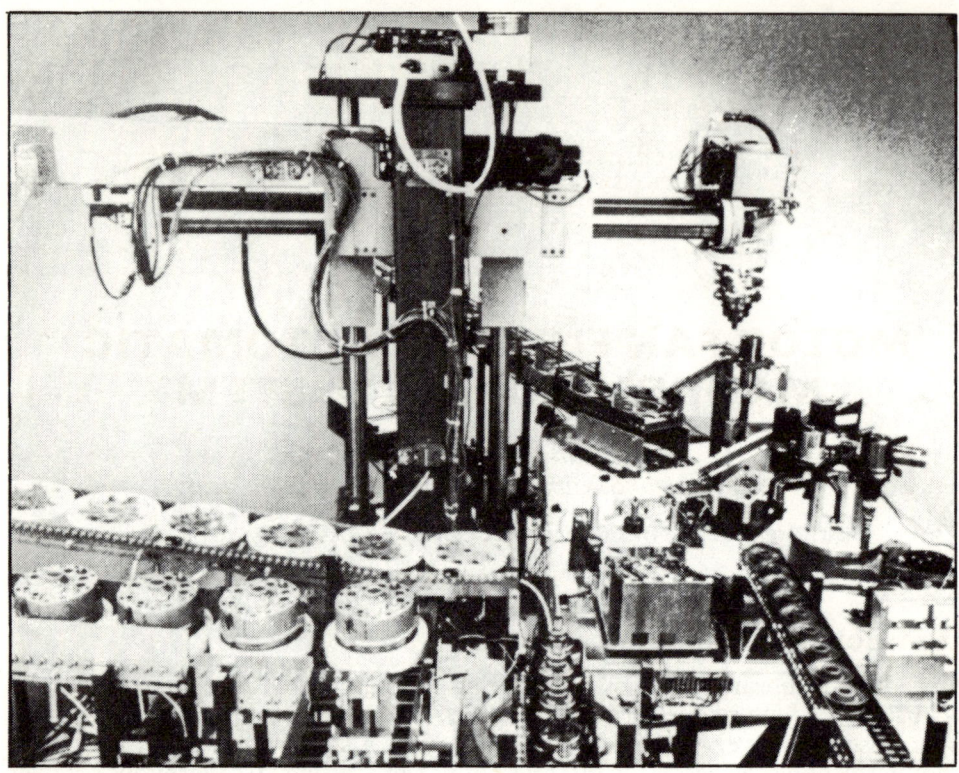

Fig. 1   The Nevins and Whitney experiment

No mention was made in this paper about the cost ratio between the general purpose part (arm) and the specific tooling, but we can assume that the latter will cost more than double the arm.

At the time of this experiment, the environment, i.e. the industrial scene, was very much different to the present, as were the automation requirements. In fact nowadays everyone is looking for total integration and automation within the factory, and so the main target is to build an unmanned factory.

When this system was conceived, the aptitude of mechanisation and manufacturing engineers, was orientated to identify the most labour intensive operations on the assembly lines and study, according to a plan of allowed investments, the most profitable operation to reduce the labour content of these operations. At that time, and still in some cases, it was possible to have manual operations and dedicated machines on the same line, so an experiment, like the one previously described, was mainly orientated to demonstrate the availability of flexible assembly means to cover the manual operations where hard automation was not applicable. The major interest now is shifting towards flexible manufacturing systems (FMS), in order to implement many additional advantages, largely overcoming the narrow standpoint of manpower reduction.

The concept of flexibility

Before showing some recent examples of highly flexible automation of assembly lines it is better to strengthen some concepts in order to face the problem from a well-grounded point of observation.

Everyone knows the dictionary meaning of the word 'flexible' – the problems will arise when attempting to apply the definition to manufacturing systems. In fact there is no way to measure the flexibility in a quantified way; only a rough evaluation can be performed.

If a flexible manufacturing system is designed which is able to perform the assembly of a defined product in a number (N) of different styles, it is necessary to explain if the system can pass from one style to another in zero time or if it is necessary to change a part of the tooling. This aspect can be easily quantified because the time necessary to retool the system, if this is the case, can be evaluated and then a percent 'down-time' figure calculated. Obviously this is not an absolute value, because it depends on the mix of production of the different styles since the tool change time is fixed[4,5]. Although after boundary limits are fixed it is possible to compare the flexibility of different FMS by comparing the percentage downtime due to tool change, this concept is similar to a comparison of blanking or moulding presses with die or mould changing.

At this point another question arises: how flexible will the system be if we want to assemble another product style for which the line has *not* been designed? The possibilities can vary from one case to another; in the worst case it is not practicably possible, in many cases there *will* be a solution and a related incremental cost. This concept gives us the key for the evaluation of performances and the degree of flexible manufacturing systems.

A flexible manufacturing system can be composed of general purpose components or dedicated components. If the system is made up of general purpose components, the value analysis of each component's function could be performed by attributing a value proportional to the function contribution to each component, if the pay back of the whole system is satisfactory all components will have the same property[6]. But in the majority of cases with general purpose components, specific tools and devices are used. In such cases specific value analysis must be carried out for all important components. Going back to the FMS evaluation it is desirable to develop a cost – performance diagram starting from the general purpose part and progressively adding the terms related to the additional function (Fig. 2). From such a diagram it is possible to evaluate the contributing parts and decide whether or not they are suitable to perform certain functions. Clearly the target to be pursued is to increase the performances, in

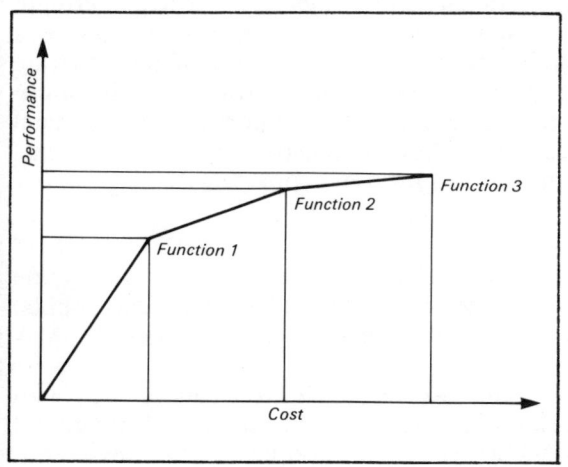

Fig. 2 Cost – performance diagram (arbitrary units)

Fig. 3   General view of the assembly line

terms of flexibility of the general purpose part in such a way that the cost necessary to increase performance through specific tooling is reduced.

To explain the concept further let us compare two different solutions of a defined problem. Suppose that in order to assemble a product it is possible to use either a single three-arm machine or three arms as stand-alone. To simplify the problem assume that the specific tooling is exactly the same – the second version will require additional parts to perform the same function. First, it is necessary to add a network of connections for the different arms which are already built-in in the multi-arm machine. Furthermore it is necessary to have a computer or a PLC programmable logic controller to coordinate the activity of the separate arms, whereas this function is already performed by the control unit of the multi-arm system. Also, it is necessary to design and build support and mechanical links for tooling and to guarantee the reciprocal position of the arms; this problem is made harder if the system is served by conveyors.

Motor fan assembly

An evident demonstration of the concept can be visualised by comparing some recent realisations of assembly lines obtained by putting together various robots, arms, etc. to the following example of a flexible assembly line using two SIGMA robots cooperating in-line[7].

The integrated line was designed using two basic SIGMA robots each with three arms and nine degrees of freedom. The layout is of the 'in-line' type, and the two robots are connected by standard conveyor modules of the power-and-

Fig. 4 The parts to be assembled

free type, specially developed by Olivetti Controllo Numerico to reduce the amount of specific design for tooling. The same concept was followed for the side conveyors necessary to connect the line to the rest of the factory, and for grippers. In this case only two basic grippers were used to obtain the different gripping functions.

The integrated line was designed in order to perform the assembly and final testing of various types of motor fans for automobile engine cooling (Figs. 3, 4) at a rate of one finished part every nine seconds. The layout of the line is shown in Fig. 5.

The electric motor body is placed on the main conveyor pallet by means of a pick-and-place unit from a dedicated permanent magnet assembly station. This pallet is designed to accept all the different parts to be assembled (Fig. 6). The pallet, running by friction along the conveyor, is clamped in position where needed for assembly operations.

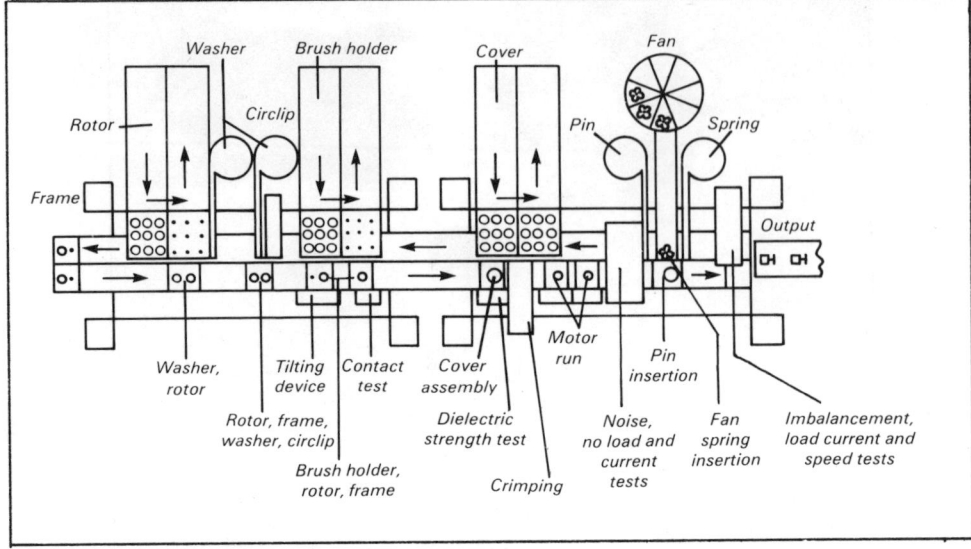

Fig. 5 Assembly line layout

Fig. 6 Main conveyor pallet

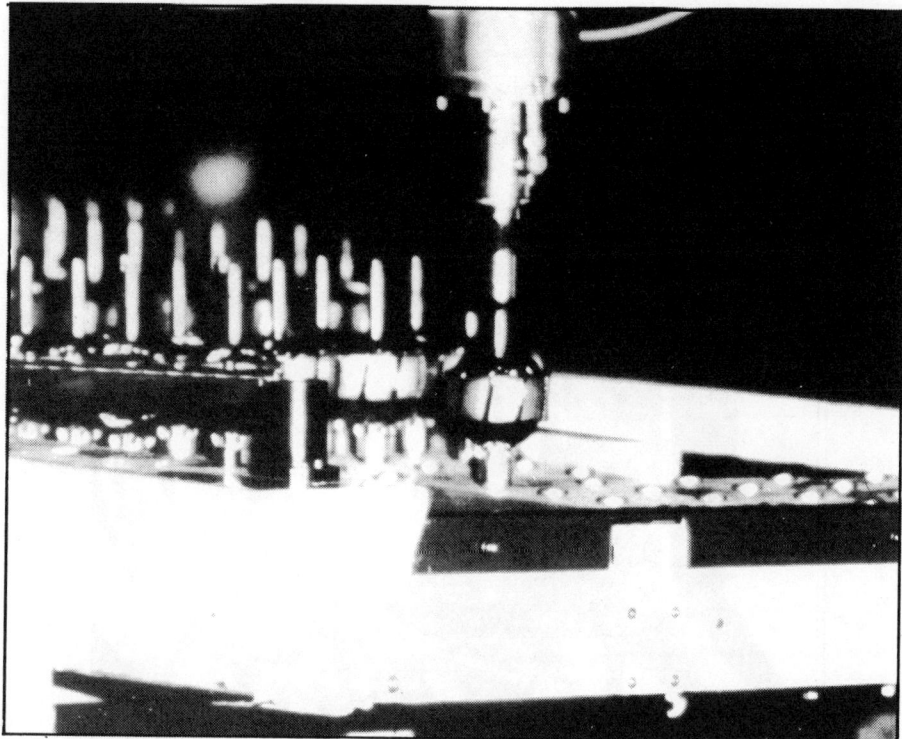

Fig. 7   Side conveyor rotor pallet

Functions of the first robot

At the first assembly station the robot's first arm takes a washer from the terminal of a hopper feeder by means of a standard two-function gripper and then inserts this washer into the shaft of the motor rotor. The rotors are taken to this assembly station on a pallet which runs along a transverse conveyor hanging on the main conveyor (Fig. 7). After the gripper has inserted the washer, it places the rotor shaft on to the motor body pallet. Gripper sensors monitor the non-conformity or the absence of a component or the wrong coupling by sending related messages to the control unit. When this operation has been completed, the pallet is released and moves asynchronously towards the next assembly station.

It is important to notice how the arm periodically performs an additional task. In fact, when the last rotor is on the transverse conveyor pallet, the arm moves the pallet from the ingoing to the outgoing branch. When the empty pallet is shifted a full one takes its place.

The second arm equipped with a standardised three-function gripper (Fig. 8) picks up a washer and a circlip from their related feeders, then moves to the main conveyor pallet, picks up the motor frame and inserts it on to the rotor shaft lying on the same pallet. A vertical transducer measures the reciprocal position of the motor frame and the rotor and calculates the position of the recess to insert the circlip; the latter is inserted by a specific finger of the gripper. All the above-mentioned functions are monitored through on–off or linear transducers.

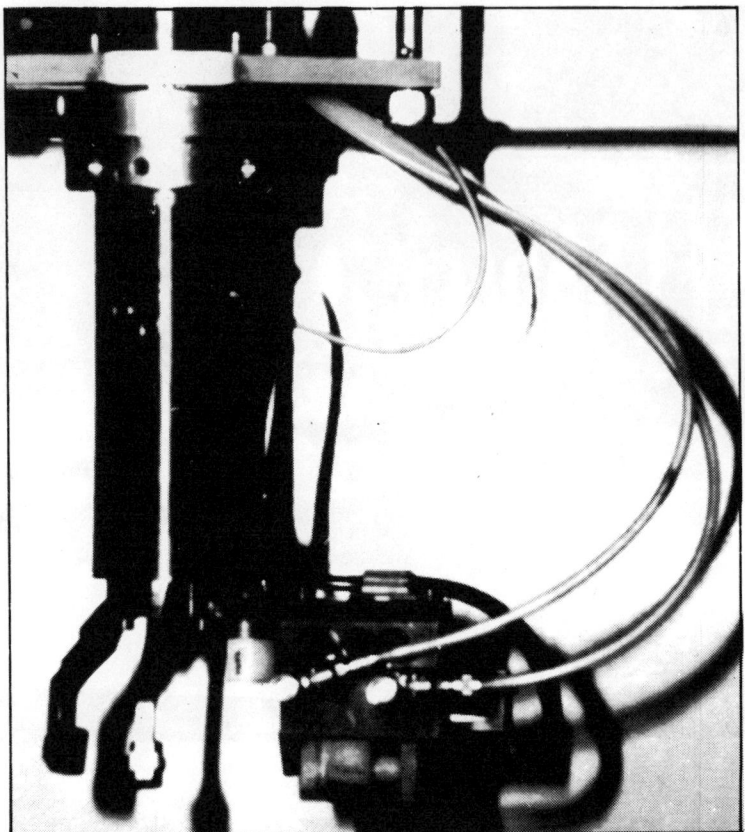

Fig. 8   Multi-function gripper

Fig. 9   Side conveyor brush holder pallet

To perform the next operation it is necessary to reverse the position of the sub-assembly. This is performed by a dedicated tilting device mounted on the side of the machine, which is designed to accept all the possible types of components to be assembled (as are all the specific tools mounted on the machine). The third arm inserts the brush holder which is taken from a side conveyor pallet (Fig. 9). After insertion, a mechanical strength test of the electrical connections is performed by pushing to a specified force and monitoring the related displacement. If this test gives a negative result the related brush holder is discarded and a new one assembled.

Functions of the second robot
At the first station of the second robot, the assembly of the motor cover is performed by taking it from a side conveyor pallet (Fig. 10). Then before crimping the cover, a high voltage dielectric strength test is performed. If this test

Fig. 10   Side conveyor cover pallet

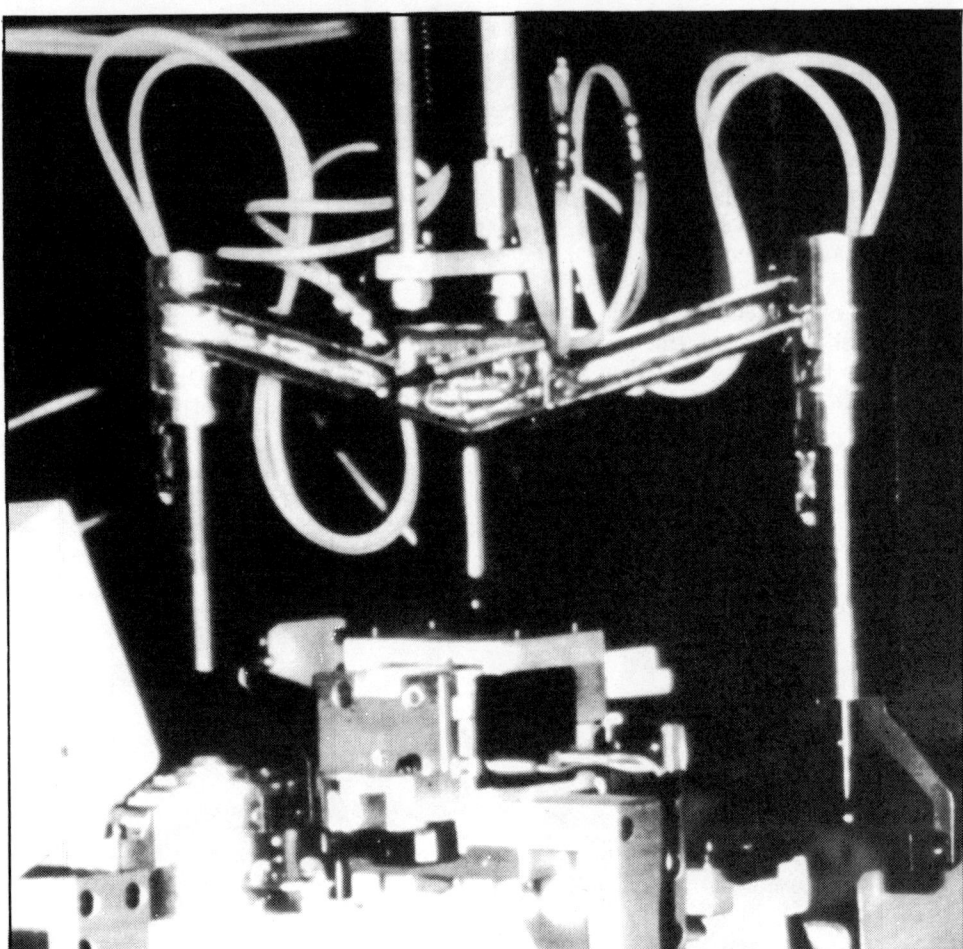

Fig. 11 The third arm has two grippers

is passed, the motor is crimped; otherwise the parts are left loose to allow disassembly for scrap recovery, if required.

The next pallet station is dedicated to 'running up' the motor for sufficient time to eliminate 'tightness'. The next station, also performed by a dedicated device, is for the measurement of motor noise and load current. To do this, a fixed point displacement gripper brings the motor into an acoustically insulated cabinet, feeding it electrical power near to the noise measuring unit.

If one of the above tests are not passed, the second arm will discard the faulty motor; otherwise it will perform the insertion of a pin in a transversal hole on the motor shaft. This pin is needed for 'driving' the plastic fan. Pin insertion is performed by bringing the motor to the pin inserter and making the shaft rotate until the pin finds the hole. The same arm then goes to the fan-feeding station and pushes the shaft into the fan recess.

The third arm has two grippers (Fig. 11) and inserts a spring to fix the fan and presents the final assembly to the testing station where the following tests are performed:

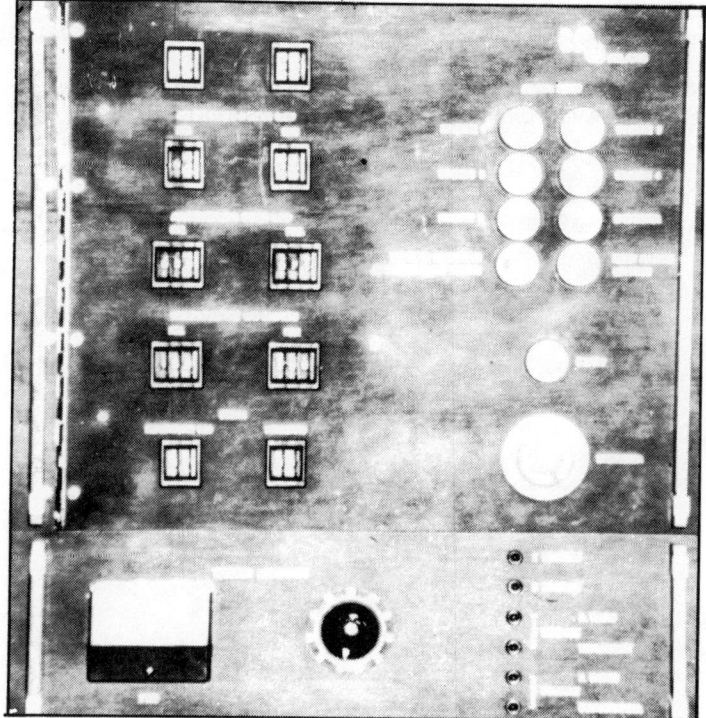

Fig. 12   Control cabinet for performance testing

☐ speed under load,
☐ mechanical imbalancement, and
☐ current under load.

All the presence and position tests are performed by the central control unit of the two robots, whereas performance tests of the product itself (i.e. current, noise, electrical strength, speed, imbalancement, etc.) are carried out by a specific control cabinet integrated with the robot control unit (Fig. 12). It is possible to set the performance ranges on the control cabinet according to the particular model or style to be produced.

References

[1] Nevins, J. L. and Whitney, D. E. Categorization and Status of Assembly Research, CSDL Report No. P-330. Charles Stark Draper Laboratory, 1976.
[2] Lynch, P. M. Economic-Technological Modelling and Design Criteria for Programmable Assembly Machines, CSDL Report No. T-625. Charles Stark Draper Laboratory, 1976.
[3] Watson, P. C. A Multidimensional System Analysis of the Assembly Process as Performed by a Manipulator, CSDL Report No. 0-364. Charles Stark Draper Laboratory, 1976.
[4] Chamberlain, R. G. Manufacturing systems and productivity. Manufacturing Engineering (July 1980). Society of Manufacturing Engineers, 1980.
[5] Del Gaudio, I., Di Maio, F. and La Monaca, E. Matching the assembly robot with the factory; 5 years experience in tooling Sigma. In, 10th International Symposium on Industrial Robots, Milan 1980.

[6] De Falco, R. and Del Gaudio, I. Comparative analysis of quality and production cost at varying automation levels. Italian Machinery Equipment (March 1976). Etas Kompass, 1976.
[7] Del Gaudio, I. Sensoriality aspects in robotized assembly process. In, Proc. 11th International Symposium on Industrial Robots, Tokyo, 1981.

AUTOMATED ASSEMBLY IN THE ELECTRICAL INDUSTRY

B. Lotter, EGO Elektrogerätebau GmbH, West Germany

First published in *Assembly Automation* (February 1984)

Abstract

In the electrical industry 50 to 75% of the total production costs of a product are in assembly. This shows that the main focus for rationalisation should be in assembly. This paper illustrates planning methods for rationalisation, describes a completed assembly line and outlines the economies obtained.

Introduction

A thermal switch, as shown in Fig. 1, consisting of 37 component parts is to be assembled at a net output of 1,000 units per hour. In order to attain economic efficiency the operating efficiency of the assembly line laid down must be at least 80%. Based on these assumptions the individual cycle time for the assembly line is calculated as follows:

$$^tT = \frac{t \times \eta}{P} = \frac{3600 \times 0.8}{1000} = 2.88 \text{ seconds}$$

where P is the production rate.

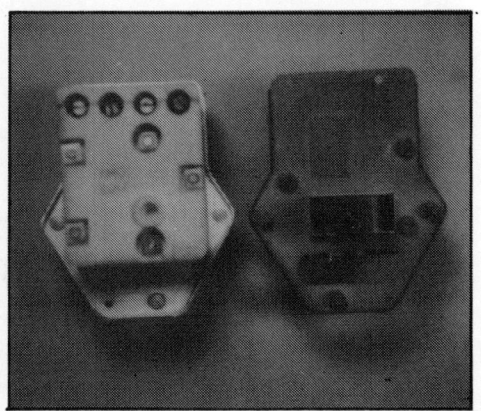 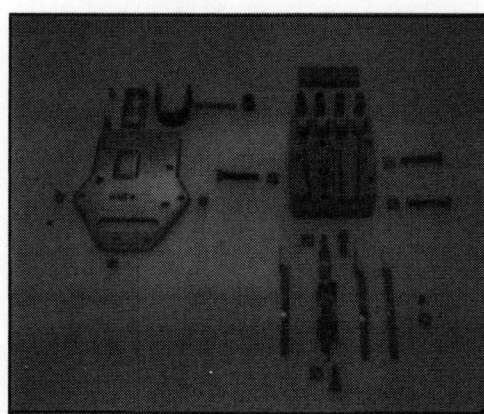

Fig. 1   Thermal switch: completed assembly and exploded view

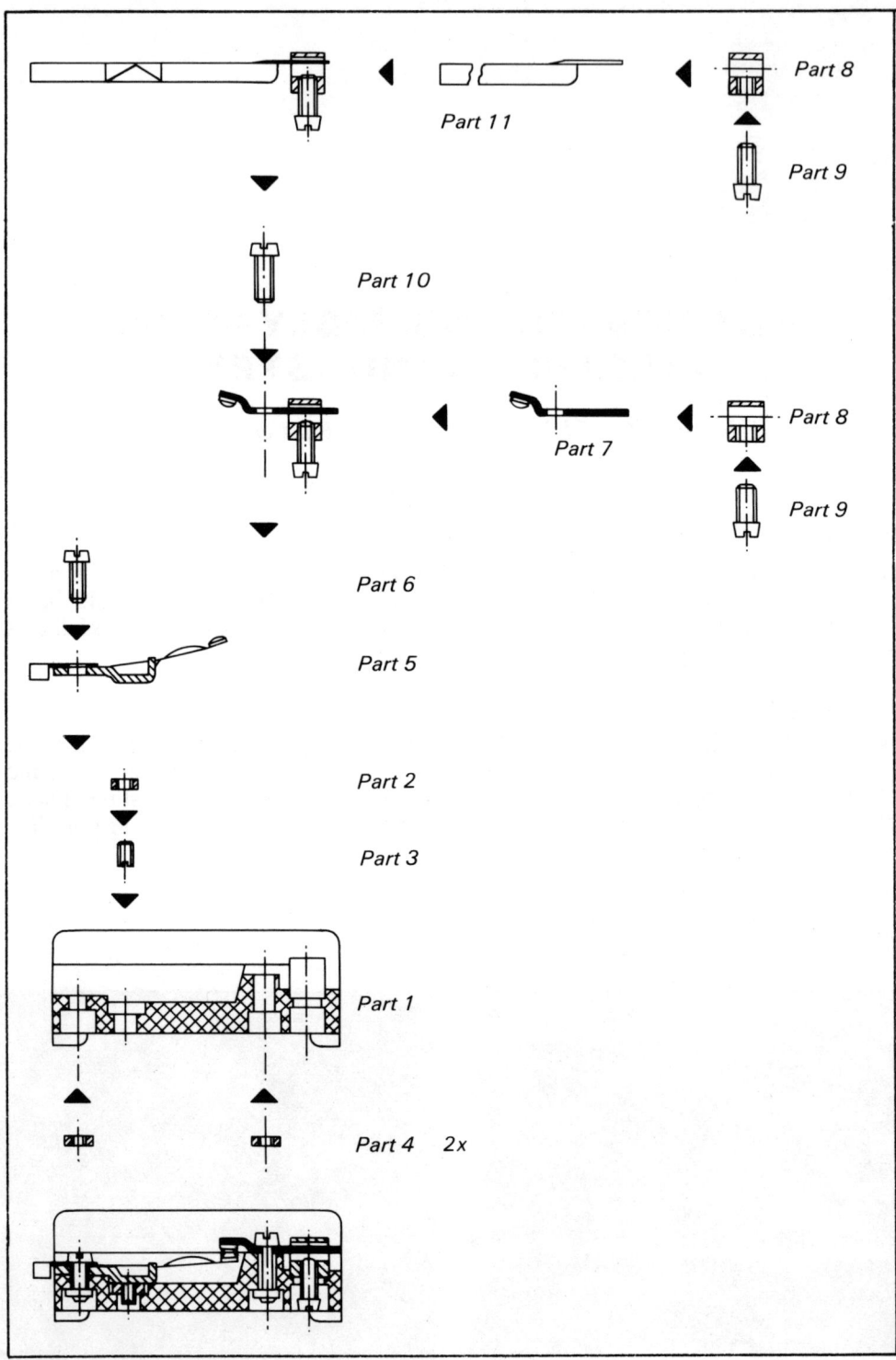

Fig. 2 Base sub-assembly

AUTOMATED ASSEMBLY IN THE ELECTRICAL INDUSTRY 77

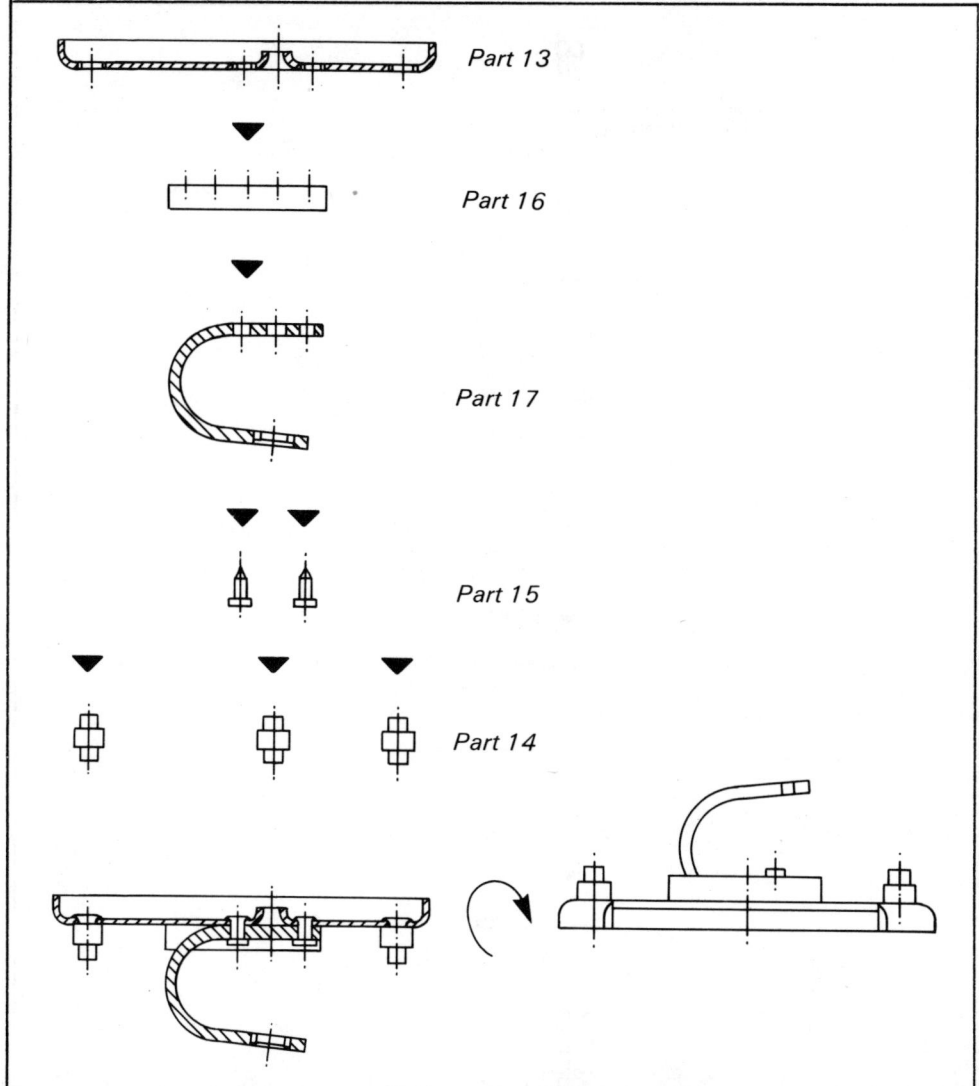

Fig. 3 Cover sub-assembly

Planning method

To determine optimum production and the most economic method a 4-stage planning method was used.

Stage 1 – ABC-analysis extended to assembly

This analysis determines the cost of a component until it reaches its operational purpose, i.e. assembly into the product. From this analysis the requirements of the quality of the component parts emerges both in their delivered condition and joining capability. In the thermal switch the result of this analysis was that the base (part 1 in Fig. 2), produced in a ceramic material, must be handled manually. Also the microswitch element (part 5) and connector strips (part 11)

78 PROGRAMMABLE ASSEMBLY

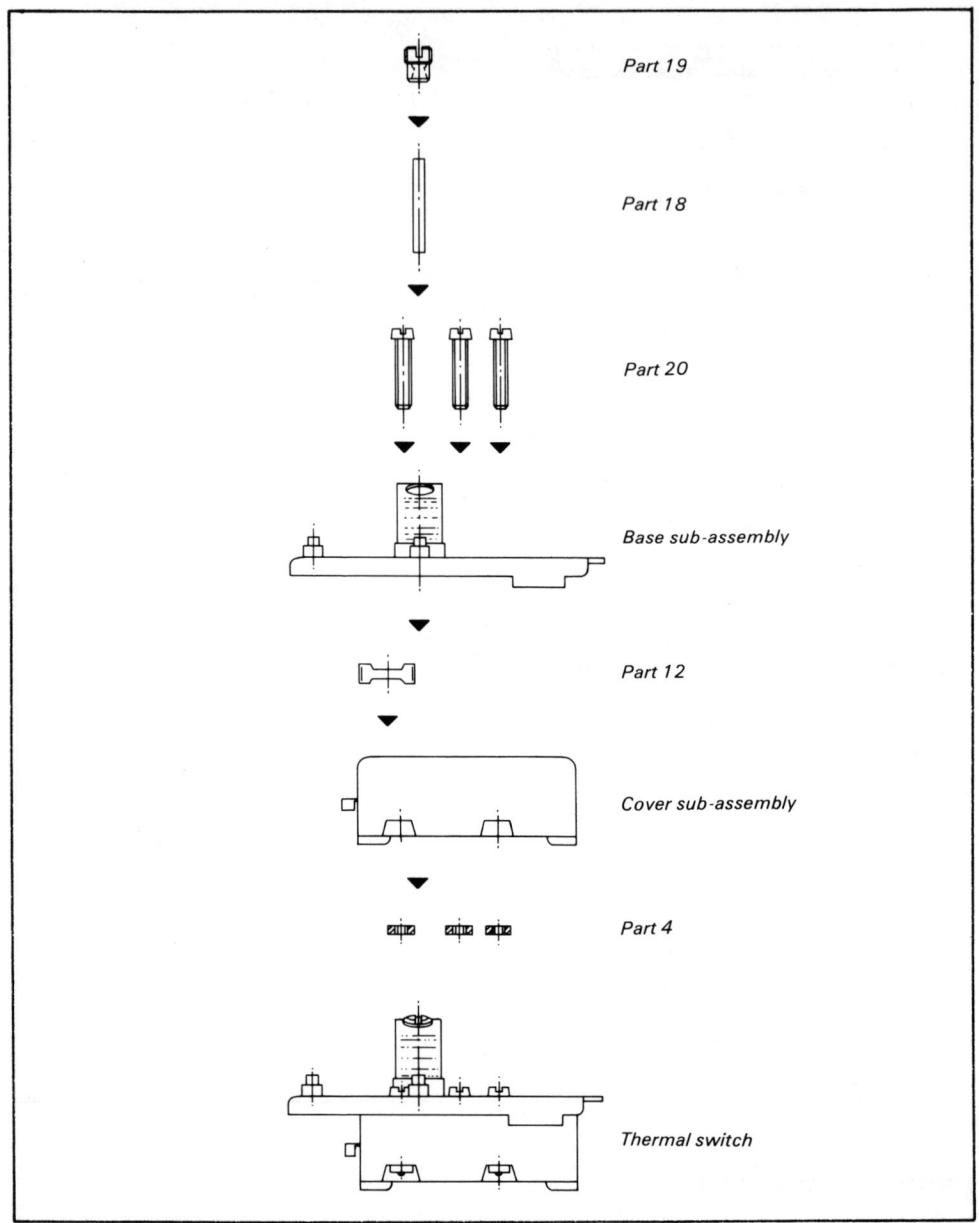

Fig. 4   Final assembly

cannot be fed automatically as they are either too sensitive or they cannot be sorted owing to their geometry. This means that these parts must be handled manually or, especially for the strips (part 11), a different production method must be found in order to achieve automatic feeding. This requires processing operations to be integrated in the assembly processes. The ABC-analysis also finds that the standard tolerances of common parts, such as screws and nuts, are not good enough to provide high availability at the automatic stations. All the other parts are capable of being sorted and joined.

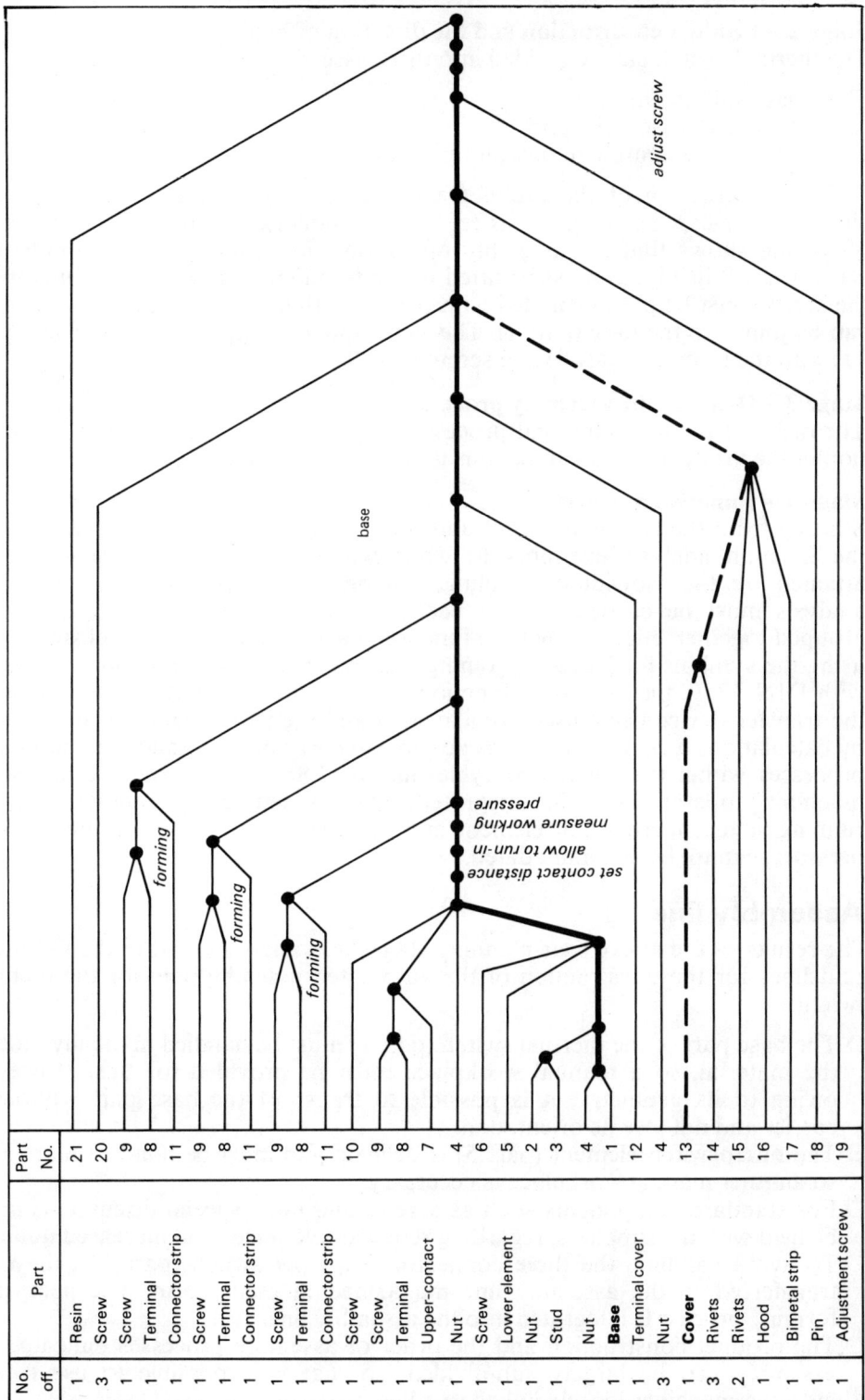

Fig. 5 Order of assembly processes – sequence of operations

Stage 2 – Product construction and the direction of joining
The thermal switch can be divided into three assembly products:

☐ the base sub-assembly,
☐ the cover sub-assembly, and
☐ the complete assembly of base, lever and a few additional parts.

The construction of the product and the resulting direction of joining are shown in Figs. 2–4. The analysis regarding product construction and direction of joining shows that pre-assembly operations are required. For example, the screw (part 9 in Fig. 2) must be fitted to the terminal (part 8); the terminal with the screw must be pre-assembled with part 7 so that the complete sub-assembly can be joined to the base (part 1). The same applies to the connector strips (part 11) with the terminal (part 8) and screw (part 9).

Stage 3 – Order of the assembly processs
The order of the assembly and processing operations is based on the construction of the product and the directions in which joining occurs (Fig. 5).

Stage 4 – Function analysis
Starting from the specified output and the resulting cycle time of 2.88 seconds, the function analysis examines to what extent this given cycle time can be attained by the individual handling, joining and machining operations. An analysis must be carried out for each component part. Some parts can be grouped together. Fig. 6 shows the functional analysis for the thermal switch. By using the symbols for handling, joining, and machining processes in accordance with DIN 3239, the individual functions of the switching and stopping times of the transfer device being used are allocated and the individual times determined by calculation. As can be seen, it is possible to carry out the handling and joining processes within the calculated cycle time of 2.88 seconds, but the necessary machining operations to be integrated, such as setting the contact distance, running-in the microswitch element and measuring and adjusting the working pressure, cannot be accommodated.

Assembly line
The results of the above four planning stages and analyses lead to the following guidelines for the construction of the automated assembly line for the thermal switch:

○ The base part of the thermal switch (part 1) must be handled manually, due to the material, so a manual workplace must be provided for this. However, owing to its geometry, it is possible to transport the base part without a carrier and not lose its orientation.
○ The microswitch element (part 5) is delicate and must be handled manually, so another manual workplace is necessary.
○ For standard components such as screws and nuts, special discussions must be held with the suppliers, regarding reduced tolerances and increased quality.
○ To avoid handling the three connector strips per switch (part 11), they are transferred to the assembly line magazined in blank form. The necessary forming process is integrated into the assembly line.
○ The product construction and the order of assembly processes indicate that assembly can be largely sub-divided, so that it is possible to use rotary indexing machines loosely linked to a line.

- The operating processes allocated to individual assembly machines must be kept as low as possible in order to obtain the highest overall availability of the individual machine.
- In order to increase and ensure the efficiency of each individual machine, each individual supply station is equipped with lead length control, shown schematically in Fig. 7. Experience shows that most faults occur during sorting of components in the vibratory conveyor. Similarly, a large number of the faults occur in the transfer between the latter and the electromagnetically driven outflow rails. Lead length control achieves a buffer effect after the conveyor by using suitable lengths of outflow rails. Faults within the vibratory conveyor and the transfer position to the outflow rails no longer directly affect the efficiency of the assembly machine, since enough good parts are buffered in the lead length to allow the removal of faults arising in the conveyor – in sufficient time so that idling does not occur in the lead length.
- Pre-assembly operations must be carried out at 'satellite' machines separate from the assembly machines, or integrated in the workpiece carrier of the individual assembly machines.
- The function analysis shows that the specified cycle time of 2.88 seconds is insufficient for a number of the working and testing processes. For those processes that take longer than 2.88 seconds duplexing must be arranged.
- Interlinking of the individual assembly machines is achieved by conveyor belts which serve as intermediate buffers between individual machines in the sense of lead length control, or lead length control of individual insertion stations.
- The cycle time of 2.88 seconds dictates rotary indexing units controlled by cams or Geneva-motions.
- Each assembly or workstation is followed by a test station with an immediate *go/no go* decision. The individual workpiece carriers are fitted with coding pins which indicate the result of the test. A single *no* signal does not shut down the machine – only three consecutive *no* signals will cause a stoppage. The cause of the breakdown is signalled simultaneously to the operator. Faulty or incomplete assemblies must be sorted out automatically at each individual machine so that only acceptable sub-assemblies are offered to the following machine.

The assembly line for the thermal switch was constructed on the basis of the above findings and guidelines. It consists of nine loosely linked assembly machines of the circular indexing type, as shown in Fig. 8.

Operation

The assembly line operates (refer to Fig. 9) as follows:

- *Machine 1* – assembles the lower microswitch element into the base. The base is loaded by the operator on to the input conveyor at a rate independent of the machine cycle. A stud and nut (parts 3 and 2), fastened together on a satellite rotary indexing machine are automatically placed into the base. A second operator places the lower microswitch element (part 5) into the base. A nut and screw (parts 4 and 6) fasten the element in place.
- *Machine 2* – fits the upper microswitch element. First, a second nut (part 4) is automatically loaded into the carrier (Fig. 9b) and then the base is placed over the nut locating on a fixed pin and an axially moveable pin (Fig. 9c) which also locates the nut. Another section of the carrier is designed to pre-assemble

82 PROGRAMMABLE ASSEMBLY

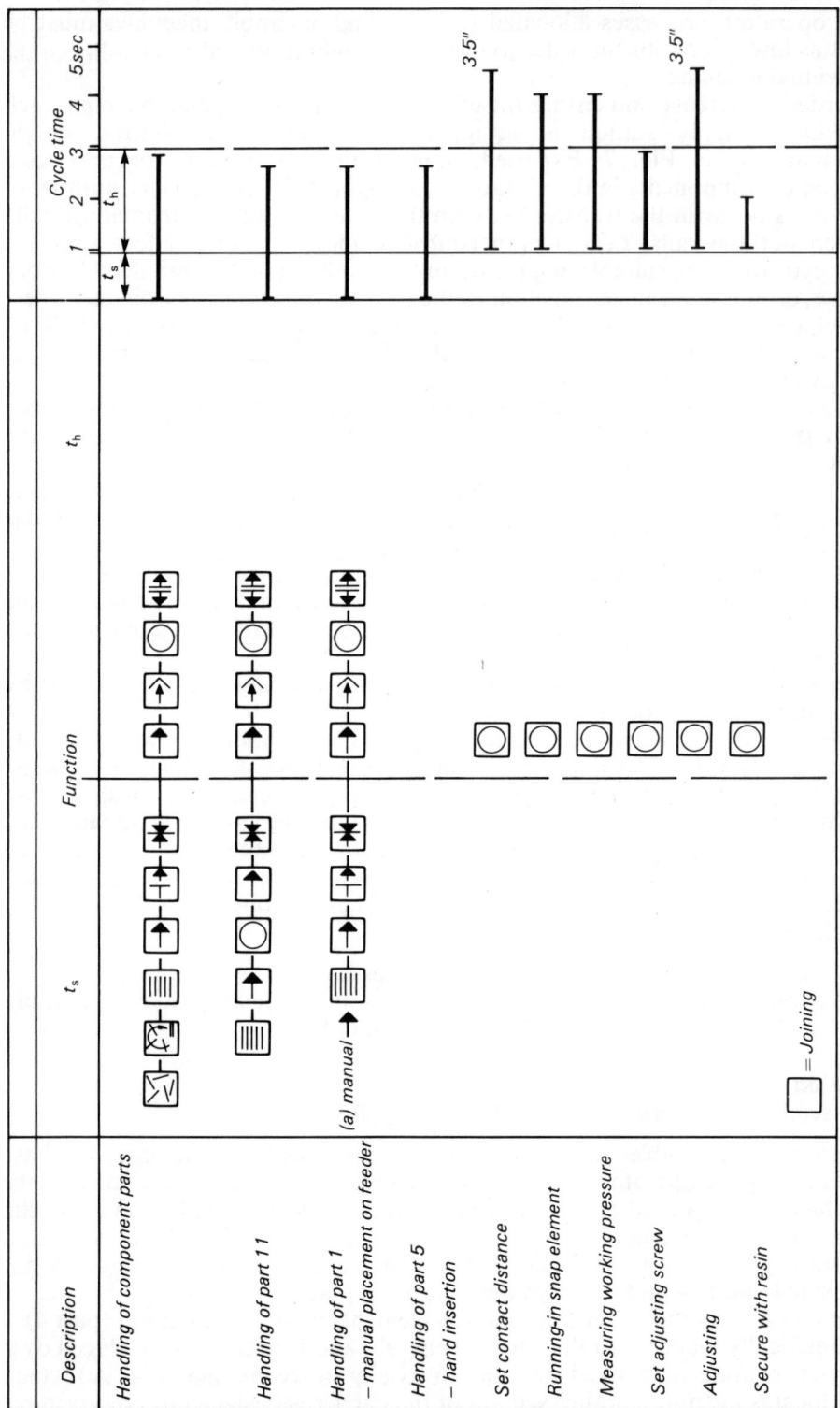

Fig. 6 Function analysis

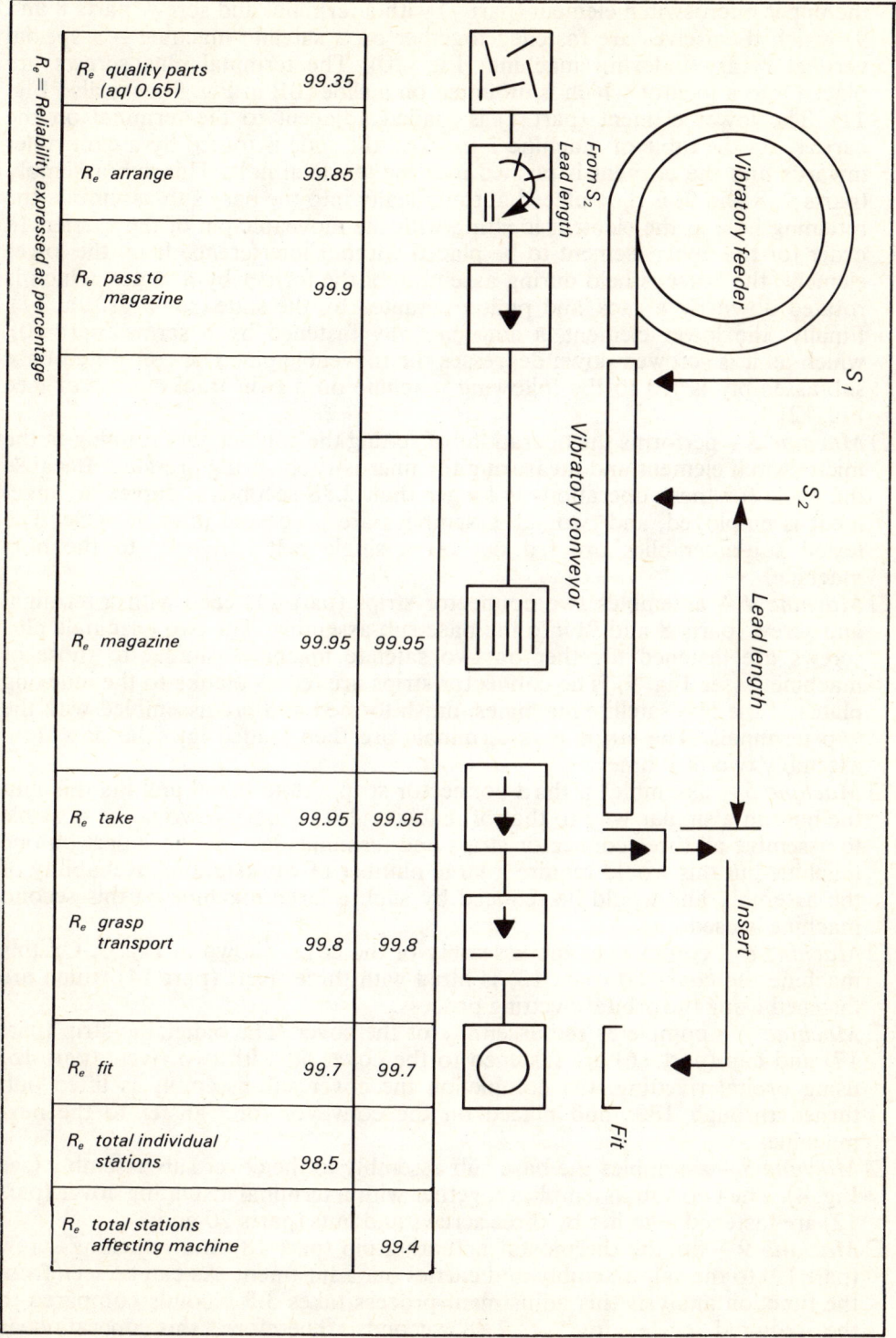

Fig. 7 Reliability of a feed-fit station with lead length (S_1–S_2 = signal generator)

the upper microswitch element (part 7) with a terminal and screw (parts 8 and 9) which themselves are fastened together on a satellite machine – a special vertical rotary indexing machine (Fig. 10). The terminal plus screws are placed into a locator which is mounted on a slide (BB in Fig. 9b, see also Fig. 11). The lower element (part 7) is loaded adjacent to the terminal on the carrier. As the table of machine 2 indexes, the slide is forced by a cam roller inwards and the element is pushed into the terminal hole. This sub-assembly (parts 7, 8 and 9) is then placed automatically into the base sub-assembly, the retaining hole in the element locating with the moveable pin of the carrier. In order for the upper element to be placed without interference from the lower element, the latter is held during assembly of the former by a finger, which is rotated down by a rack and pinion actuated by the slide (see Figs. 9a, 11). Finally, the lower element is automatically fastened by a screw (part 10), which as it is screwed down depresses the moveable pin. The completed base sub-assembly is fed to the following machine on a twin track conveyor (see Fig. 12).

☐ *Machine 3* – performs the operations of setting the contact gap, running-in the microswitch element and measuring the microswitch spring pressure. Because the cycle for these operations is longer than 2.88 seconds a duplex arrangement is employed, and two sub-assemblies are processed in each cycle. The tested sub-assemblies are fed out on a single belt conveyor to the next machine.

☐ *Machine 4* – assembles two connector strips (part 11) each with a terminal and screw (parts 8 and 9) into the base sub-assembly. The two terminals plus screws are fastened together on two satellite machines similar to those of machine 2 (see Fig. 9). The connector strips are fed as blanks to the indexing plates of the two satellite machines, finish formed and pre-assembled with the two terminals. The strips plus terminals are then loaded into the base sub-assembly two-at-a-time.

☐ *Machine 5* – assembles a third connector strip and terminal and fits this into the base in a similar way to that of machine 4. In theory it would be possible to assemble all three connector strips and terminals into the base part on one machine but this would require a large number of workstations. Reliability of the assembly line would be reduced by such a large machine so this second machine is used.

☐ *Machine 6* – commences the assembly of the cover shown in Fig. 3. On this machine the cover lid (part 13) is fitted with three rivets (part 14) which are fastened using the orbital rivetting process.

☐ *Machine 7* – completes the assembly of the cover. The bimetallic strip (part 17) and cap (part 16) are fastened to the cover lid with two rivets (part 15) using orbital rivetting. On completion the cover sub-assembly is lifted out, turned through 180° and placed on the conveyor for transfer to the next machine.

☐ *Machine 8* – assembles the base sub-assembly to the cover sub-assembly (see Fig. 4). The two sub-assemblies together with a terminal insulating cover (part 12) are fastened together by three screws and nuts (parts 20 and 4).

☐ *Machine 9* – fits the thermostat actuating pin (part 18) and adjusting screw (part 19) to the sub-assembly and carries out adjustment. As can be seen from the function analysis this adjustment process takes 3.5 seconds compared to the required cycle time of 2.88 seconds. Therefore, this operation is duplicated. When the adjustment has been completed the adjusting screw (part 19) is secured by applying a resin and the finished assembly is automatically unloaded.

Economics

In order to assess an investment the economic efficiency of the available production methods has to be investigated. It is necessary to assign to each of the alternative methods the same relevant costs asssuming the same output and the same methods of calculation. A method using machine-hour rates, or alternatively a place cost investigation, is suitable for this task.

The investigation of the described assembly line for the thermal switch was based on the comparison of two production methods:

○ *Method 1* – manual assembly in accordance with MTM on individual assembly station with partial loose interlinking.
○ *Method 2* – mechanised assembly.

The costings for the comparison are itemised in Table 1.

Table 1   Costings

Operating costs
    planned quantity = 16,000/16h/day = 1,000 pieces/h.

calculable depreciation $(K_A) = \dfrac{\text{replacement value}}{\text{machine life } T_N \text{ (4 years)}}$

calculable interest $(K_Z) = \dfrac{\text{replacement value}}{2} \times \text{interest rate (10\%)}$

space cost $(K_R) = m^2 \times$ DM per annum (120)

energy cost $(K_E) =$ kWh $\times$ (DM 0.20) $\times T_N$
                    compressed air m$^3 \times$ (DM 0.08) $\times T_N$

maintenance cost $(K_I) = 10\%$ of replacement value

machine life $(T_N) = 220$ days $\times$ 16h $= 3,520$h

machine hour rate $(K_{MH}) = K_{MH} = \dfrac{K_A + K_Z + K_R + K_E + K_I}{T_N}$

Personnel expenses per hour

production personnel   = number of personnel × wage + 110% of incidental wages costs

non-productive personnel   = number of personnel × wage + 110% of incidental
(charge-hand, setter)   wage costs

allowances (shift) $= \dfrac{\text{shift allowances per day}}{16\text{h}} =$ average per hour

supervision (foreman) $= \dfrac{\text{wage + incidental wage cost}}{172.33\text{h}}$

The replacement value with a useful life of four years and an assumed inflation rate of 5% per annum is calculated as purchasing price × factor of 1.215

For manual assembly (method 1), planning in accordance with MTM on individual assembly places with partial loose interlinking gave a standard time of 3.9 minutes per unit. Assuming an average efficiency of 135% this gives an

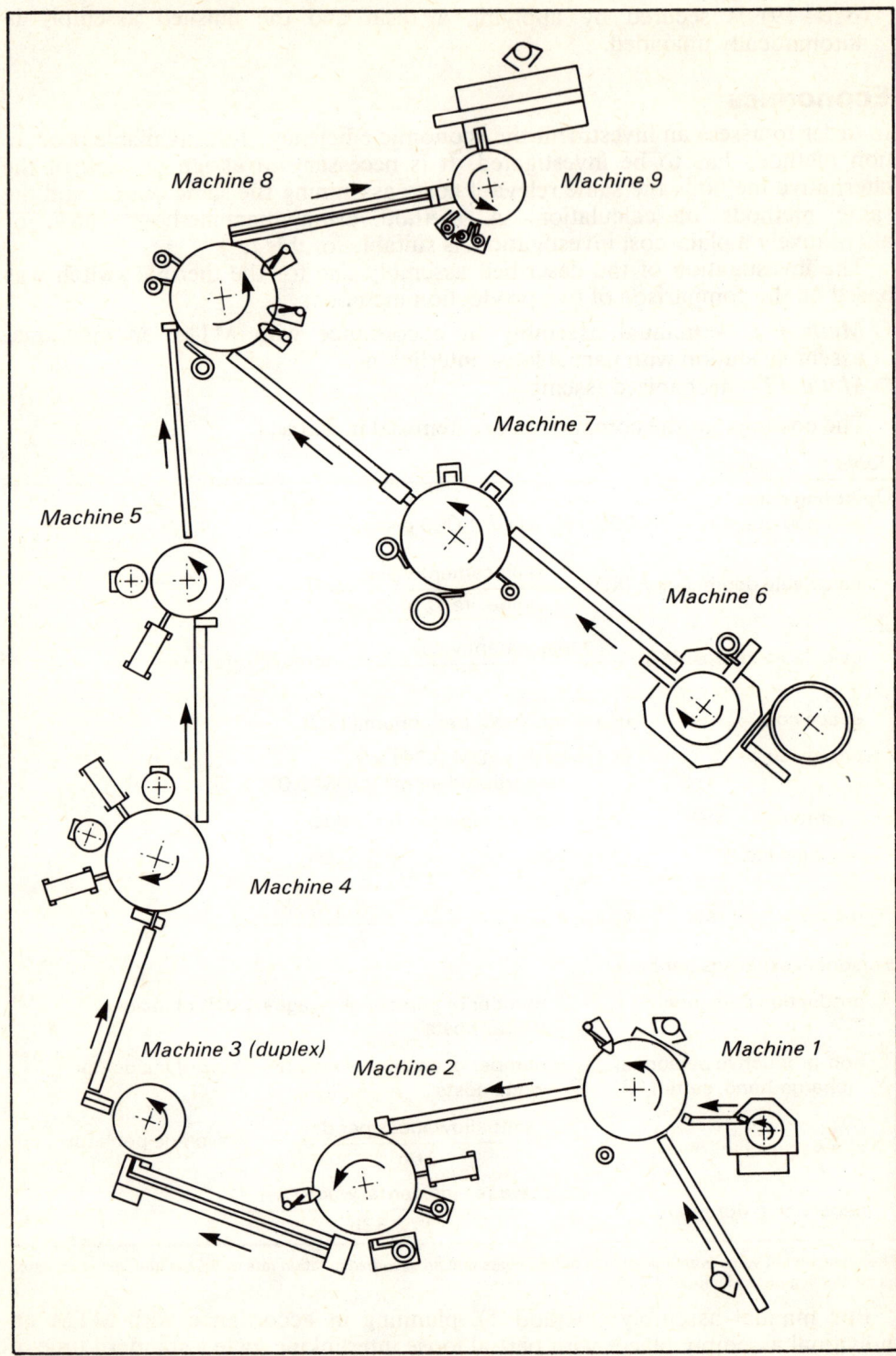

Fig. 8   Layout of assembly line

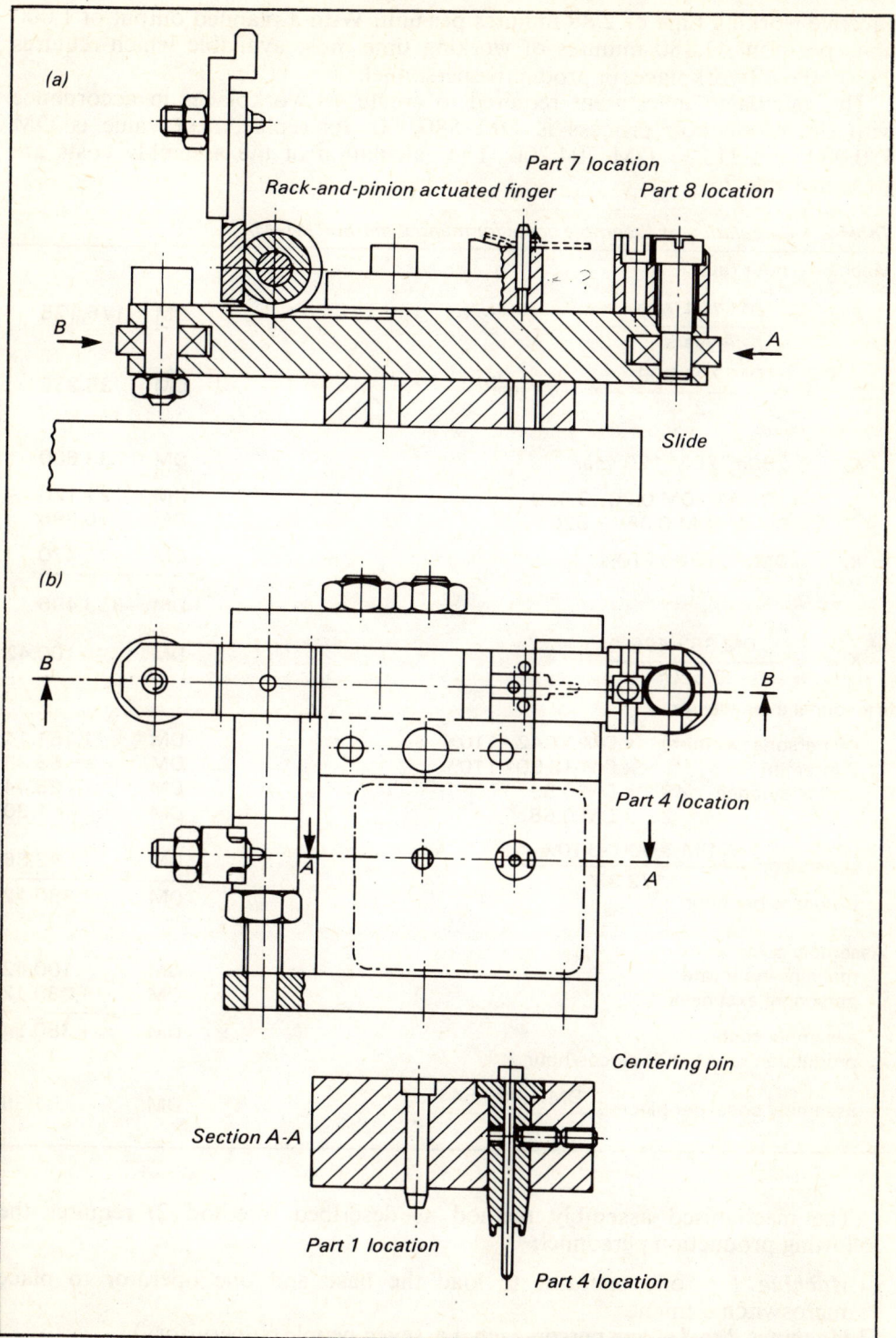

Fig. 9   Work carrier design for machine 2: (a) section B-B, (b) plan, (c) section A-A

effective working time of 2.88 minutes per unit. With a planned output of 1,000 units per hour, 2,880 minutes of working time must available which requires $2880/60 = 48$ workplaces or productive personnel.

The calculated investment required to create 48 workplaces in accordance with the production process is DM 580,000. Its replacement value is DM $580,000 \times 1.215 = $ DM 704,700. The calculation of the assembly costs are given in Table 2.

Table 2 Calculation of assembly costs for manual method (MTM)

Machine – hour rate

$$K_A = \frac{\text{DM } 704,700}{4 \text{ years}} \qquad\qquad \text{DM} \quad 176,175$$

$$K_Z = \frac{\text{DM } 704,700}{2} \times 10\%/\text{year} \qquad \text{DM} \quad 35,235$$

$K_R = 280\text{m}^2 \times \text{DM } 120/\text{year}$ DM   33,600

$K_E \begin{cases} = 30 \text{ kWh} \times \text{DM } 0.20 \times 3,520 \\ = 60 \text{ m}^3 \times \text{DM } 0.08 \times 3,520 \end{cases}$ DM   21,120 / DM   16,896

$K_I = \text{DM } 704,700 \times 10\%$ DM   70,470

                                                  DM   353,496

$$K_{MH} = \frac{\text{DM } 353,496}{3,520^h} \qquad\qquad \text{DM} \quad 100.42$$

Personnel expenses/hour

49 persons (women)	× DM 11.42+110%	DM	1,151.14
2 foremen	× DM 13.90+110%	DM	58.38
shift allowance 48	× DM 0.53	DM	25.44
2	× DM 0.65	DM	1.30

$$\text{supervisor} \quad \frac{\text{DM } 3,600 + 110\%}{172.33^h} \qquad \text{DM} \quad 43.86$$

expenses per hour   DM   1,280.12

Assembly costs

machine-hour rate	DM	100.42
personnel expenses	DM	1,280.12
assembly costs	DM	1,380.54
production rate 1,000 pieces/hour		

$$\text{assembly costs per piece} \quad \frac{\text{DM } 1,380.54}{1,000} \qquad \text{DM} \quad 1.38$$

The mechanised assembly method as described (method 2) requires the following production personnel:

- *Machine 1* – one operator to load the base and one operator to place microswitch element.
- *Machines 2 to 8* – one person each, i.e. seven people (supervisory).
- *Machine 9* – one person (supervising) and one person (removal and packing).

This gives a total of 11 operators for method 2.

The procurement cost for the automatic system is DM 1,950,000, which has a replacement value of DM 1,950,000 × 1.215 = DM 2,369,250. The calculation of the assembly costs for this method are shown in Table 3.

Table 3 Calculation of assembly costs for automated assembly method

Machine-hour rate

$K_A = \dfrac{DM\ 2,369,250}{4\ years}$ DM   592,312.50

$K_Z = \dfrac{DM\ 2,369,250}{2} \times 10\%/year$ DM   118,462.50

$K_R = 160 m^2 \times DM\ 120.0/year$ DM   19,200.00

K_E = 40 kWh × DM 0.20 × 3,520   DM   28,160.00
    = 130 m³ × DM 0.08 × 3,520   DM   36,608.00

K_I = DM 2,369,250 × 10%   DM   236,925.00

     DM 1,031,668.00

$K_{MH} = \dfrac{DM\ 1,031,668.00}{3,520^h}$ DM   293.08

Personnel expenses/hour

11 persons (women)   × DM 11.42 + 110%   DM   263.80
2 set-up men   × DM 15.20 + 110%   DM   63.84
shift allowance   11   × DM 0.53   DM   5.83
     2   × DM 0.71   DM   1.42

supervisor $\dfrac{DM\ 3,600 + 110\%}{172.33^h}$ DM   43.86

expenses per hour   DM   378.75

Assembly costs

machine-hour rate   DM   293.08
personnel expenses   DM   378.75

assembly costs   DM   671.83
production rate 1,000 pieces/hour by 80%

assembly costs per piece $\dfrac{DM\ 671.83}{1,000}$ DM   0.672

A comparison of the two manufacturing methods gives:

○ assembly costs with manual production = DM 1.380 per unit,
○ assembly costs with mechanised production = DM 0.672 per unit.
○ thus the saving with the mechanised method = DM 0.708 per unit.

This means an annual saving at 16,000 units per day × 220 days × DM 0.708 of DM 2,492,160.

The amortisation period of the assembly line for the thermal switch is calculated as follows:

$$\dfrac{capital\ investment}{saving + annual\ depreciation} = \dfrac{DM\ 1,950,000}{DM\ 2,592,160 + DM\ 592,312} = 0.63\ years$$

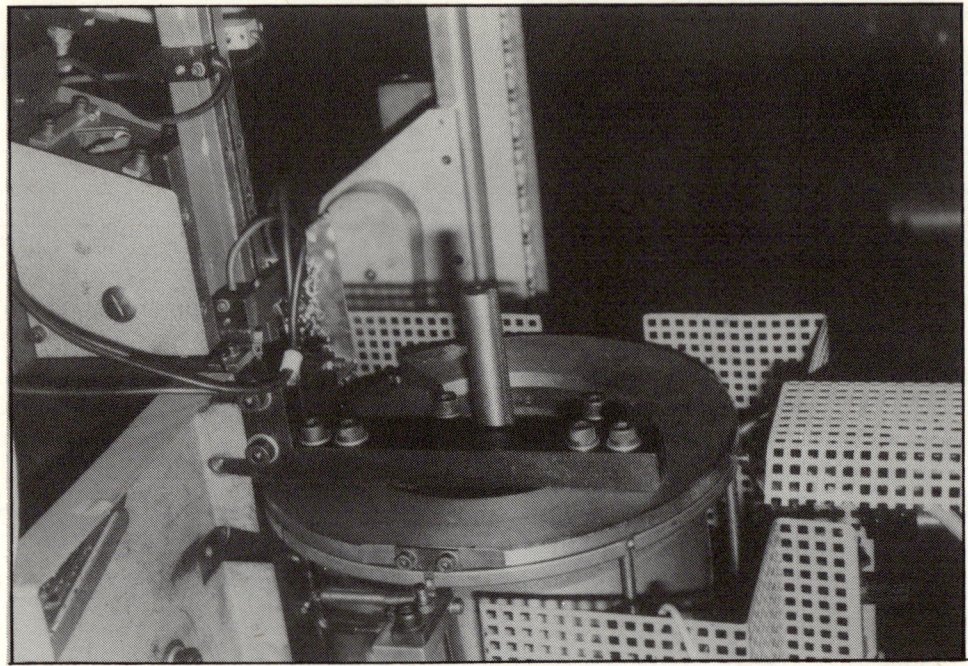

Fig. 10   Satellite assembly machine for fastening terminal and screw (parts 8 and 9)

Fig. 11   Work carrier for machine 2

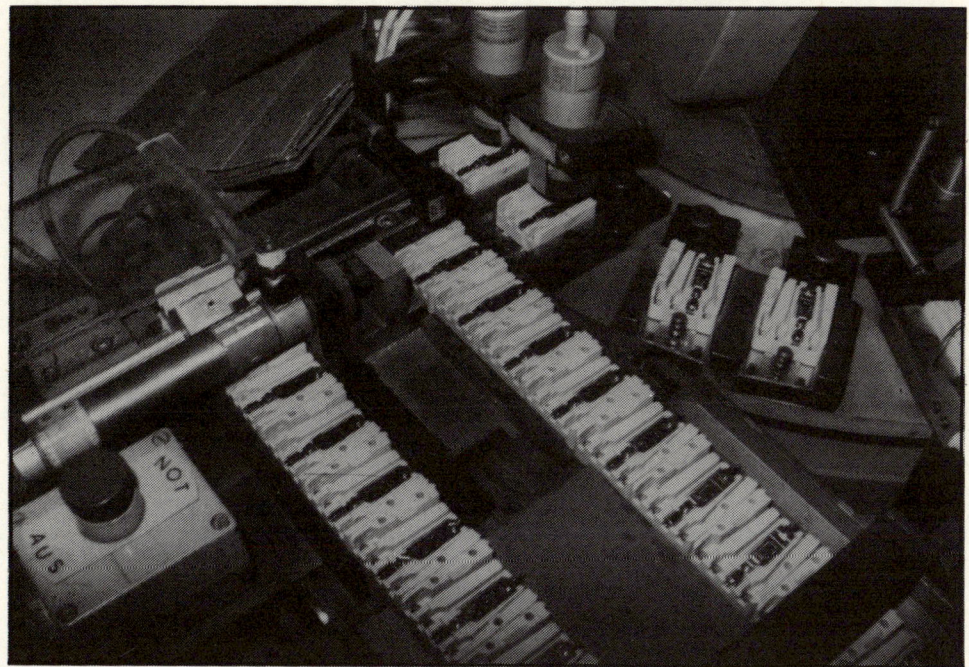

Fig. 12   Duplex loading station on machine 3

The short amortisation period is due to two-shift operations. With single-shift operation the K_{MH} is increased from DM 293.08 to DM 586.16, so that the saving in assembly costs is reduced from DM 0.708 to DM 0.422 per unit. So the annual saving is 16,000 × 220 × 0.422 = *DM 1,485,440*.

Thus the amortisation period for single shift operation is:

$$\frac{\text{capital investment}}{\text{saving + depreciation per annum}} = \frac{\text{DM } 1{,}950{,}000}{\text{DM } 1{,}485{,}440 + \text{DM } 592{,}312} = \textit{0.94 years}$$

These calculations show that with a high capital investment it is necessary to operate more than one shift in order to reach a short amortisation period.

VERSATILE ASSEMBLY BY DEDICATED PROGRAMMABILITY

W. B. Heginbotham, D. Gatehouse, D. Law and G. Wakefield, PERA, UK

First presented at the 2nd International Conference on Assembly Automation, 18–21 May 1981, Brighton, UK

Abstract

Versatile programmable assembly systems have the best chance of success when dealing with particular product families. Different versatility requirements are defined and the characteristics of a programmable assembly machine currently in use are described and evaluated.

Introduction

Assembly by robot for small batch production situations with indefinite product variability still lies within the realms of science fiction; small batch assembly is only possible within limits determined by the rule – 'can one deceive the machine into thinking that components are the same when, in fact, they are not. This can also be interpreted as a general rule for the application of industrial robots to variable product situations and is a simple and useful axiom.

In order to establish a firm basis on which to proceed, it is useful to restate the different types of machine versatility required to achieve a solution to the problem presented by programmability[1,2]:

V_F = internal positional versatility, i.e. the ability to select different positions within a machine's known universe.

V_P = external positional versatility, i.e. the capability of coping with random positional variability of peripheral equipment and components.

V_M = manipulative versatility, i.e. the ability to reproduce complex curve tracing functions quickly.

V_T = time versatility, i.e. the capability of coping with asynchronous time variability in the robot's peripheral environment.

V_C = product variability (normally a gripping problem).

The machine considered in this paper was designed to partly assemble the components for a 'ring binder' for loose-leaf inserts. Assemblies consisted of:

94 PROGRAMMABLE ASSEMBLY

☐ A flat steel strip of variable length.
☐ Varying numbers of ring binder half-loops rivetted to the steel strip with programmable spacings.

These component parts and two different assemblies are shown in Fig. 1.

As far as versatility requirements are concerned, the needs were for V_F and V_C as previously defined. The need for V_P and V_T were engineered out by using:

○ A hand-loaded adjustable magazine input of strips.
○ A vibrating bowl feeder for half-loops (all standard).

There was no requirement for V_M.

The method used throughout the factory before the application of the machinery described here was as follows. The operator selected a strip which had been pre-punched with the correct number and spacing of holes for the particular batch from a vibratory bowl feeder into a mechanism which presented a half-ring to a fixture under the press stroke. The operator then placed each strip by hand onto successively fed half-rings and operated the press to rivet the end over. The automatic system of presenting the half-rings to the operator was controlled and operated by pneumatics. A good operator could carry out up to 22,000 rivetting operations per 8-hour shift, the average performance being 17,000 operations per 8-hour shift. There were about 100 such workplaces throughout the factory.

Specification for the automatic machine

Since labour was difficult to obtain, it provided a stimulus to design an automated machine. It was intended to have one operator per four machines, supplying them with steel strips and rings, instead of one operator per machine

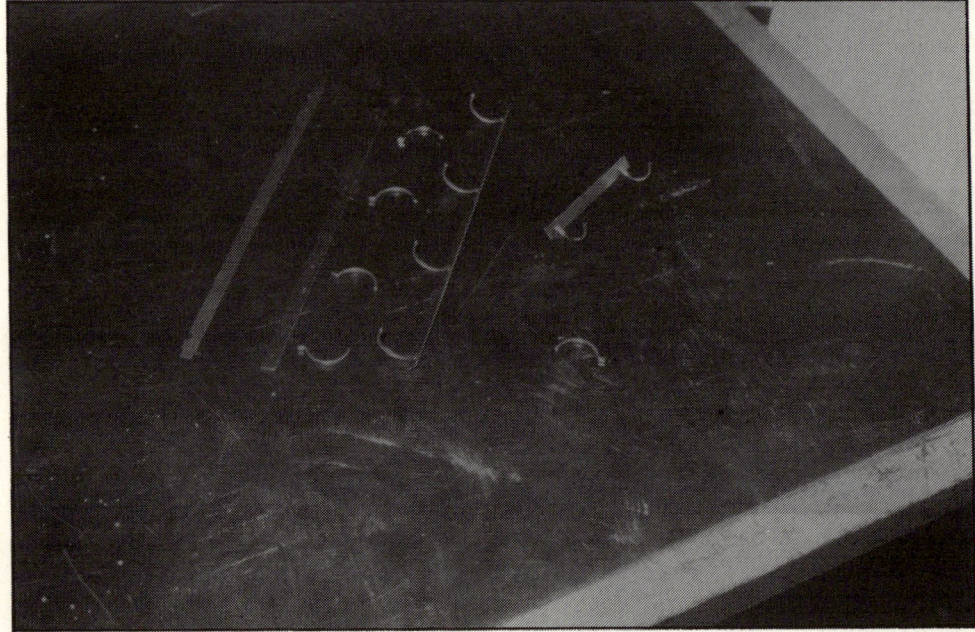

Fig. 1   Typical loose-leaf binder components and assemblies

as before. The strips were to be indexed through the press automatically. A maximum of eight rings (changed to five on subsequent machines) was to be rivetted, and the position of each one was to be programmable, so that when a batch was finished the machine could easily be prepared for the next batch.

The machine could be described as a dedicated programmable machine because it could only make batches of ring binders but with many permutations and combinations of loop numbers and steel strip sizes.

The clearance between the rivet projections on the half loop and the holes in the strip was 0.5mm along the length of the strip, thus requiring that the indexing of the strip in relation to the magazine should be ideally $\pm$ 0.1mm. Two versions of the machine were eventually built, and about 20 copies of the second machine have been built for production use. Two revisions were made to the specification of the first machine. Firstly, to feed the metal strips from the opposite side of the press to the rings, which allowed more overlap in the movement of the grippers, and therefore speeded up the process. This configuration had originally been rejected by the customer because it differed from the manually operated machine. Secondly, the maximum number of rings that could be rivetted was reduced from 8 to 5. Other minor changes and improvements were made.

General description of the production prototype automatic machine

Three views of the machine are shown in Fig. 2 (a, b, c).

Magazine system. The magazine system is shown in Fig. 2(a), where it can be seen that the steel strips (A) rest by gravity on projections (B) from the creep drive chain (C) and are lined up by the adjustable blade (D).

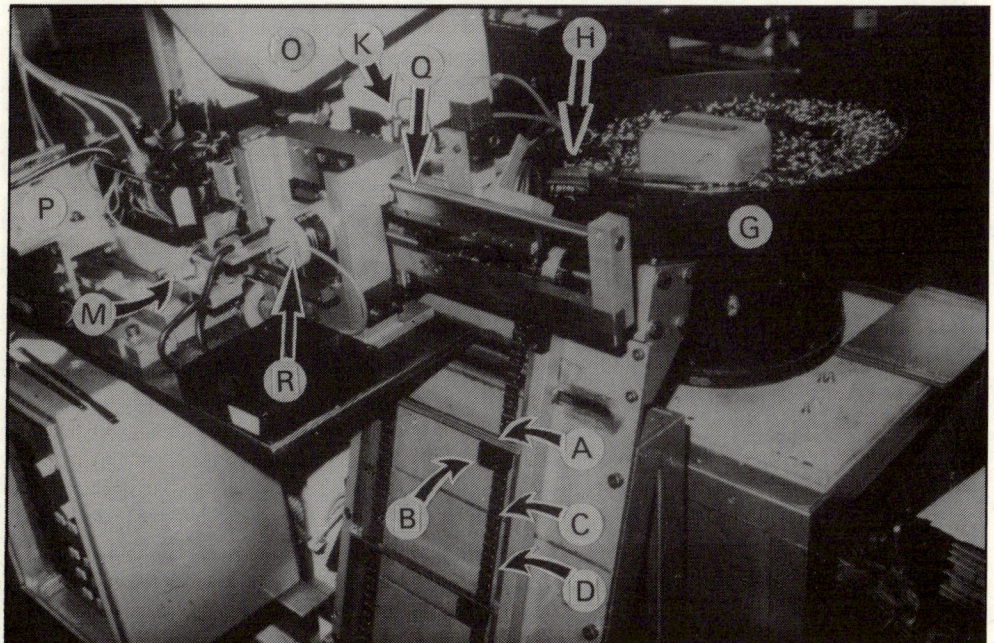

Fig. 2(a) Input magazine system for steel strips and vibratory bowl feeder for half-rings

96 PROGRAMMABLE ASSEMBLY

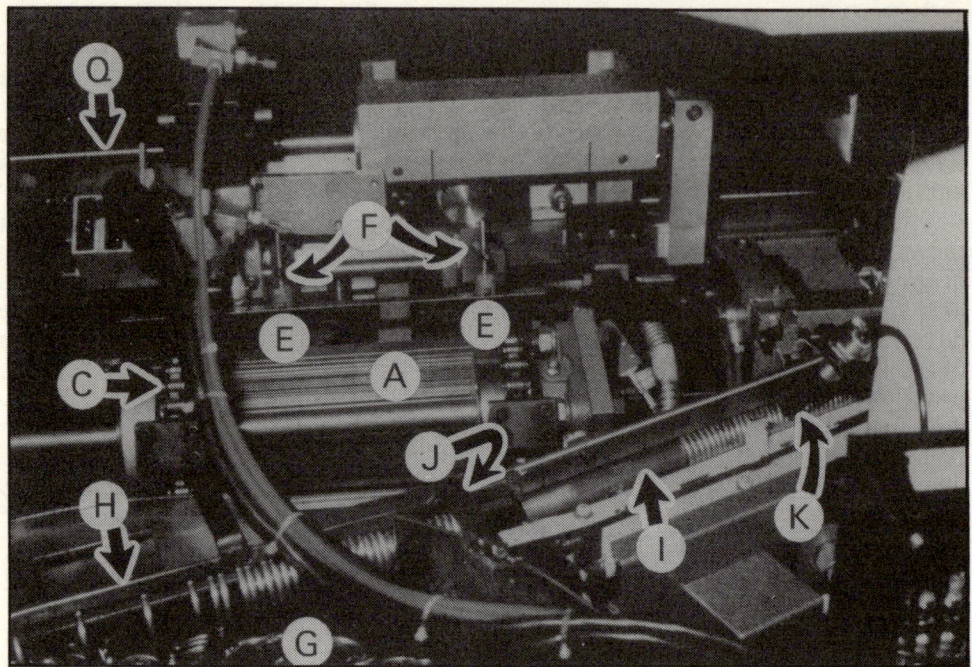

Fig. 2(b)   Strip 'pick-up' station and magazine feed of rings into the press holding fixture

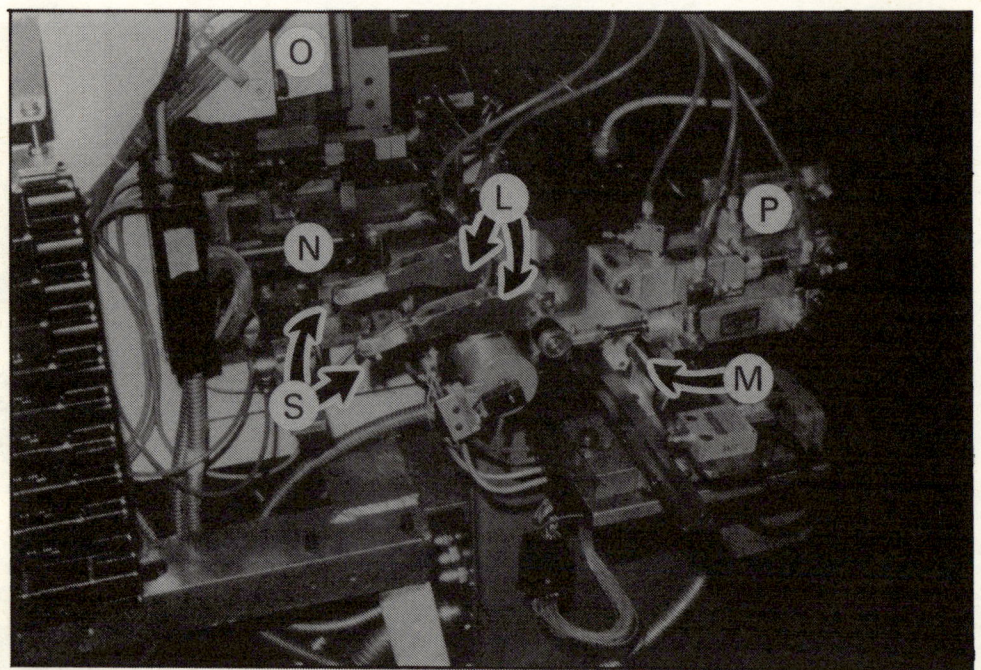

Fig. 2(c)   Output station of machine

Pick-and-place device No. 1. The strips are picked up by the permanent magnets (E) on the end of a pair of swinging arms (F) (Fig. 2b). These swing arms working in parallel form part of a pick-and-place motion. Sideways transfer of these units to complete the pick-and-place motion is provided by mounting them (Fig. 2a) on a slide (Q) driven by stepping motor (R) through a rack and pinion system. This pick-and-place unit is subsequently referred to as 'pick-and-place unit No. 1'.

Thus, the system:

☐ Ensures the accurate alignment and position of each strip in relation to the transfer carriage on which pick-and-place system No. 1 is mounted.
☐ Lifts the strip clear of the magazine stack.
☐ Allows the strip to be moved sideways and placed over the first ring for rivetting. When the first hole is located correctly, a pressure finger is applied, thus holding the strip/ring assembly in place so that the magnet pick-up arms (F) can release themselves against this holding pressure and return to the strip magazine for further pick up.

Half-ring feed system. The vibratory bowl feeder (G) has a 'slit-and-hang' rejection type orienting device (H) (Fig. 2a, b). The half-rings are then transported via a linear vibrating rail feeder (I) into a breech mechanism which holds the ring in the correct position for clenching. The intermediate storage magazine formed by the linear vibrating rail feeder (I) is fed intermittently from the bowl according to signals from proximity sensors (J) and (K) [(J) is below the track], forming upper and lower level gauges to control the store, so that:

○ The store is never empty.
○ The queue of parts cannot interfere with the orienting and feed system by 'backing-up' into the orienting zone.

Pick-and-place device No. 2. Double grippers (S) are carried on swinging arms (L) mounted on slide (M) forming No. 2 pick-and-place device, (Fig. 2c). After the first rivetting operation has taken place, this device is driven to a position opposite the strip, which will be in position (N) overhanging the press (O). (L) is rotated to a horizontal pick-up position to grip the strips. A mechanical gripper was essential for this process to preclude the possibility of slipping while the strip was being indexed to successive rivetting positions. The grippers (L), therefore, lift the strip after each rivetting operation and are indexed sideways with the carriage (P) under the influence of a stepping motor, a rack-and-pinion drive and its allied control system to place holes over successive rings. After the appropriate number of rivetting and indexing operations has taken place, finished assemblies are transferred sideways to a delivery station and dropped into a storage bin.

Gripper unit. The gripper design is of interest because there is not much room available in the daylight of the press. The grippers were effectively designed like pliers but with the jaws of the pliers rotated through 90°. Actuation of the gripper was by small pneumatic cylinders.

Programming. The switchblock (I) controls the distance that pick-and-place device No. 1 travels into the jaw of the press (i.e. according to the position of the first hole). The other switchblocks (II) allow seven distances to be set, thus defining the target positions of the strip for up to seven rings after the strip has passed to pick-and-place unit No. 2 (Fig. 3).

98 PROGRAMMABLE ASSEMBLY

Fig. 3 Layout of control console for the production prototype machine

Control system. The machine is controlled by four simple cycles, which are interlocked. A cycle comprises four, six or eight steps that occur in sequence. Each step corresponds to an unambiguous mechanical configuration of the machine, e.g. pick-and-place unit No. 1 is positioned over the strip magazine. Microswitches sense the positions of the various parts of the machine (e.g. (T) shows those on the press clutch operating lever, Fig. 4). Generally, each degree of freedom has three possible positions, either at one end of its travel or the other end, or somewhere in between. The latter position is transitory, during which the machine position is not known.

The press is controlled by an eight step sequence – pick-and-place unit No. 1 also by an eight step sequence, and pick-and-place unit No. 2 by the combination of four and six step sequences. The use of these sequences divides the operations of the machine into identifiable mechanical subsystems, and allows the use of standard modular electronic sequence control boards[3].

The machine must proceed in an orderly manner: pick-and-place unit No. 1 takes a strip, puts it onto a ring in the press fixture, the press then operates, clenches the ring at the bottom of its stroke, retracts and stops. Pick-and-place No. 2 takes hold the strip and the strip is indexed along by a predetermined amount controlled by the setting of the appropriate bcd switch memory. Obviously, the action of the various cycles is synchronised in some way. The machine can only proceed from one state to another when certain conditions are met, for instance, the pick-and-place devices are only allowed to go under the

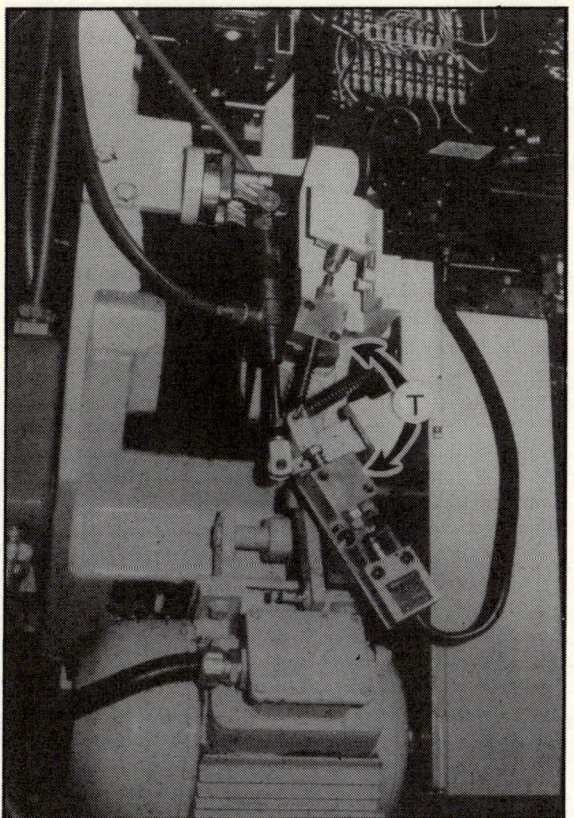

Fig. 4   Limit switches

press when the press is open and stationary. The cycle for pick-and-place No. 1 as shown in Fig. 5 illustrates the principles described above.

Note, that there are three 'actuators' which are controlled by the sequence and whose positions determine the state of the sequence. Only one of these corresponds directly to a mechanical movement of the grippers, that being up/down. The other two are 'auxiliary actuators', numbers 2 and 4, that are essentially memory devices. The motion of the pick-and-place device to and from the press is under the influence of a stepping motor, and so is not conveniently described as a three state actuator (on, off, indeterminate). With reference to Fig. 5, it can be seen that states 1 and 5 both issue a start command to the stepping motor control board, and then wait for the return signal that the motor has reached its target position. It may not be immediately apparent why a total of eight states is necessary. Consider the up/down movements of the pick-and-place device: this device is lowered to pick up a strip and corresponds to the transition from state 1 to state 2, then it lifts (3 to 4) and lowers over a ring (5 to 6) and retracts (7 to 0). Each of these transitions must be separated by the operation of another actuator (in this case an auxiliary actuator with no mechanical function) because, for example, if an 'up' command is given immediately after the machine has reached 'down', then an ambiguity exists because the machine control system cannot distinguish between the end of the 'down' stroke and the beginning of the 'up' stroke. Therefore, the two auxiliary actuators are needed to eliminate the ambiguity; hence the need for an eight step sequence.

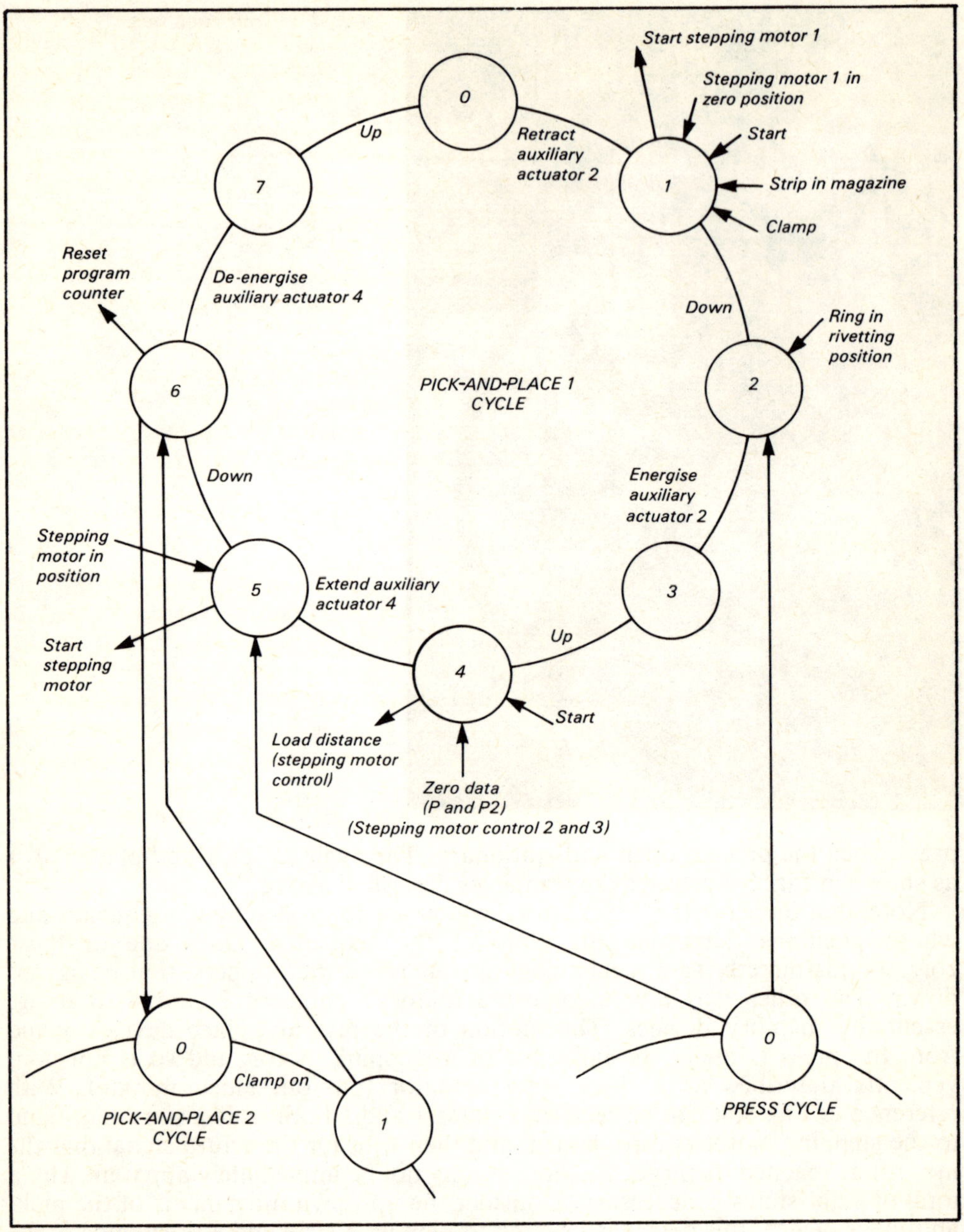

Fig. 5 Pick-and-place unit no. 1 control sequence

Following the device through its cycle, it operates as follows: Starting at state 0, auxiliary actuator 2 is retracted and the machine moves to state 1. A message is sent out to start the motor, and four signals indicating the overall state of the machine are awaited, i.e. that the stepping motor has reached the zero position, that there is a strip in the magazine, that the strip clamp in the press jaws is open and that a start signal is present.

The grippers move down under the influence of a pneumatic cylinder, and the device moves to state 2. A check is now made to ensure that a ring is in the rivetting position, and that the press is ready. Auxiliary actuator 4 is then energised and the machine moves to state 3. The grippers move up, giving state 4. A message from the bcd switch memory, defining the distance the strip must be moved, is accepted by the motor controller. Auxiliary actuator 2 is extended, giving state 5. A message is then sent out to start the stepping motor, and a return message that it has reached the target position is awaited. There is also an interlock with the press. The strip is lowered, giving state 6, which requests the strip clamp and waits for confirmation that it is closed. Auxiliary actuator 4 is de-energised, giving state 7, and the grippers move up, pulling themselves away from the strip to give state 0.

Machine performance. The machine was able to rivet rings at the rate of 12,000 per 8-hour shift, as compared with the average rate of 17,000 for a manual operator. However, it was envisaged that one operator would be able to cope with four of the new machines, keeping them loaded with steel strips and half-rings.

Conclusion

The time taken to rivet three rings to a strip was 7.2 seconds. It would be difficult to achieve this performance with a general purpose assembly robot because of the greater inertia of its arm. The use of two pick-and-place devices also speeds the process considerably, by having one returning to its start position whilst the other is working. Whilst being a machine dedicated to producing half-strips for ring binders, it is programmable so as to cope with all likely product variations, i.e. between two and eight rings, at variable positions along the strip adjustable to the nearest 0.1mm. It had been recognised from the start of the project that faster assembly times could be realised by feeding strips and rings from opposite sides of the press. The customer, however, was anxious to retain the general layout of the manually operated machine. This restriction was later relaxed, allowing a 17% increase in speed.

Due to the use of microswitches on all motions to provide interlocks for the machine sequence, serious 'self-inflicted' damage (by the machine running on after some fault had occurred) was guarded against. The machine was found to be very reliable when put into production.

Acknowledgement

Grateful thanks are due to Bensons International Ltd (Stroud, UK) for permission to publish the details of the programmable assembly machine.

References

[1] Heginbotham, W. B., Pugh, A., Gatehouse, D. W. and Law, D. A versatile variable mission assembly machine. In, Proceedings of the 3rd CIRT, 6th ISIR Conference, Nottingham, 24–26 March 1976, pp. 53–70.

[2] Heginbotham, W. B. Programmable Automation. Chartered Mechanical Engineer, November 1977, pp. 78–81.

[3] Ashley, J. R., Heginbotham, W. B. and Pugh, A. Developments in programmable assembly devices. In, Proceedings of the 1st International Conference on Robots, IITRI, Chicago, April 1970.

Chapter 3
DESIGN FOR ASSEMBLY

It is acknowledged that the most critical factor in automating the assembly process is the product's design. Design-for-assembly techniques and how they are applied are discussed in the four papers of this chapter.

DESIGN OF DATA PROCESSING EQUIPMENT FOR AUTOMATED ASSEMBLY

F. L. Bracken and G. E. Insolia, IBM Corporation, USA

Abstract

The basic approach in designing for simplified assembly and minimum parts is to (a) list every known function, (b) list every known part and sub-assembly in an exploded view, (c) categorise dimension in rank of importance and (d) give consideration to field replaceable units. A robotically assembled printer is considered.

Introduction

Producing low volumes of data processing equipment mandates that a manufacturing method must be flexible and produce more than one product. Selection of products, processes and designs which will provide compatibility through the fabrication and assembly stages must then be carried out. Designing products for ease of manufacturing is an absolute necessity if we are to minimise numbers of parts and achieve machine assembly or easy manual assembly.

It has been our experience that designing for ease of assembly does not necessarily lead to minimising part numbers. Nor does the minimum parts design assure that you will achieve machine assembly or easy manual assembly. We must pay careful attention to both factors if benefits are to be realised.

Basic design approach

Almost every design which is being undertaken in industry today is one that has evolved through many generations. They are based almost always on the 'best' parts, sub-assemblies or ideas from a previous design. In most instances where this has happened, the product has many parts some of whose function may not be essential to the end function.

Designing for simplified assembly and minimum parts may require something of a revolution. The designer should step back and consider what functions are required and just what is to be accomplished.

The basic approach is straightforward. First,

LIST EVERY KNOWN FUNCTION

These can be few or many, depending on what the individual designer wants to define as basic functions. These may be a single element or a major sub-assembly.

The next step is to list every known part and sub-assembly in an exploded view.

EXPLODED VIEW

At this point, there is no need for dimensions or part detail, only an awareness of every detail.

The next and one of the most important considerations is to

CATEGORISE DIMENSION IN RANK OF IMPORTANCE

It should now be easy to

PICK A POINT TO BEGIN THE DESIGN

This point can be in the product or sub-assembly, pallet, or benchtop.

Every engineer and designer knows that the most critical dimension should be contained within a single element if at all possible. This reduces tolerance build-up between parts and the likelihood of a need for adjustments. Tolerances on the dimensions then can be held to the accuracy of machining or manufacturing equipment tolerances. Applying the basic rules of the University of Massachusetts design for assembly and parts reducing program (Boothroyd, et al, 1982) will result in a minimum number of parts. If it is impossible to contain dimensions in a single part, then a division should be considered in light of the effect of such a division.

Field replaceable units (FRU) may determine how something is designed for replacement. The ease by which products can be repaired is becoming a major factor in the overall cost of a product.

GIVE CONSIDERATION TO FIELD REPLACEABLE UNITS

Proper consideration can lead also to fewer parts and greater ease of assembly.

COVER CONSIDERATION

Properly designed covers can provide many more uses to the customer than just cosmetics. If designed properly, covers can lend to structural integrity, as well as binding and holding parts in place within the assembly. In optimal use, they can suffice as card holders, and component securing features.

One use for covers that should be avoided is as a place to mount indicators, switches, thermal indicators, and so on, which may require extensive cabling. Try mounting such hardware so that it can be accessed through the case, reducing cables to and from covers. Cable layout should be one of the first details considered in the design. Remember, the design stage is the time when proper cable and FRU planning can reduce intermediate connections and thus cut maintenance and service time.

SEQUENTIAL ASSEMBLY

Any part not having a rightful place in the sequence must be dealt with by other means. Five possible alternatives are:

☐ Combine the part with another part and handle as a sub-assembly.
☐ Change planned design.
☐ Use assembly aids which will assist in manufacture and which can be removed from product.
☐ Design assembly aid in product.
☐ Do it manually as part of planned process.

In all cases, the ultimate system will be realised only by considering the component, product, process, assembly, test and maintenance of the product at the design stage.

Robotically assembled printer

The base of the printer concept vehicle is used as the starting point for our stacked assembly. Features are designed in the base to accept and hold parts and sub-assemblies in place until secured by the cover or other parts or sub-assemblies. Assembly aids are incorporated into the base also. One such feature is a block to support a drive motor. This feature could have been left out of the design, requiring an assembly aid to be used by the robot and then removed after the motor has been secured to the frame. Other features of this base will be highlighted as the features are defined for other parts that assemble into the base.

Fig. 1 shows the print mechanism sub-assembly. If the exploded view of the print mechanism is studied, it becomes apparent that the frame has been used as

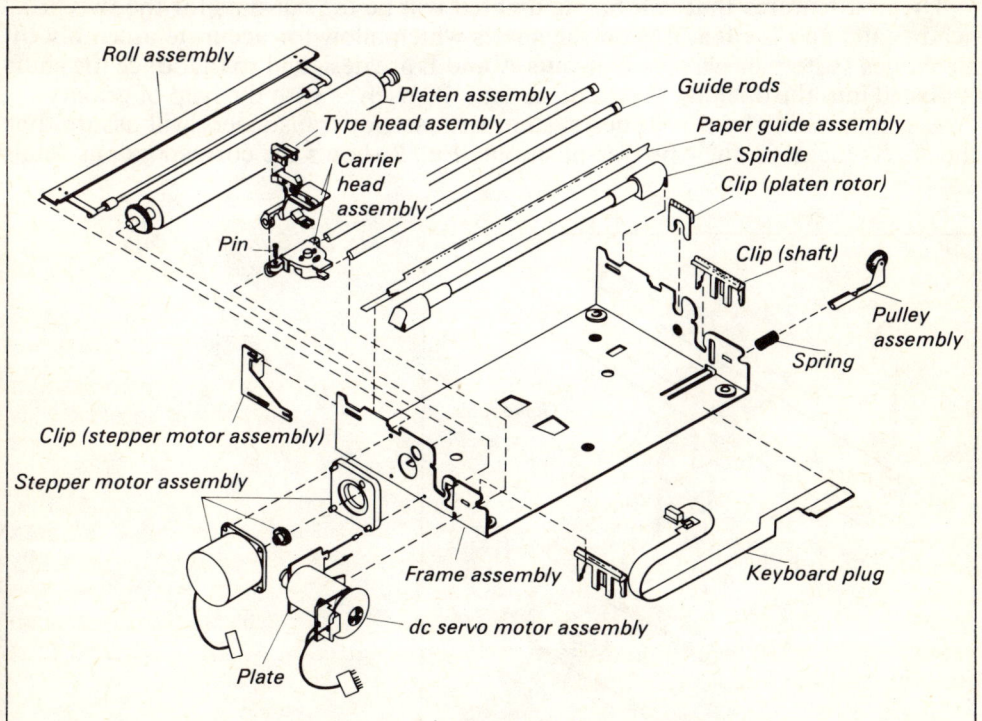

Fig. 1   Print mechanism sub-assembly

the starting point for the stacked design of this sub-assembly. Each part has been engineered and designed for sequential assembly. Features designed in the frame also account for the method used to fit the mechanical sub-assembly into the envelope of the final product.

The investigation of the features of each part and how they relate to each other and to the frame is provided here in detail.

Note the left upright of the frame in Fig. 1. There are a minimum of four critical dimensions. The first is the height of the centre line of the platen shaft from the centre line of the shaft carrying the print mechanism. The horizontal distance between them is also important because the print mechanism rotates about the supporting shaft. The only other dimension of interest for this discussion will be the need for parallelism over the entire length of the frame.

The dimensions are highlighted for two very important reasons:

○ If the critical dimensions of a mechanism can be contained within a single element, then the accuracy of the manufacturing process will be the only contributing factor in each feature's relationship to another.
○ It is important to reduce the number of required adjustments which may be impossible or very expensive to do in assembly. It will also reduce the number of parts in most cases since allowing for adjustment usually requires the use of threaded fasteners, tools, fixtures, and additional parts.

It is, therefore, possible to design into a mechanism requirements for an adjustment that would not have been needed if proper attention had been given to the original design. An example of this will follow.

The first features that will be highlighted will be the cut-outs for the two print head shafts and the features on the shafts which allow for accurate assembly by a robotics system or person. Cut-outs A and B are designed so that once the shaft is placed into the opening they tend to seat themselves with the help of gravity.

A gentle nudge by a robotics system with sensor transducers will assure that the shafts locate in their proper positions. Fig. 2 shows the corresponding detail

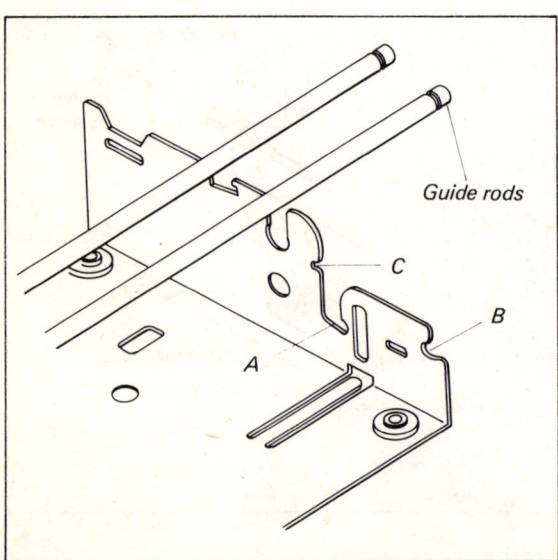

Fig. 2   Guided rod detail

DESIGN OF DATA PROCESSING EQUIPMENT 109

Fig. 3   Exploded view of head and rods

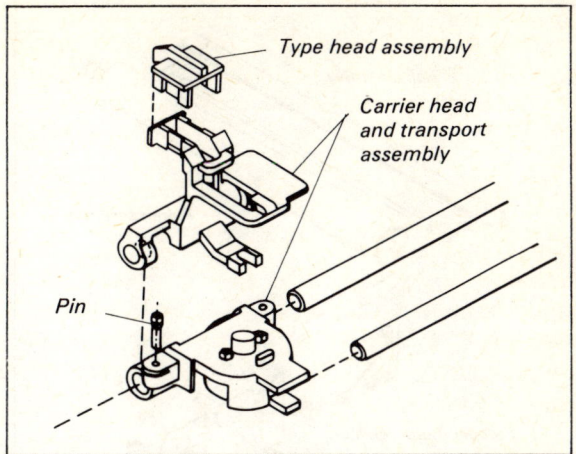

for shaft positioning (marked A and B). The two grooves in the ends of the shaft serve multiple purposes: first, they function to locate the shafts within the pallet to assure positive location for the robot to grasp the parts; second, they assure that the parts cannot move in horizontal direction once they are snapped in place by a retainer.

The parts on the two transverse rods are the print element carrier. These must move relative to one another on the rods between the uprights. There is a problem getting the rod through the two parts and into the uprights with only one robotic manipulator available for assembly.

Fig. 3 shows the head and rods in an exploded view. A plastic dowel is placed through the parts to hold them in proper position. The robotics system then picks up the transverse rod that will carry the head back and forth, inserts the rod (pushing out the plastic dowel) and places all three parts into the mechanism.

The paper guide is used to direct the paper around the platen and up past the print head mechanism. Feature C of Fig. 2 is used to hold the top of the paper guide into the upright.

Fig. 4 shows the paper guide. The unique aspect of this part is the spring

Fig. 4   Details of the paper guide

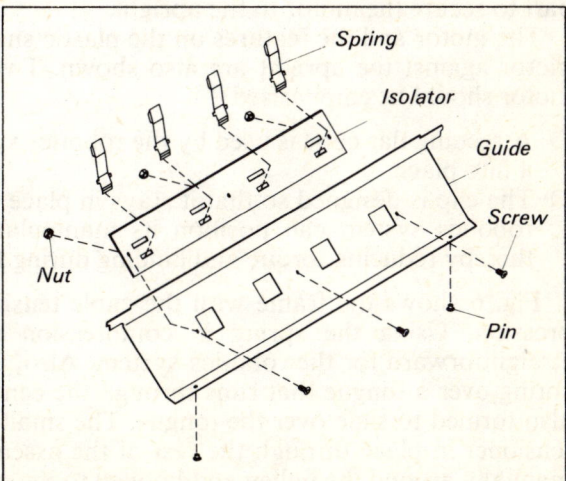

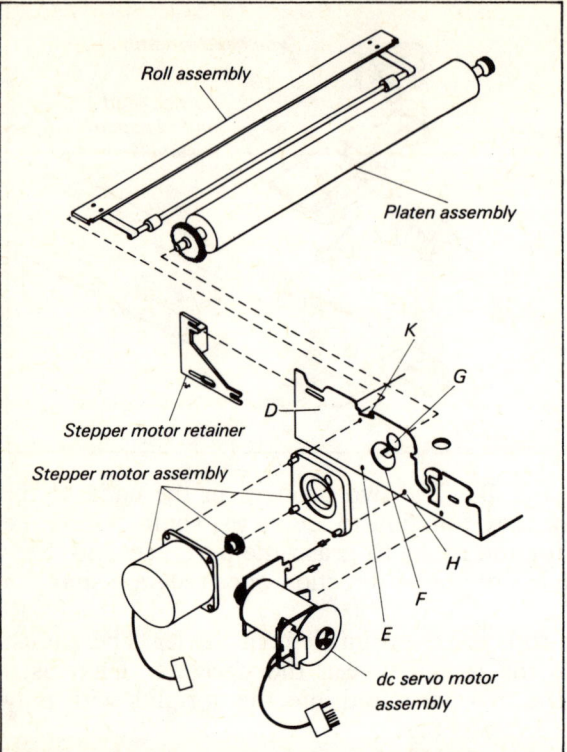

Fig. 5 Detail of platen motor mounting

properties of the material; these aid in the placement and retention of the paper guide in the mechanism. Cut-outs in the base accept the locking studs riveted into the paper guide. To clear the lip of the studs, the robot pushes against the paper guide deflecting the metal and travels in horizontal direction from one end of the guide to the other while applying pressure to snap both studs in the base.

In Fig. 5, D, E, F, G, and H define where the platen motor will mount. The motor with the three pins which were specifically designed for mounting to the frame are shown. The studs have grooves that will mate with the locking plastic part to secure the motor to the upright.

The motor and the features on the plastic snap that are designed to wedge the motor against the upright are also shown. Two aspects to the assembly of the motor should be emphasised:

○ A rectangular cube is used by the robotics system to handle the clip and force it into place.
○ The clip is designed so that it stays in place after initial placement so that the robotics system can position its manipulator to push directly on the clip, thereby reducing torque and binding during final insertion.

Fig. 6 shows the frame with the cable tensioner. The spring is used in compression. Using the spring in compression makes this assembly step very straightforward for the robotics system. Also, the frame is cut out to accept the spring over a tongue that runs through the centre of the spring. The tensioner is also formed to slide over the tongue. The small feature A is designed to hold the tensioner in place through the rest of the assembly line until the cable is placed manually around the pulley and hooked to the print mechanism.

Fig. 6 Frame with cable tensioner

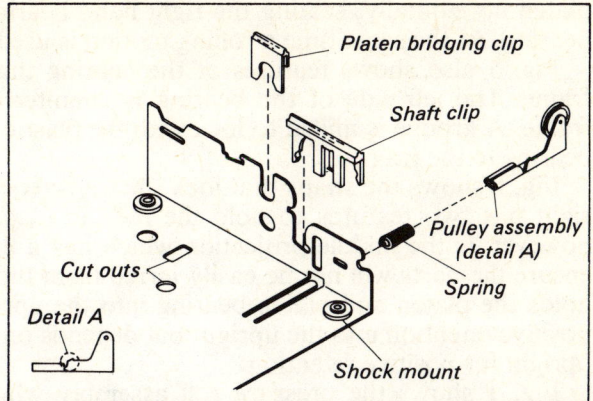

The task of stringing the cable from the helix gear through the frame and around the tensioner and then fastening it to the print mechanism offers a real challenge to automation. The solution was to eliminate the cable by going into direct drive, eliminating the need for helix gear, motor and the cable. The head could then be driven back and forth on a helix shaft and corresponding nut on the print mechanism.

Fig. 7 is the platen. The left end must be placed into a bearing which is already in the upright and the gear must be meshed with the gear on the shaft of the motor. The right end has a bearing attached which must be placed in the right upright.

A special tool is required for assembling the platen into the mechanism (Fig. 8). The tool is required due to the close clearance between the paper guide and the platen. A block at the top of the tool is designed to allow the manipulator to grasp the tool. To insert the platen, the robotics system picks up the tool and forces it over the barrel of the platen. The platen is dropped vertically into the upright but offset horizontally to the right. The robotics system then moves the

Fig. 7 Platen details

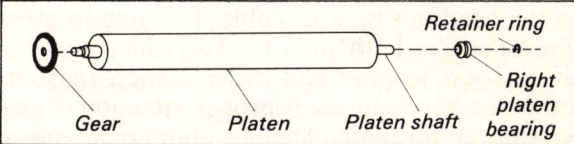

Fig. 8 Platen assembly tool

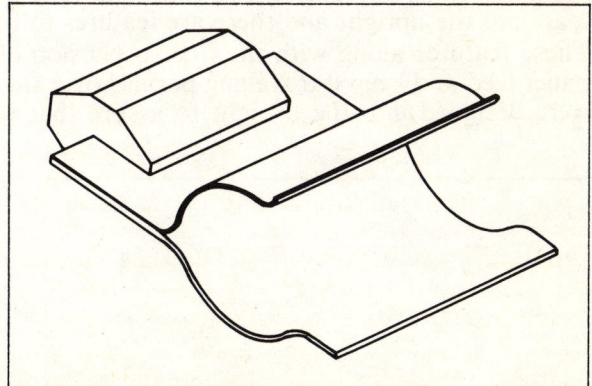

platen horizontally, seating the right hand bearing and inserting the shaft in left bearing. At the same time a rolling motion is used to engage the gears.

Fig. 5 also shows features of the bearing that allow it to be inserted in the frame. The left side of the bearing is chamferred to lead the bearing into the frame. A groove is milled to lock with the plastic fastener which in turn locks the bearing to the frame.

Fig. 6 shows the snaps that lock the transverse shaft in place. The snap on the right has two features to hold the rods in place. The most important feature, however, is the middle projection which has a lip that snaps into the upright to ensure the parts will not be easily jarred from their position. The snap on the left holds the platen and platen bearing into the upright. Presently it does not have positive retention into the upright but depends on features of the snap, platen and upright for positive retention.

Fig. 1 shows the pressure roll assembly which, in some ways, is the most interesting part and incorporates techniques for using machine features to hold insert, hold in place and provide function within the machine. The small rolls rotate around the shaft. The shaft is held by loops in the leaf springs at either end of the assembly. The leaf springs are fastened to the flat bar by screws or spot welds. The small notch at either end of the flat bar is designed to allow the upright frame to nestle into the slot. This holds the assembly in place horizontally once it is placed into the mechanism.

In Fig. 5, K is the feature that holds the flat bar. The insertion is accomplished in two steps. The assembly is placed on top of the platen and into the K feature. The robotic manipulator is rotated to allow pushing down on the assembly while one grasps the back of the flat bar and moves it toward the platen. The spring deflects and the flat bar is slid under the lip in the upright. At the same time, the rolls go down over the centre of an arc on the platen, pulling the bar tight into the upright and holding it in place. Pressure must be applied to remove the assembly.

The dc servo motor (Fig. 1) is designed to hook over the top of the frame with the tongue protruding into the base to give stability in the horizontal position. The weight of the motor hangs on the cable to depress a spring and keep tension on the head mechanism cable. It is important to note that the cables have been lengthened on both motors. To reduce part numbers and assembly steps, it is always best to plan and order cables, fans, etc., with a cable long enough to reach the drive source. If proper attention is paid to design of structure, they can be removed for maintenance without removing any parts.

Fig. 9 shows the plastic rod that holds the paper. The ends are designed to snap into the upright and there are features to feed both 8.5in. and 11in. paper. These features along with the free suspension of paper on the shaft control the paper feed and keep it travelling parallel over the printing surface. These features were designed into the upright to assure that the best possible dimensions for

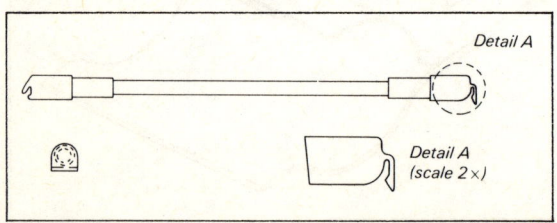

Fig. 9   Plastic rod to hold platen

parallelism to the platen could be gained. These dimensions also help determine how well the paper feeds.

Now that the mechanical parts have been defined and highlighted, the system envelope will be further illustrated. The base of the printer was used as the starting point of design for the total vehicle assembly. The features of the base will now be highlighted as the rest of the machine is described and sequentially assembled.

The first part of the machine to be assembled into the base is the frame. It is shock mounted at the four corners to reduce mechanical noise and vibrations that would be transmitted to the base and cover. The mounting posts are incorporated into the base of the frame. The second feature is the four post and snap features that hold the keyboard.

If such features would prohibit moulding or make it too difficult to manufacture, then part of the snap feature could be in the keyboard. A keyboard that fits into the uprights has not been designed for this mechanism. It is sufficient at this time to point out what needs to be done to make the keyboard assemble easily into the product.

Since electrical connection is required between the keyboard and the logic, this vehicle layout requires a cable. If planned properly, plugging in the cable could be a manual operation that takes place at the proper time in the sequential process. It is likely that the cable could be placed on the base first with features designed to hold the cable in place and prevent it from interfering with the rest of the assembly. Connecting to the keyboard and logic would be a final manual operation.

The other alternative would be to reconfigure the logic cards and board so that direct plugging could result when the keyboard is placed into the vehicle. Fig. 1 shows the keyboard plug. It is apparent that automated assembly will not be practical as designed.

The assembly of the logic board into the base of the machine is accomplished by designing the base to accept the board. Changes to the board were not required to accomplish machine assembly.

The power supply would require additional design changes to fully realise machine assembly without human intervention. The on-off switch placed on the cover would require a cable as it is presently designed. It is possible that this manual operation could be eliminated with the correct selection of hardware.

The connection between the power supply and the logic board has been designed to eliminate the need for cables. The power supply was designed for two possibilities: one method would be to add a 90° connector on the board and plug with the tabs or the edge of the power supply; another approach is to place an AMP feed-through connector on the power supply and plug the pins of the board directly into the power supply. This step is difficult for automated assembly because of the number of pins involved. Also, the connector has no lead-in to facilitate assembly.

The cooling fan in the power supply is assembled much the same as the drive motor in the print mechanism. The exception being that a plastic snap is not used to hold the motor in place. Instead, the cover of the power supply has been designed with features to bind the motor downward holding it in place. This eliminates the need for an additional plastic snap.

The logic cards have not been modified. They are standard size logic cards, placed in dummy socket to provide a positive location for the robotics system to pick up. They can be inserted into the standard board repeatedly.

114 PROGRAMMABLE ASSEMBLY

Signal cables from the logic to the different motor, keyboard, end of forms, etc., are still manual operations. The cable that pulls the print element back and forth is a manual assembly step.

Robotically assembled power supply

The general rules of designing for automation are very easily adopted for use in designing power supplies. Eliminating wires and threaded fasteners, combining several parts into one, and providing for a sequential method of assembly will all lead to a product which is sufficiently simple for automated manufacture.

To further ease the assembly of threaded fasteners they may be manufactured with a conical or bullet-nosed tip. These tips help in centering the screw prior to running it and compensate for slight misalignment.

Simplification of discrete components which are printed circuit card mounted is of great importance when designing for automation. The following are several concepts for components in our power supply which have simple and effective installation procedures which may be performed by a robotics system.

The first component to be considered is a diode (Fig. 10). Presently, the diode is assembled to a heat sink with a nut and star washer. The heat sink is assembled in turn to a printed circuit card with a screw, two washers, and a nut which is then flow soldered. The stamped eyelet of the diode has a wire soldered to it. The wire's other end is soldered to the printed circuit card.

The preferred concept eliminates all screws, nuts and washers by screwing the diode directly into a threaded heat sink and forming the heat sink so that it may be placed directly on a printed circuit card and soldered. The heat sink is formed with a 90° angle and a small tab on the undrilled side.

The tab serves to make the electrical connection from the threaded portion of the diode to the printed circuit card and to allow the remainder of the heat sink to lie directly on the printed circuit card for mechanical support. The tab may be pre-tinned to allow for reflow soldering.

The lead of the diode may also be pre-tinned and reflowed in the same manner. The lead is formed with two different diameters (Fig. 11) so that the larger diameter will act as a mechanical support and spacer for mounting the

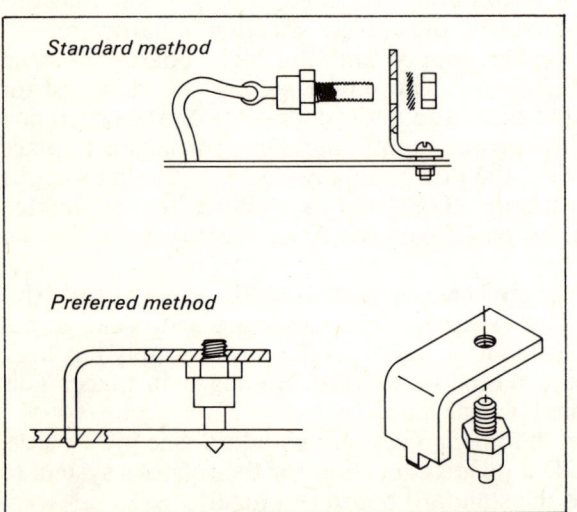

Fig. 10 Standard and preferred methods for diode assembly

Fig. 11 The diode lead is formed from two different diameters

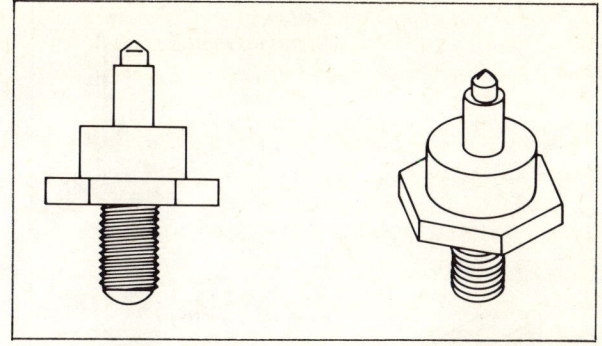

heat sink/diode assembly. The tip of the lead is formed in a cone or bullet nose to aid in location and the tab of the heat sink is tapered for the same reason. If the heat sink tab and diode lead are not pre-tinned they may require wave soldering to make all connections.

Power transistors

The present method of mounting power transistors to printed circuit cards is to use two screws assembled from the bottom of the printed circuit card, six washers, and two nuts which are assembled from the top (Fig. 12). Shown in the preferred method are two special screws which will eliminate all of the other nuts and washers. The screws are brass with a bright tin surface treatment and the lower six millimetres of thread removed (prior to tinning).

The heat sink is drilled and tapped so that the power transistor may be assembled to it using only the screws and then the entire sub-assembly will be installed to the printed circuit card. The threaded portion of the screws will extend below the bottom of the heat sink and rest on top of the printed circuit card to act as spacers for air flow beneath the heat sink. The screw tips and transistor leads may be either reflow soldered or wave soldered to complete all connections to the card.

Fig. 12 Standard and preferred methods of mounting power transistors to printed circuit cards

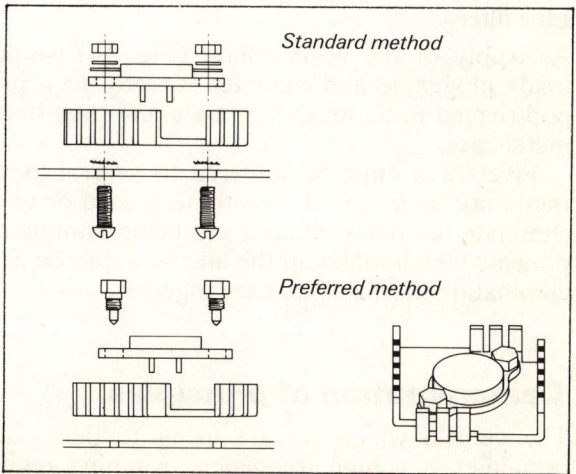

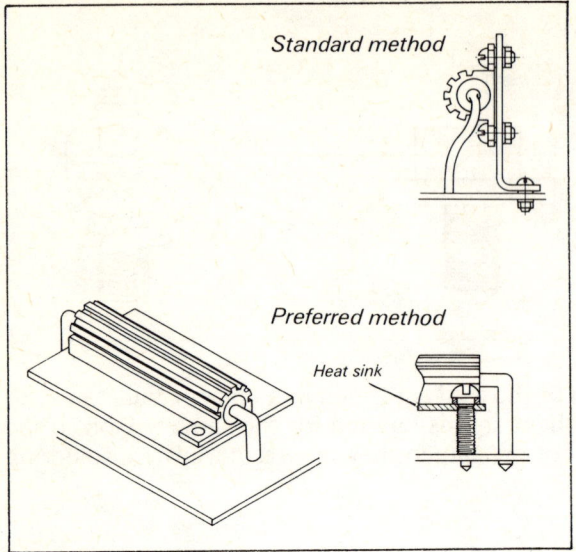

Fig. 13 Standard and preferred methods for mounting load resistors to printed circuit cards

Load resistors

The same pre-tinned brass screw concept may be used for mounting load resistors to printed circuit cards. Several screws, washers, nuts, and wires are eliminated as well as four hand soldering operations. The resistor leads are formed at right angles and with tapered tips to simplify the installation (Fig. 13). The heat sink is drilled and tapped so that the resistor/heatsink sub-assembly may be prepared and then dropped into the printed circuit card. All mechanical and electrical connections will be made by either reflow soldering or wave soldering the screws and leads.

Another method of attaching these load resistors to a printed circuit card is to use clinch nuts on the heat sink instead of tapping the drilled holes. The clinch nuts serve as a spacer between the heat sink and the card and also eliminate the need for a special screw. All connections would once again be made by either wave or reflow soldering.

Line filter

Assembly of the ac line filter (Fig. 14) would be greatly simplified if it were made pluggable and mounted directly to a printed circuit card. It is presently pop-riveted to an insulating plate and then that sub-assembly is pop-riveted to a metal case.

Five wires must be soldered to several locations. Forming the leads so that they may be mounted directly to a card or to a card mounted connector would eliminate ten parts. Studies are being conducted to determine whether the components which make up the line filter can be arranged directly on a card thereby eliminating the shielding case together.

Demonstration of principles

The vehicle which we are using to demonstrate our design for automation principles is a high frequency switching regulator from an existing processor

DESIGN OF DATA PROCESSING EQUIPMENT 117

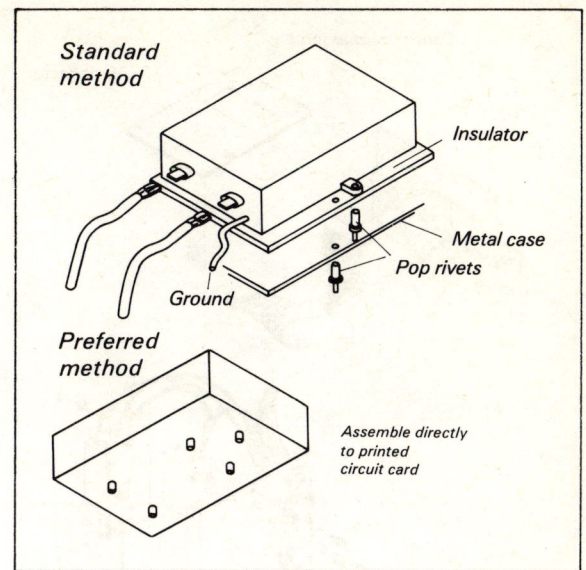

Fig. 14 Standard and preferred methods for line filters

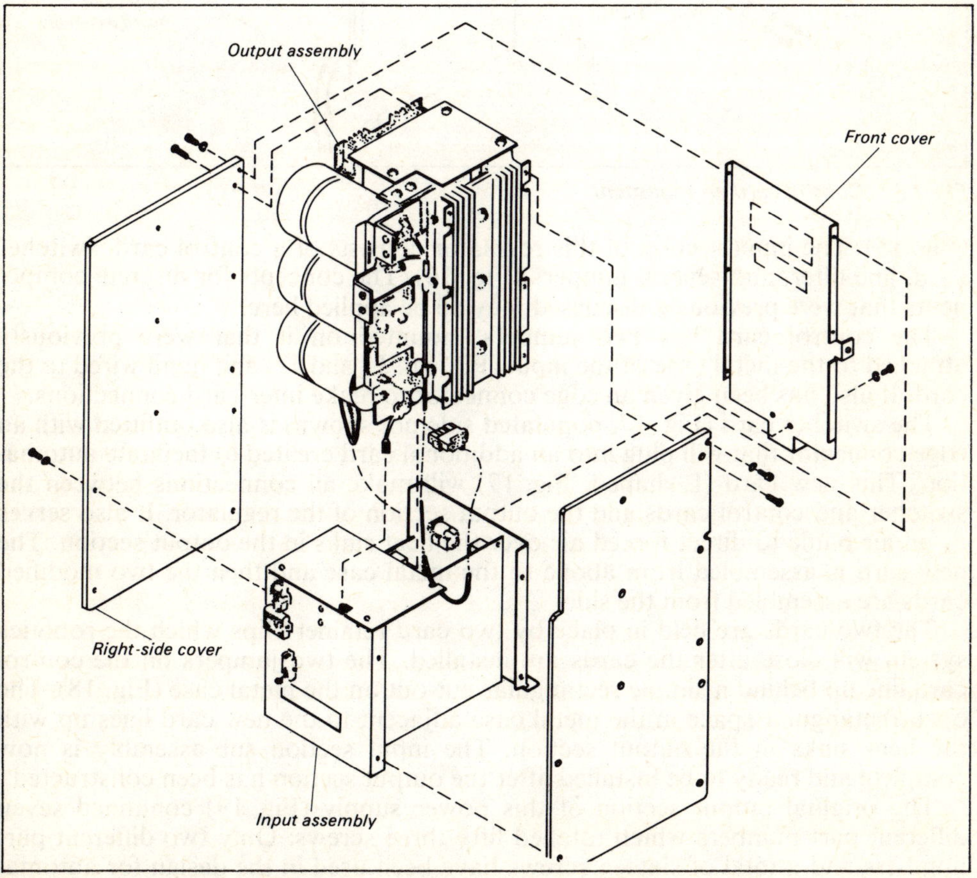

Fig. 15 High-frequency switching regulator

118 PROGRAMMABLE ASSEMBLY

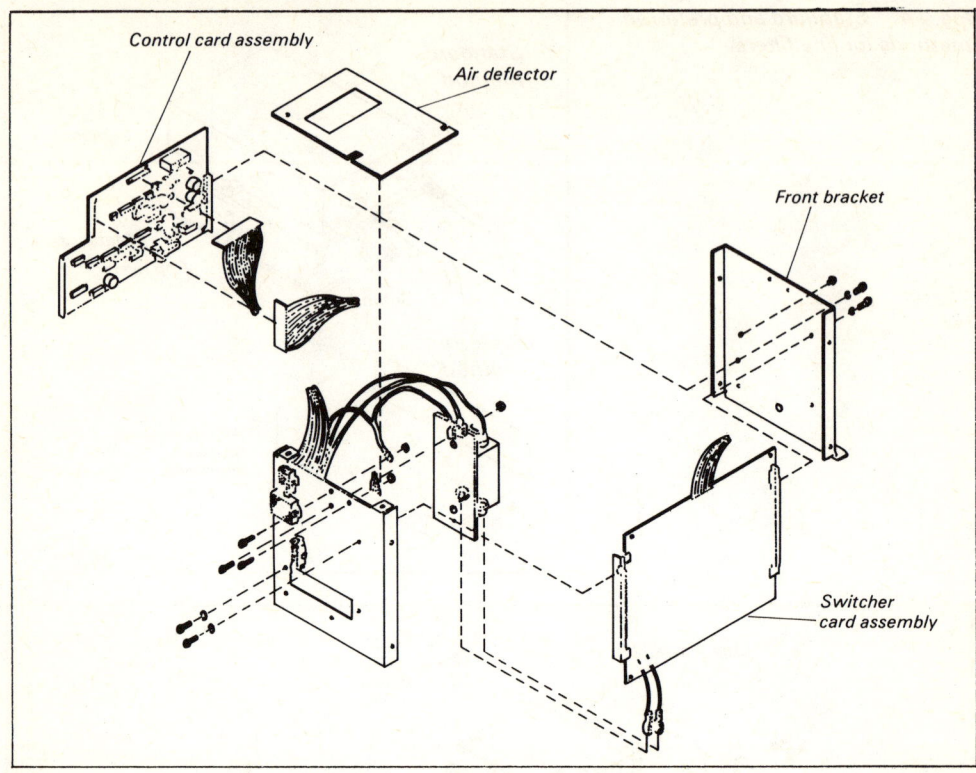

Fig. 16   Control card arrangement

(Fig. 15). The input section of this regulator consists of a control card, switcher card, line filter, and several jumpers (Fig. 16). The concepts for discrete components that were previously discussed may all be applied here.

The control card has two jumpers mounted on it that were previously attached to the metal case of the input (Fig. 16, J1 and J2) and hand wired to the card. It also has been given an edge connector to make inter-card connections.

The switcher card (Fig. 17, populated side not shown) is also outfitted with an edge connector that will plug into an additional card created to facilitate automation. This new card (L-shaped, Fig. 17) will make all connections between the switcher and control cards and the output section of the regulator. It also serves as an air baffle to direct forced air over the heat sinks in the output section. The new card is assembled from above to the metal case and then the two modified cards are assembled from the side.

The two cards are held in place by two card retainer clips which the robotics system will close after the cards are installed. The two jumpers on the control card line up behind a simple rectangular cut-out on the metal case (Fig. 18). The open rectangular space in the metal case adjacent to the new card lines up with the heat sinks in the output section. The input section sub-assembly is now complete and ready to be installed after the output section has been constructed.

The original output section of this power supply (Fig. 19) contained seven different part numbers which totalled fifty-three screws. Only two different part numbers and a total of sixteen screws have been used in the design for automation (Fig. 20). Twelve of these screws are necessary for making electrical con-

DESIGN OF DATA PROCESSING EQUIPMENT   119

Fig. 17   Switcher card assembly

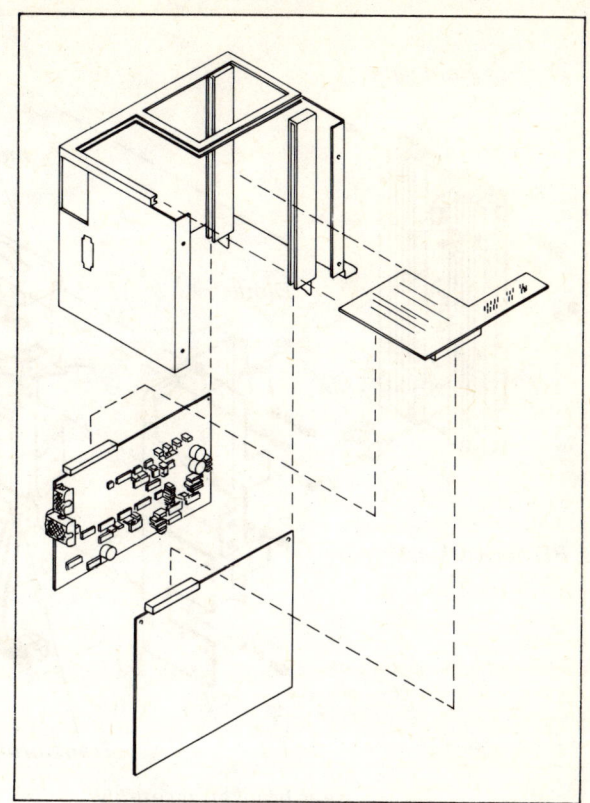

Fig. 18   Metal case configuration

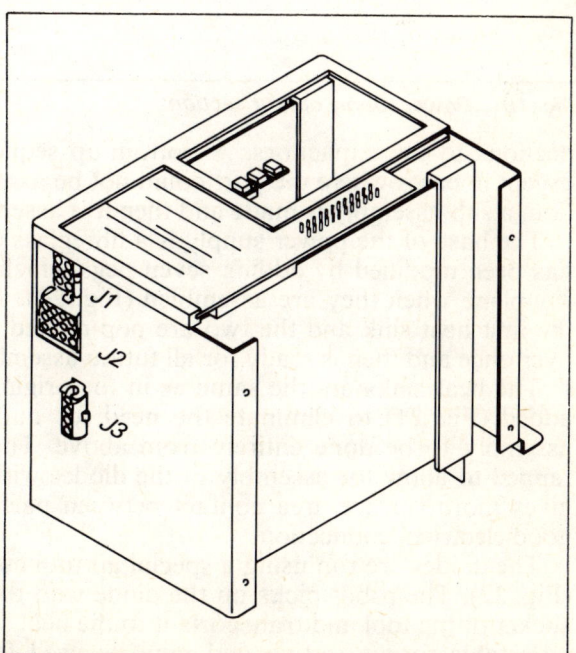

120 PROGRAMMABLE ASSEMBLY

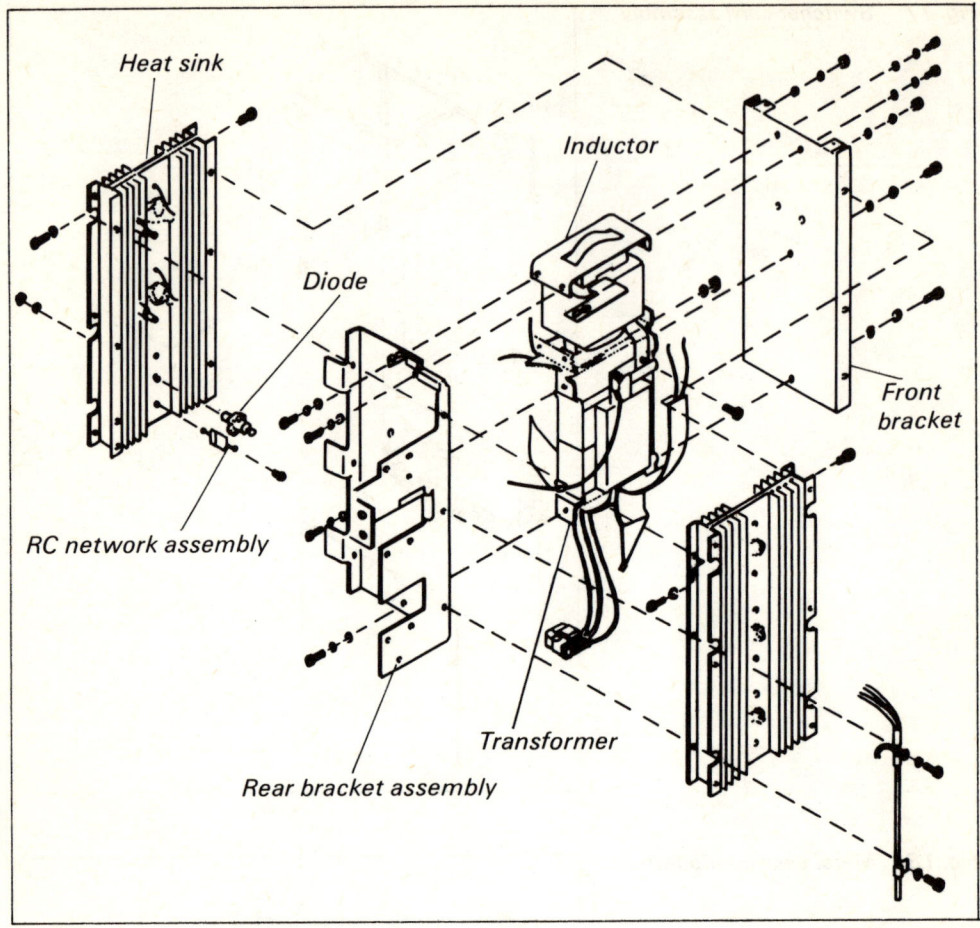

Fig. 19 Power supply output section

nections to the capacitors. A bottom-up sequence of assembly has been used except in cases where the part could not be assembled from above. In that situation, a sub-assembly is made and then it is assembled from above.

The base of the power supply is a fibreglass cover from the original supply. It has been modified by adding seven pegs which will locate the capacitors in the x-y plane when they are assembled (Fig. 20). The glass fibre plate is located on the first heat sink and the two are pop-riveted together. The assembly is turned over once and then is ready for all future assembly procedures.

The heat sinks are the same as in the original supply. Clinch nuts have been added (Fig. 21) to eliminate the need for nuts and washers and to allow the assembly to be done entirely from above. The heat sink has been drilled and tapped to allow for assembly of the diodes without nuts and washers. This also gives more surface area contact between parts for better heat dissipation and good electrical connection.

The diodes are run using a special air tool manipulated by the robotics system (Fig. 22). The robot picks up the diode with the aid of a rubber grommet in the socket of the tool and transports it to the heat sink for assembly. The tool has an adjustable torque setting and may be used for various jobs by changing the

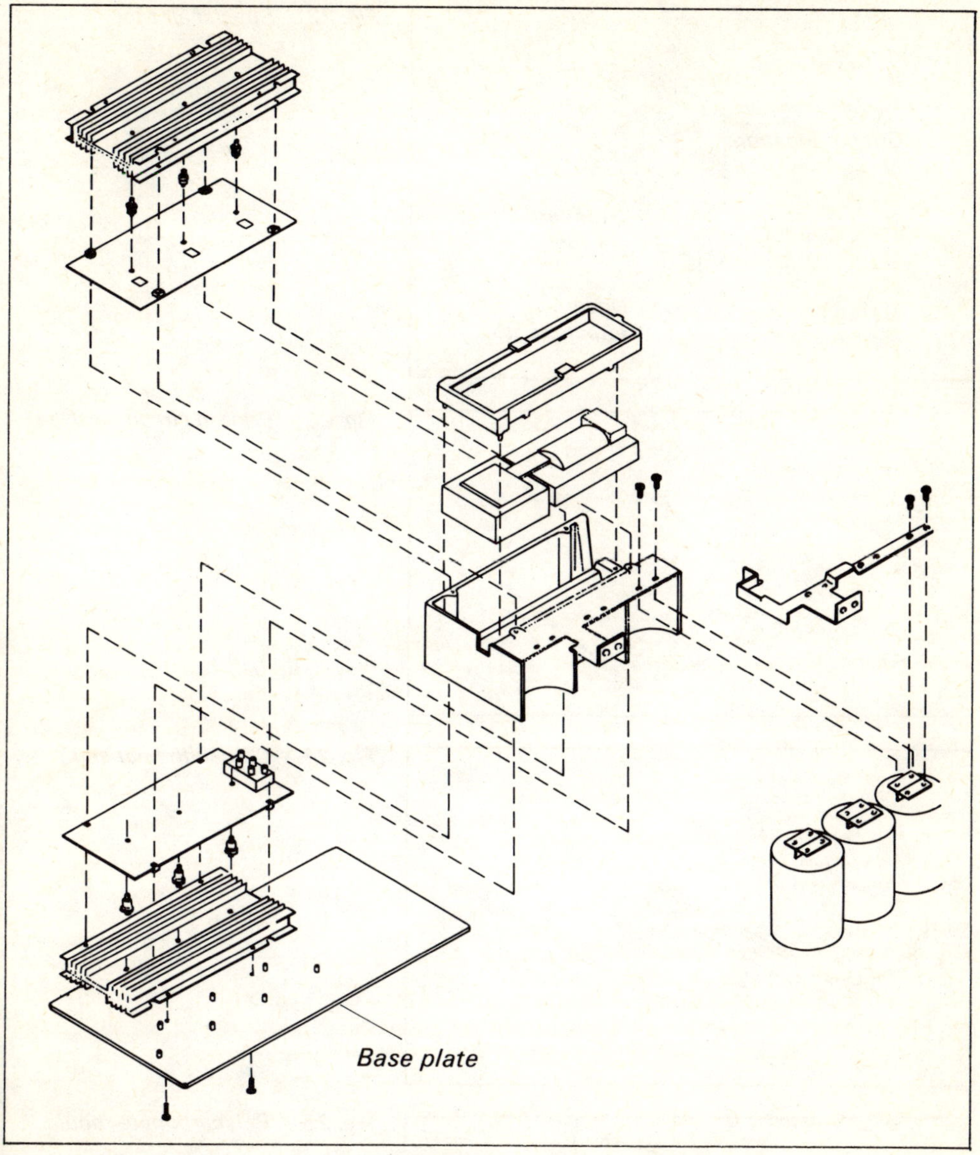

DESIGN OF DATA PROCESSING EQUIPMENT 121

Base plate

Fig. 20   Power supply designed for automation

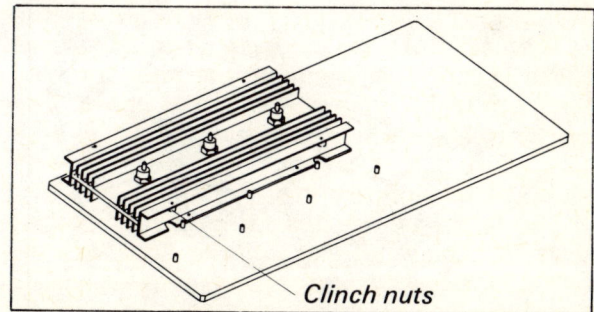

Fig. 21   Clinch nuts on the heat sink eliminate the need for nuts and bolts

Clinch nuts

122   PROGRAMMABLE ASSEMBLY

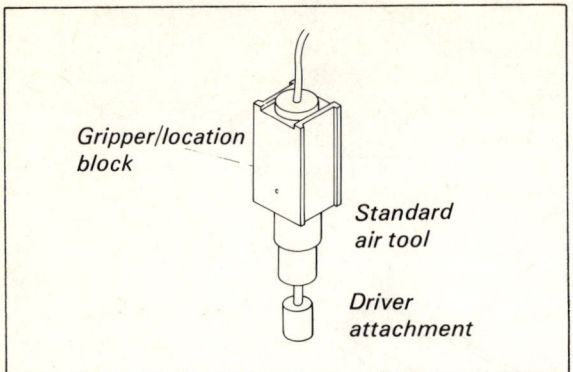

Fig. 22   Special air tool

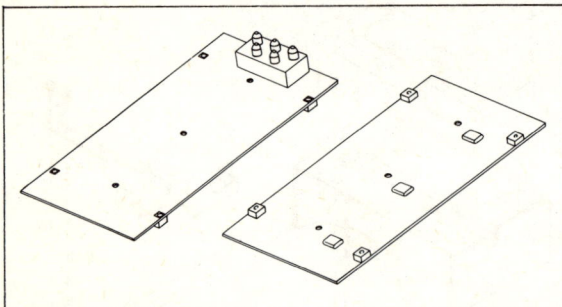

Fig. 23   Printed circuit card

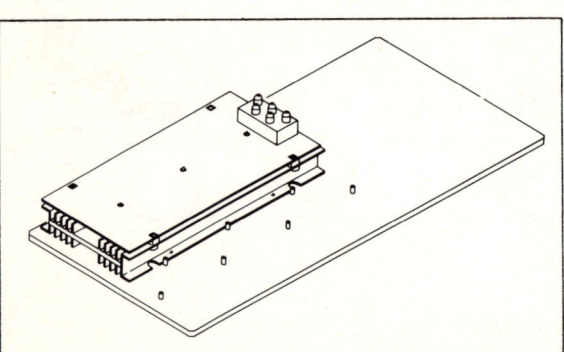

Fig. 24   Card with heat sink

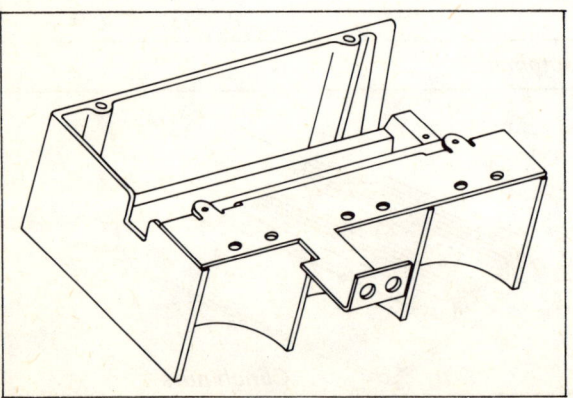

Fig. 25   Polycarbonate mould

driver head. The leads of the diodes are pre-tinned so that they may be reflow soldered using the hot gas process mentioned earlier.

The connections from these three diodes to the transformer are made through a 5oz copper printed circuit card. The card has an RC network assembled to it for each diode and a connector for contacting the transformer (Fig. 23). Studies are being done to develop a high current connector with a low insertion force for this application. The card also has four spacers to keep it a specific distance from the diode cases. The card is located on the heat sink/diode sub-assembly and the diodes are reflow soldered to make the electrical connections (Fig. 24). This design for the heat sink/diode/printed circuit card sub-assembly eliminates six-hand soldering operations. There is an identical heat sink/diode/pcc sub-assembly which is built at this time and assembled to the power supply later.

The next piece to be assembled is a polycarbonate mould which takes the place of two brackets, a deflector plate and numerous fasteners (Fig. 25). The mould has four conducting cylinders incorporated into it as well as a bus bar-connector. The cylinders serve as electrical connections between the two heat sinks. The bottom of the mould is formed to nest the magnetics in their proper position for connection to the printed circuit cards. A secondary coating of conductive material such as zinc or nickel will enhance the shielding characteristics of the mould.

The present method of connection from the transformer to the diodes is through hand-soldered wire leads (Fig. 26). By placing the transformer leads into a pluggable package the assembly time and difficulty are greatly decreased. The transformer and choke will be assembled into their own one piece package with pluggable leads at the supplier (Fig. 27) to further facilitate automated assembly.

Fig. 26 Standard and preferred methods of connection for the transformer

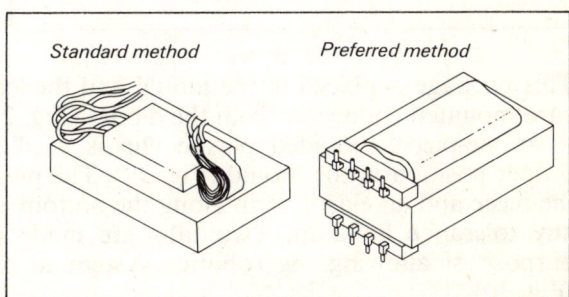

Fig. 27 Magnetic package

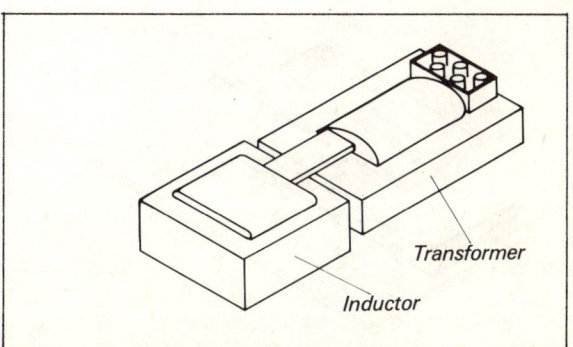

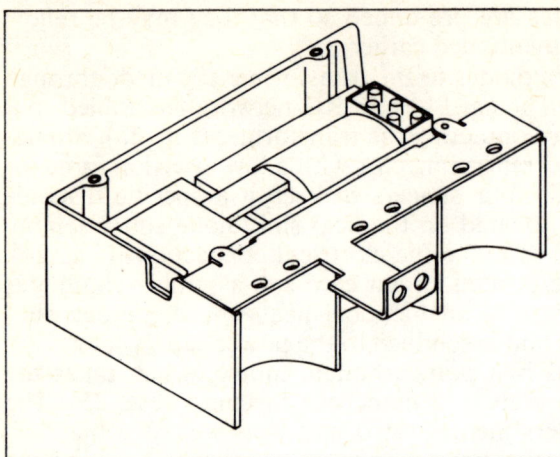

Fig. 28   Magnetics in mould

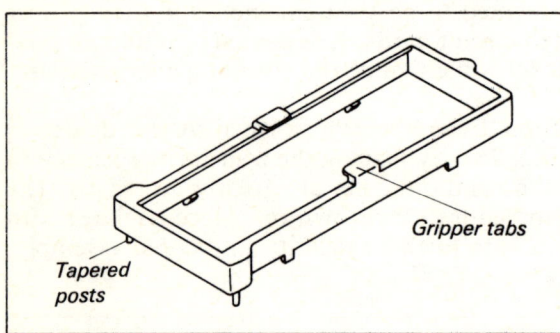

Fig. 29   Retaining clip

Gripper tabs

Tapered posts

This package is placed in the mould and the leads are ready for location on the card mounted connector from the diodes (Fig. 28).

A one-piece, moulded plastic clip is used to retain the magnetics in their proper position in the mould (Fig. 29). The posts are tapered to help in aligning the piece and an elastic strip along the bottom edge will compress to account for any tolerance build-up. Two tabs are made part of the piece solely for the purpose of allowing the robotics system to pick it up and locate it properly (Fig. 30).

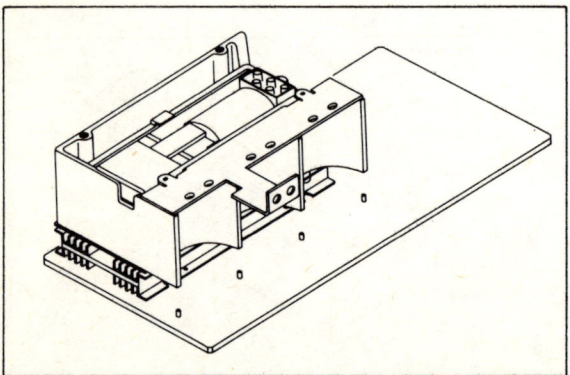

Fig. 30   Assembly of one-piece moulded plastic clip

DESIGN OF DATA PROCESSING EQUIPMENT 125

Fig. 31   Capacitor

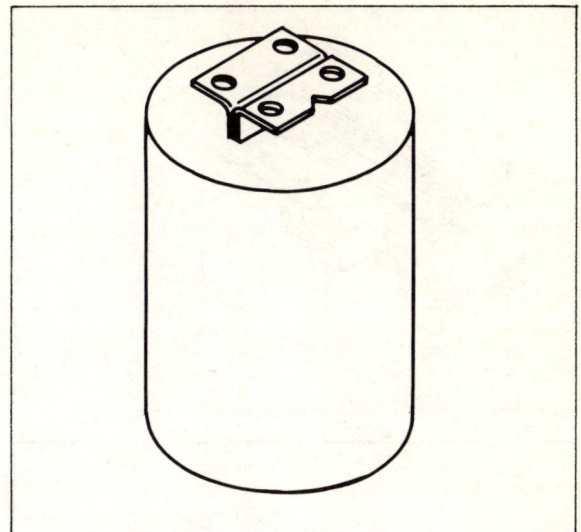

Fig. 32   Capacitors in assembly

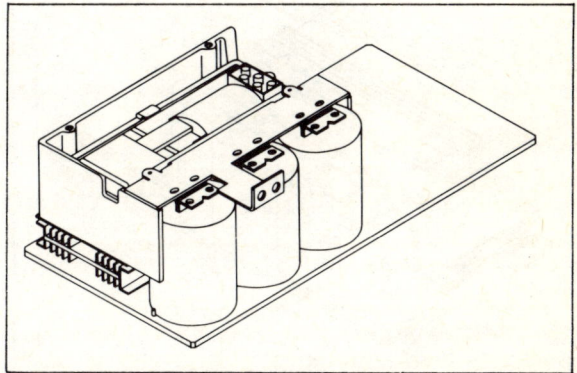

The capacitors used in this concept are the same ones used in the original supply. A notch cut into the negative lead (Fig. 31) will allow the robot to align the capacitors at the correct rotation and the studs on the glass fibre base will locate then in the *x-y* plane. Once the three capacitors have been located on the base plate (Fig. 32) the bus bar/connector may be assembled.

Fig. 33   Bus/connector in assembly

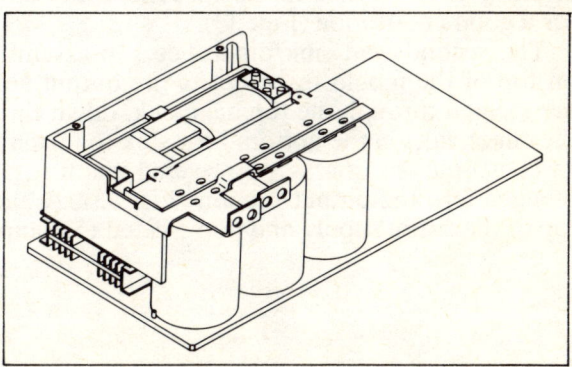

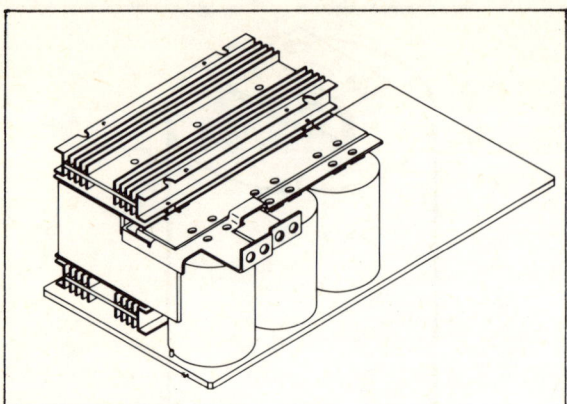

Fig. 34 Heat sink/diode/pcc sub-assembly

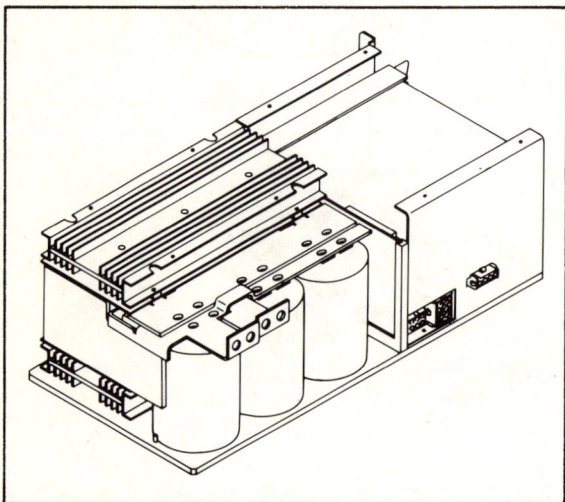

Fig. 35 Final assembly

The bus bar/connector takes the place of three parts and six screws that held the parts together. It makes the connections from the inductor and capacitors to the world outside of this regulator. In the original design, the capacitors were rotated 90° relative to how they are depicted here. This caused the old bus bar to be much more complicated than it needed to be. The addition of short oval slots in place of the drilled holes will relieve any strict tolerances while still allowing for a good connection (Fig. 33).

The second heat sink/diode/pcc sub-assembly is turned over and assembled on top of the mould to complete the output section (Fig. 34). Four long screws are located through the top heat sink, down through the mould and screwed into the clinch nuts in the bottom heat sink to fasten the assembly.

The input section is pop-riveted on the glass fibre base plate after being plugged into the output section (Fig. 35). A glass fibre cover is then located on top of the entire supply and pop-riveted to complete the assembly.

PRODUCT DESIGN FOR AUTOMATIC ASSEMBLY

T. Lund and S. Kähler
Danish Technology Ltd., Denmark

First presented at the 2nd European Conference on Automated Manufacturing/Automan '83,
16–19 May 1983, Birmingham, UK

Abstract

The key to success in automatic assembly centres on the design of the product being assembled. So when planning for automatic assembly one has to take a careful look at product design for automatic assembly. In this paper the authors present strategy and a number of guide rules with the aim of optimising the design of the product as well as the automatic assembly system through parallel development of product and production technology. This strategy maximises the chances of success in an automated product assembly programme.

Introduction

Product design is the first of many steps in the manufacturing process. All of the opportunities for and limitations on efficiency in manufacturing and possibilities for rationalisation of assembly are established at the product design level.

The technical development of assembly equipment is relatively slow, and the efforts to create universal machines are not yet convincing. Thus the task of designing a new product for automated assembly is often combined with the task of designing new and specialised assembly equipment.

These tasks have to be performed simultaneously, with considerable demands on the designer to foresee the consequences of his decisions:

☐ He must design the product so as to attain high quality.
☐ He must know which design parameters determine the quality of the assembly (in the broadest sense).
☐ He must, if he is to be able to deal with this very complex task, be acquainted with the general principles of designing for ease of assembly.

Rationalisation of assembly

The task of rationalisation of assembly must be looked upon as total, in other words as an optimisation of a whole product and production system. Many factors play a part in this complex optimisation, but there are four main goals which must be emphasised:

○ Improvement of the effectiveness of assembly, i.e. increased productivity in relation to manpower and investment resources.
○ Improvement of product quality, i.e. improved product value from the buyer's standpoint in relation to the product's price.
○ Improvement of the assembly system's profitability, i.e. increased utilisation of the equipment.
○ Improvement of working environment within the assembly system.

Production systems are normally conservative; that is, changes in product or system create both forseeable and unforseeable problems – and therefore should be avoided. A re-organisation of assembly should not be regarded as an end in itself, but should be used as a link in a total rationalisation using the four goals mentioned above.

Design rationalisation

The designer determines the structure of the product; that is, its component construction and the way in which these components are joined, in addition to determining the detailed design of each of the components.

This will normally result in a fairly precise production process and a correspondingly precise assembly process. If the designer proposes another product structure the number and type of processes and assemblies will be altered. If he proposes a different component design, still other process and assembly methods will have to be applied.

The most radical assembly improvements lie in selecting and designing product alternatives in which certain assemblies can be disposed of, or greatly simplified. In other words: *Assembly can first and foremost be rationalised by changing the products so that assembly becomes superfluous or at least simplified.*

Product oriented rationalisation

If we assume the starting point is in the production system, particularly in the assembly system, then opportunities for rationalisation lie in exploiting an optimal assembly system.

In his design of the product, the designer determines which type of assembly system is feasible, in addition to establishing how the system will function through his specification of the component's quality.

A product cannot be regarded in isolation when we are discussing assembly problems. A product is normally divided into a series of product variants; certain sub-systems in the product can appear in other products and certain components can be applied in various sub-systems or can be produced because of group-technological similarities with other components.

Thus design for automatic assembly can be said to be the process of achieving the insertion of a product into a well-structured product, building element and component programme.

When should design for assembly be applied?

The answer is not 'always'. The designer will normally concentrate first and foremost on getting the product to function within the economic limitations laid down. Time is at a premium; as a result the most important activity in the closing phase of design is to get the product detailed so it can be in production as soon as possible – in other words getting the drawings finished. Assembly deliberation can easily become a minor part of a large hectic process. The result being a non-optimal product from the assembly point of view.

Modes of operation

Products with optimal assembly are developed today by means of a design process consisting of many steps. The finished, marketed product is given a quality 'lift', focusing on assembly.

An alternative – better, but seldom used – mode of operation is to attach greater significance to assembly deliberations in the early phases of design, in order that the product's structure and design is geared to an optimal assembly process. Such a process will normally require parallel development of product and production system, with special emphasis on the assembly system.

Areas of application

A constructive application with a view to rationalisation of assembly can be applied to three areas or on three levels:

☐ Product assortment.
☐ The product (structure, building blocks).
☐ The components.

Such application, whether it be purely revision of a product or a new design, can occur in the following ways:

○ Creation of design degrees of freedom, so that alternatives containing good assembly oriented characteristics can be created.
○ Application of the principles of design for assembly, primarily elimination and secondly principles of improvement.

Design for assembly

Product assortment

The overall goals for optimising the assembly may be expressed in the following terms:

☐ Constant, high product quality.
☐ High productivity.
☐ High utilisation.
☐ High quality working environment.

By analysing these goals, one may determine the factors contributing to their attainment. High utilisation for instance is influenced by the size of the production series assembled by the same equipment, which means that you have to follow the design principle: 'avoid variations' or 'make sure that variants can be assembled in the same way'.

The goal of 'high productivity' is influenced by the number and length of stoppages in the use of the equipment. Thus one must either design a system to accept many variations in components or must ensure components with few variations.

Table 1. Product assortment principles in 'design for assembly'

- Mechanical assembly requires quality control
- Automatic assembly requires high quality
- Assembly is (also) a management problem
- Design for mechanical assembly
- Design for automated assembly
- Design for automated, flexible assembly
- Manual assembly has many virtues
- Eliminate assembly
- Design for standard equipment
- Design for special equipment
- Avoid variations in the product
- Make sure that variants can be assembled in the same way
- Automation may result in improvements in the working environment
- Avoid 'dangerous' assembly methods

So one has to decide upon an assembly and design policy and take decisions about the product mix in order to obtain the maximum effect of rationalisation. The principles to be considered are listed in Table 1.

It is evident that these principles do not unanimously point to a solution. In a certain situation a principle may often lead to good results, in another it may be wrong. One has to choose which principle to follow.

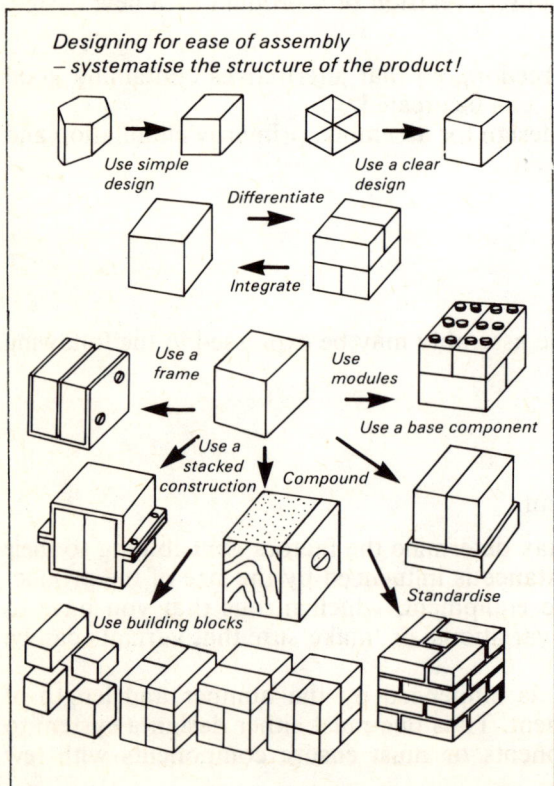

Fig. 1 Survey of structural principles for 'design for assembly'

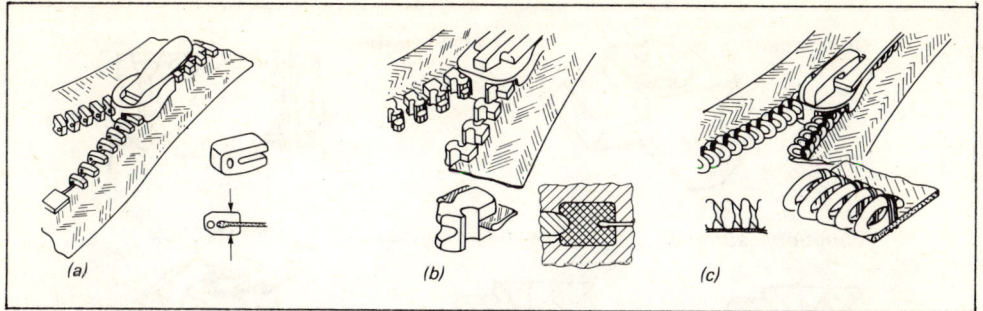

Fig. 2 Three different production methods for zip fasteners. (a) Metal zip fastener elements produced by pressure casting and mounted one by one on a band. (b) Band layed in an injection moulding machine and moulding each element around the band. (c) Zip is formed from plastic cord, which is bent and sewn into shape as the teeth of the fastener

The product (structure, building blocks)

The fundamental structure of the product in principle determines its components, but normally a much more detailed structure (a greater number of machine parts) is chosen in order to satisfy different demands. The designer works on two levels: a fundamental structure level, where techniques and solutions are logically connected, and a quantitative structure level, where he takes decisions on distances, tolerances, positioning in space and division into machine parts. In this way the function of each component is defined and the detailed design may be carried out.

This structural design actively has many possibilities for alternatives and important principles of design for assembly may be applied.

Two main design principles should be noted, 'simplicity' and 'clarity' as means for obtaining optimum solutions, few parts, few assemblies and simple assemblies.

By analysing different products, one recognises different structural principles, and in particular solutions created by 'integration' or 'differentiation', the latter implying the use of more components. Both principles may be applied in design for assembly and may lead to good results. A survey of structural principles is shown in Fig. 1. As an example Fig. 2 shows the effect of integration for the production of zip fasteners.

The components

The structure of the product determines the basic design of the components and the principles of the assembly. This means that the choice of product structure fixes most of the assembly problems. But one still has to think carefully about the detailed design of the components, in order to facilitate the operations of the assembly system.

Detailed design of a component means specifying the following basic properties: shape, material, dimension, surface quality and tolerances.

The assembly quality will depend on one or more of these basic properties, the most important is the shape, because certain surfaces on the component are utilised in the assembly process itself.

These surfaces, which limit a component, have a variety of tasks. The function of the component is realised by the *functional surfaces*. Among these, some surfaces, the *connecting surfaces*, relate to other components by touching them. The non-functional surfaces may be called *free surfaces*.

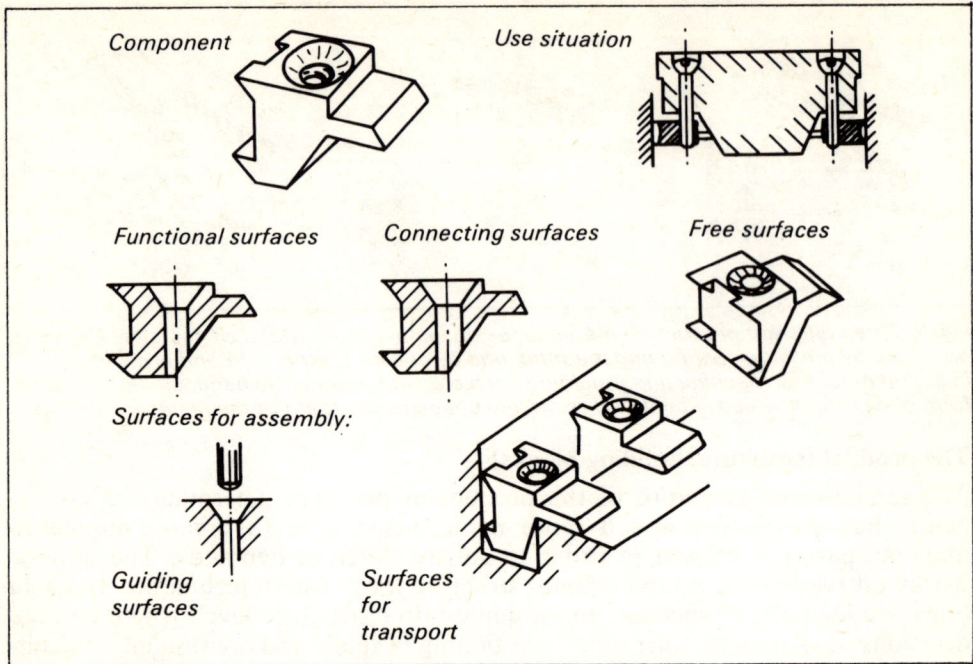

Fig. 3 Clamping piece from a switch

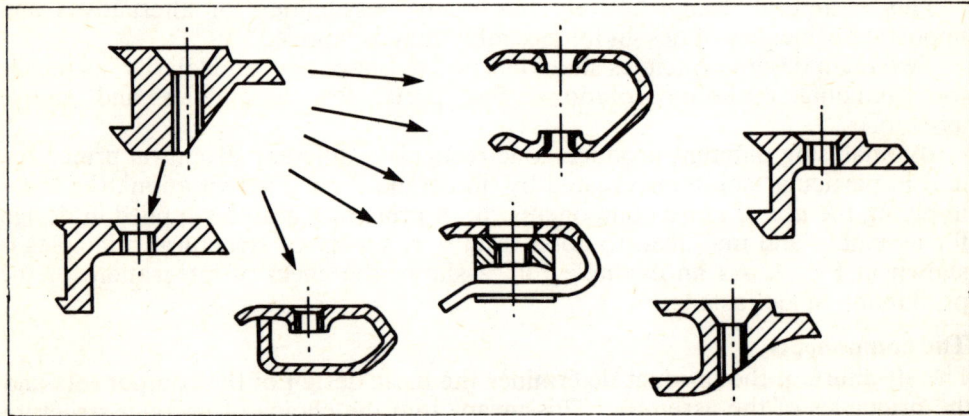

Fig. 4 Alternative designs for the clamping piece shown in Fig. 3. The functional surfaces are unaltered, but new component shapes are indicated

Assembly surfaces are surfaces used in the assembly system for orientation, transport, positioning and guiding. Designing the component for assembly implies utilising the functional surfaces and manipulating or changing the free surfaces, to create good conditions for the assembly operations (see Figs. 3 and 4).

The dialogue between the designer and the production engineer is focused on the main conditions and the design possibilities, and is simplified if one uses sketches showing the functional surfaces and the free surfaces, for example in colours. This technique is important, especially with more complex components.

Table 2. Survey of main principles for the designing of components for automatic assembly

> Avoid assembly operations:
> – integrate component
> – utilise integrating production methods
>
> Avoid orientation operations:
> – use magazines
> – use components connected in bands (tapes)
> – integrate the production of components into the assembly
>
> Facilitate the orientation operations:
> – avoid clamping or hooking
> – put special faces on the component for orientation
> – avoid components of low quality
> – make the components symmetrical
> – or make it clearly asymmetrical
>
> Facilitate transport:
> – design the component for easy transport
> – design a base component
>
> Facilitate a simple pattern of movements:
> – make all joins simple
> – put special faces on the component for guiding purposes
>
> Choose the right method for joining components together:
> – avoid joins
> – avoid separate connecting elements
> – use integrating production methods

The main operations which have to be taken into account when designing the components are: orientation, transport, connection and joining together.

Table 2 gives a survey of the main principles for designing good components.

Strategy for design assembly

Several principles have been discussed which may be adopted in designing for assembly, usable on different levels in a design project, reaching from overall goal-setting and planning of the assortment of products via structuring of the product to design of the product's components.

The question is now how to perform 'design for assembly', that is to say, how to use the principles in practice.

Assembly problems are traditionally dealt with in the production preparation area, but it is too late to consider the processes and the assembly when all drawings and parts lists are finished. The designer has already consciously or unconsciously chosen and fixed the assembly method. Thus one must use a design strategy where assembly problems are dealt with at the same time as the structural design and the detailed design of the product.

The design process may take the form shown in Fig. 5. Note, that the central part consists of the important design principles 'to integrate – to differentiate' in accordance with the principles in Fig. 1.

In this scheme it is possible to see all the levels, where there are possibilities of creating alternative solutions. Alternatives are necessary for finding the optimum solution. The quality of a single solution cannot be verified.

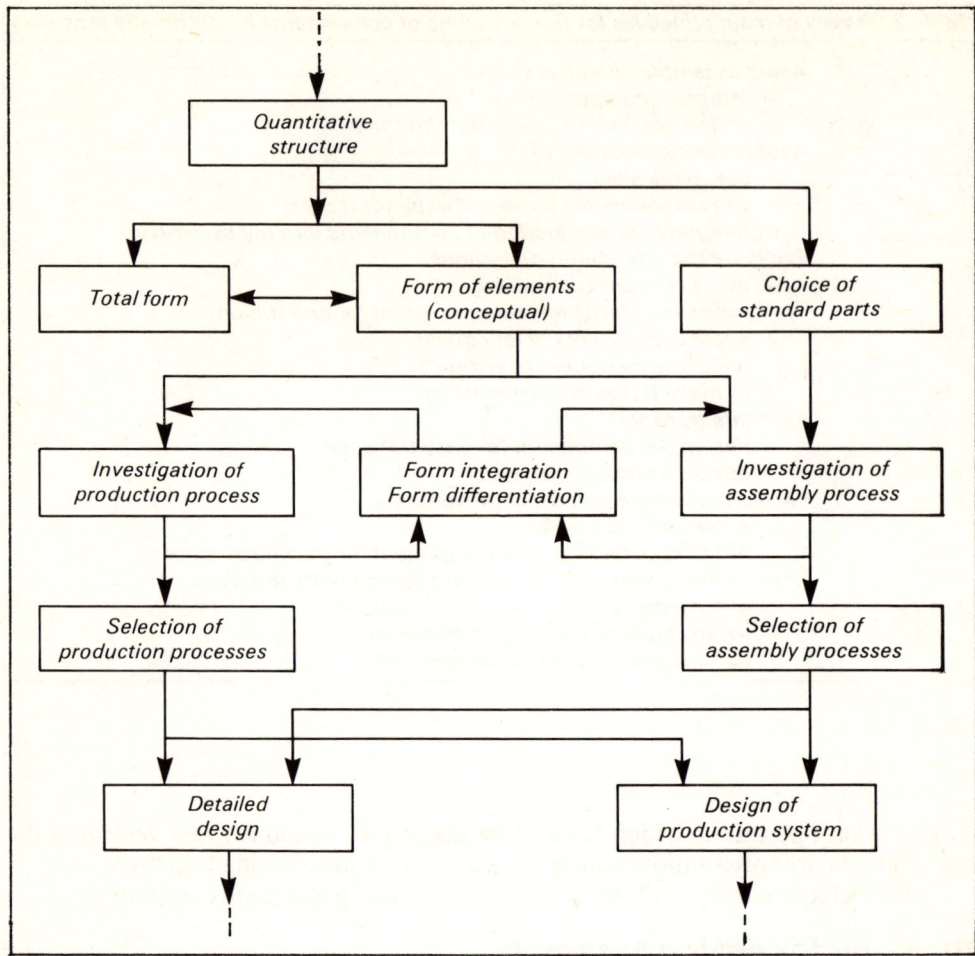

Fig. 5 Idealised procedure for integrated design of product and production (assembly) system

Case Study: The ECG electrode

This example refers to a development project for the Danish company Simonsen & Weel A/S (S & W). The project was managed by one of the authors, Thomas Lund, while working for IPU (Institute for Product Development). The project is an example of the integrated development of a product, a production technology and a production system.

The background for the project is that S & W produces and sells electronic equipment for patient surveillance (Fig. 6). A so-called ECG electrode is used as the pickup for the small electric currents produced by the functioning of the heart. When the project was started, S & W was buying its electrodes from an American firm, and produced the electronic equipment itself. In order to increase its profit margin, S & W decided to produce ECG electrodes itself. The product which S & W wished to produce is shown in Fig. 7. As S & W wanted us to deliver a fully automatic equipment for production of this product, we carried out a rapid analysis of the proposal.

Fig. 6 S & W's latest apparatus for patient monitoring

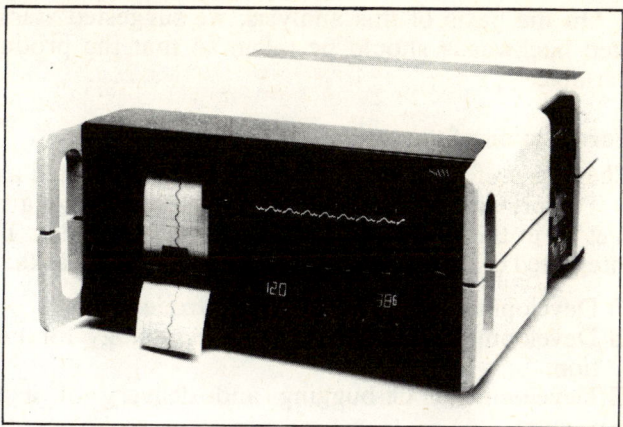

Analysis of the proposed product

An analysis of the product (Fig. 6) gave the result:

○ The properties of the product as a measuring instrument would probably be excellent.
○ Investments could be made in the way that the product is used. For example, it could be filled with electrolytic paste during production, so that the nurse does not need to do this.
○ The cost of materials in the product could probably be reduced.
○ The product's form is based on experience with manual production and is not well-suited to fully automatic production. The many variants, for example, would be very troublesome.

Fig. 7 The product as originally proposed. Note that the requirement as to variations concerns the wire and plug and not the actual electrode. Variants can therefore be completely avoided

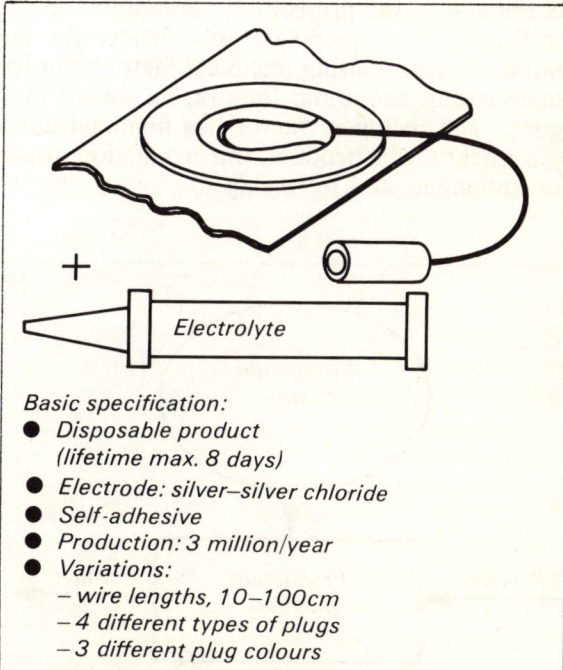

Basic specification:
- Disposable product (lifetime max. 8 days)
- Electrode: silver–silver chloride
- Self-adhesive
- Production: 3 million/year
- Variations:
  - wire lengths, 10–100cm
  - 4 different types of plugs
  - 3 different plug colours

On the basis of this analysis, we suggested starting up a project, but that a step backwards should be taken so that the product's design could become a variable.

Formulation of the task

The object of the project was therefore formulated as follows:

The principal aim of the project is to develop a new commercial opening for S & W in the field of disposable ECG electrodes. This aim is to be achieved by integrated (parallel) evolution of the following tasks:

☐ Development of a new ECG electrode.
☐ Development of a production technology for fully automatic mass production.
☐ Development, de-bugging and delivery of a fully automatic production system.

This formulation gives the necessary *constructive degrees of freedom* for an attempt to *optimise* S & W's commercial possibilities.

A very simple model of the project is shown in Fig. 8. Note that the project really consists of two projects, which are carried out simultaneously: a 'horizontal' project, which has as its outcome the product, and a 'vertical' project, its outcome the production system. The two projects meet in the production process. Thus the development of this process plays a central role in the overall project. The two sub-projects are, as it happens, of very different types. One of them is to result in a product which is to be mass-produced. The other is a project whose result is a product (a machine) to be produced on a one-off basis.

Strategy and financial considerations

As the aim of the project is to create the most profitable commercial possibility for S & W, it is necessary to consider the development project not just as a matter of manipulating technical factors, but to a considerable extent also one of manipulating economic factors, as shown in Fig. 9. As can be seen from the figure, each technical factor has financial consequences. Only by simultaneous adjustment of the product, the production process and the technical properties of the equipment and by trying to convert the technical properties into economic

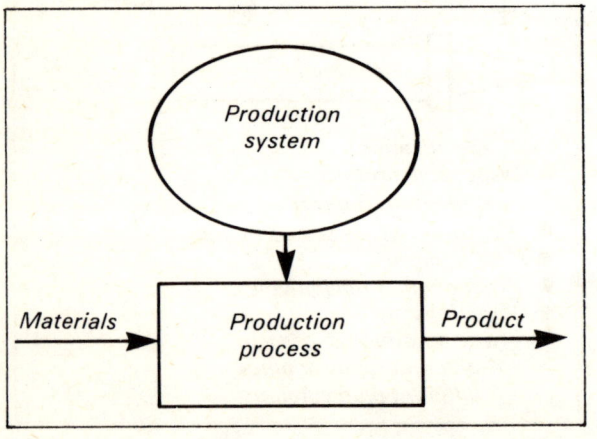

Fig. 8 A simple version of a project model

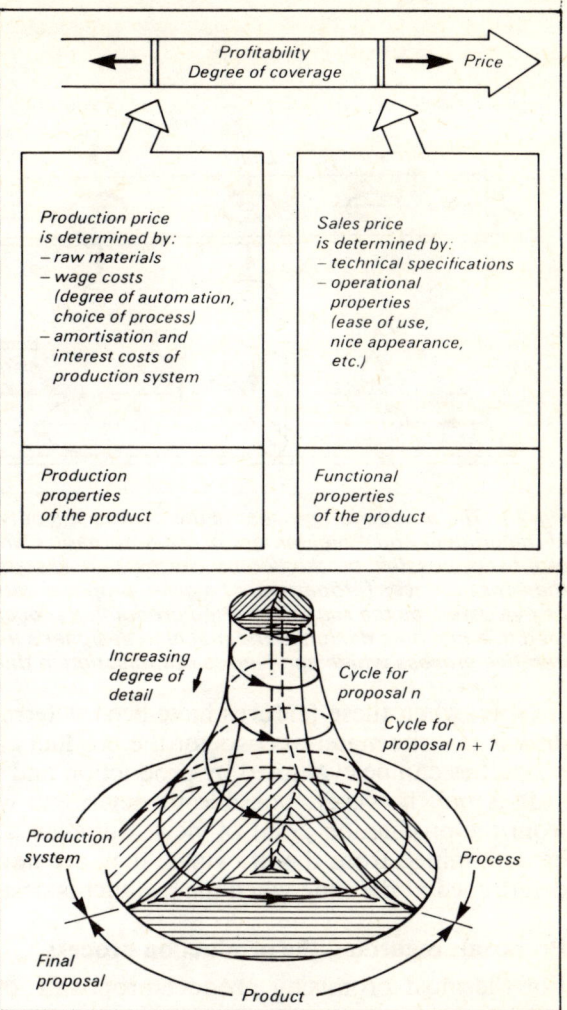

Fig. 9 The aim of the project is to create the most profitable commercial possibilities, i.e. to maximise the degree of coverage. This is done by manipulating the properties of the product, so as to obtain the highest possible sales price for the lowest possible production price. Note that both the product's production properties and its functional properties can be manipulated. All too often, only the functional properties (determined by the market) are altered. Then the management accept the production properties which the product happens to get ('The production department will have to solve that one')

Fig. 10 During the project, a large number of different proposals were examined. Each proposal is a combination of the design of the product, in conjunction with production technology and an outline for a production system. After a number of cycles, the relations between the three variables begin to be stabilised, so that an optimum, balanced combination is attained

consequences, is it possible to achieve the establishment of the most profitable commercial possibility.

The project strategy is an iterative process, as illustrated in Fig. 10. A large number of alternatives are investigated. Each alternative consists of proposals for the design of the product, the production technology and the production system. The individual cycles of this process are characterised by continual rejection of proposals which have shown themselves to be less advantageous, together with a continually increasing degree of detail.

Proposals regarding the product

A proposal for a product, as it normally appears from the drawings, is principally characterised by:

☐ The design (structure).
☐ The form of the components.
☐ The number of components.
☐ The choice of materials.

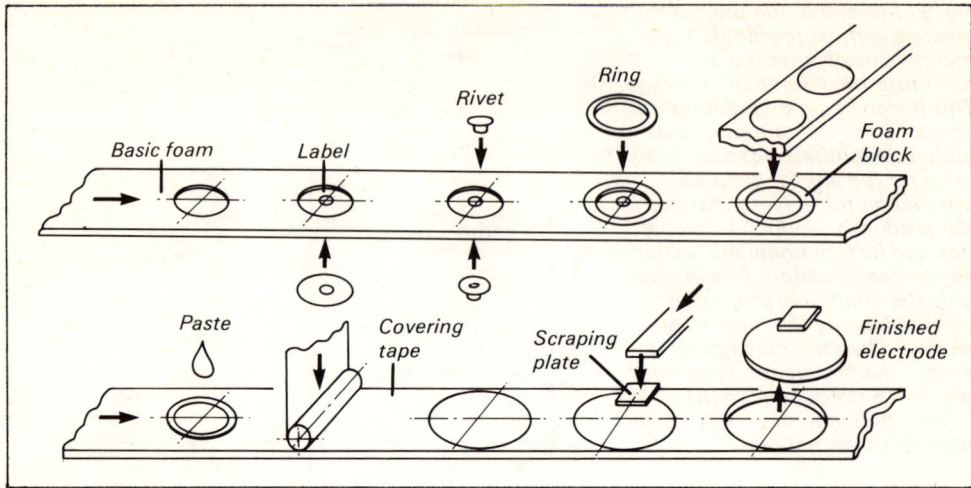

Fig. 11 The production process is the central link between the two projects: development of the product and development of the automatic plant. For a long series of product proposals, possible production processes were sketched as shown in the figure. In this way, the consequences (properties) of a given proposal with respect to production technique can be evaluated, as the roughly drafted production process can easily and rapidly be transformed to a machine design in the machine designer's imagination or on his sketch-pad. An iterative process which can lead to optimisation in thus initiated

Note, when these features have been determined — for example in the form of drawings — the material costs for the product can be calculated and its functional properties can be evaluated by production and testing of prototypes.

It is *not*, however, possible to determine whether its properties are optimal from the production point of view. But even if the production properties cannot be determined from the drawings, they are nevertheless more or less completely determined at the time when the product is designed (drawn).

Proposals regarding the production process

For the most promising product proposals, one or more production processes were proposed, each characterised by:

○ The manufacturing processes for the individual components.
○ Assembly processes.
○ Checking processes.
○ The processing sequence.
○ The form of the raw materials as delivered (as individual components, in coils, etc.).
○ The pattern of movement of tools and other active elements.
○ The timing of the processes.
○ Buffer storage.

The various proposals were documented/sketched as shown in Fig. 11, and we were now well on the way to being able to evaluate the properties of the product with respect to its production system.

Proposals regarding the production system (machine and operator)

For the most promising production processes, proposals were made for alternative systems (both machines and operators), which could realise the production processes in question. Each system was characterised by:

☐ The operator's tasks.
☐ The machine's tasks.

As it rather rapidly became apparent that it would be both technically possible and economically advantageous to have fully automatic production, the operator's tasks were reduced to:

○ Supplying raw materials.
○ Removal of the finished products and packaging.
○ Fault correction.
○ Supervision.

The machine was characterised by:

☐ Its frame design.
☐ The design of the tools.
☐ The control system.
☐ The checking system.
☐ The feeding and orientation system for raw materials.
☐ The storage system.
☐ Its appearance.
☐ The way in which it is operated.
☐ The cycle time.
☐ Its productivity.
☐ Its flexibility (readjustment to other tasks).
☐ Its size.
☐ Its weight.

These properties determine the performance (capacity, quantity) of the system, and the cost of the machine and the direct wage costs for carrying out the actual production.

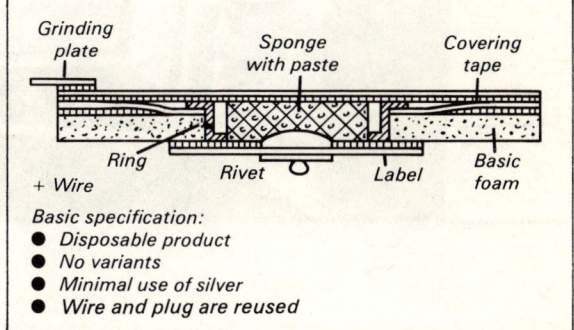

Fig. 12 The resulting product is made up of nine components. There are no variants of the electrode. The requirement that there should be variants is satisfied by the wire (which is the part in which variations are desired) which connects the electrode and the measuring apparatus being a reusable product, which is not connected to the disposable product (the electrode) until it is to be used (snap lock connection). This offers advantages with respect to both production technique and economy (it eliminates the need for materials for the wire and plug)

Fig. 13 The machine has a capacity of 3 million units per year. Emphasis has been placed on an attractive appearance, even though it is a special machine of which only a single example is to be manufactured. If the operator is proud of his machine, it will be looked after more carefully. In this way a pleasant working environment and high productivity are obtained simultaneously

It was then possible to evaluate which combination of product design, production technology and production system could provide S & W with the best commercial possibilities.

The chosen combination of product design and production system is shown in Figs. 12 and 13.

Conclusion

Between 70 and 80% of the total production costs for a product are determined by the designer. Therefore he plays a very important role when the assembly process is to be rationalised.

This paper has focussed on some of the design principles which can be used in the design process, and has a strategy for design for assembly on three levels:

☐ Product assortment.
☐ The product (structure, building blocks).
☐ The components.

It is our experience that a grasp of the principles to be followed in design for assembly can be a great help in the design process. But management of the design process, in order to obtain a design which can be assembled easily, is unfortunately difficult. Thus in the future it will be important to obtain an overall view of the situation and to use the methods illustrated.

ASSEMBLY ORIENTED DESIGN

W. Eversheim and W. Müller, Technical University Aachen, West Germany

First presented at the 3rd International Conference on Assembly Automation,
25–27 May 1982, Boeblingen, West Germany

Abstract

Adapting the design of a product to match the technical characteristics of assembly machines is one of the essential conditions for automation in assembly. An important portion of all assembly operations is the handling of assembly parts and connecting elements. By means of guidelines for the handling oriented design of parts the automation of handling operations can be facilitated during the assembly process.

Introduction

The present situation in assembly is — in contrast to the machining of parts — marked by the fact, that up to now not much rationalisation has been carried out[1,2]. Smaller production volumes makes the planning expenditure and the necessary investment too high to introduce rationalisation into assembly operations.

One of the reasons for this is because of the complexity and precision of the assembly process which makes for greater difficulty associated with the delivery of material, the quality control, the control of the assembly process and the storage of parts. In addition there has been a lack of useful techniques for the creation of good organisational conditions and also some technical equipment vital to the automatic assembly process was not available[3]. By means of the latest developments in handling techniques it will, however, become possible by introducing automated devices to rationalise the conditions in the assembly area in the near future.

Experience with conventional assembly machines shows however, that— even with regard to an increasing technical standard of assembly devices — the assembly oriented design of products is an essential precondition for an economic automation[4]. By using assembly-oriented design methods the design department is capable of reducing the demand on automation systems in the assembly area[5].

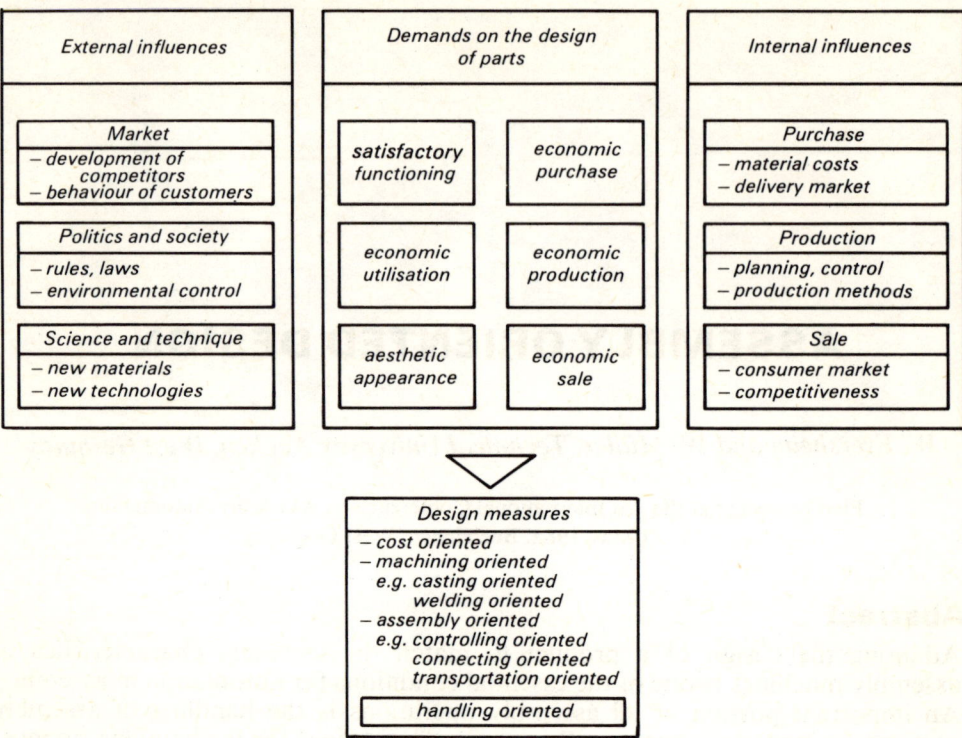

Fig. 1 Demands on the design of product

Demands of the automated assembly process on product design

Assembly is the final part of the production process. And up to now the demands of the purchasing department (e.g. the use of marketable parts) as well as the demands of parts production (e.g. the moulding oriented design) traditionally have been considered during the design of a product. In addition to these influences which come from the inside of the factory, there are a number of external influences, such as the customer's requirements, which also have consequences for the shape of the product (Fig. 1).

The designer's difficulty now is to consider all these multiple demands in his product. In doing this the assembly oriented design is only one way for him to design a product.

Alternatives for an assembly oriented design

The principal alternatives for a designer to realise a design oriented to the demands of an automated assembly, can be divided into three groups:

☐ structuring of the product,
☐ standardising of the parts,
☐ assembly oriented design.

The structuring of the product and the standardisation of the parts are directed at the rational division of assembly tasks and the improvement of the

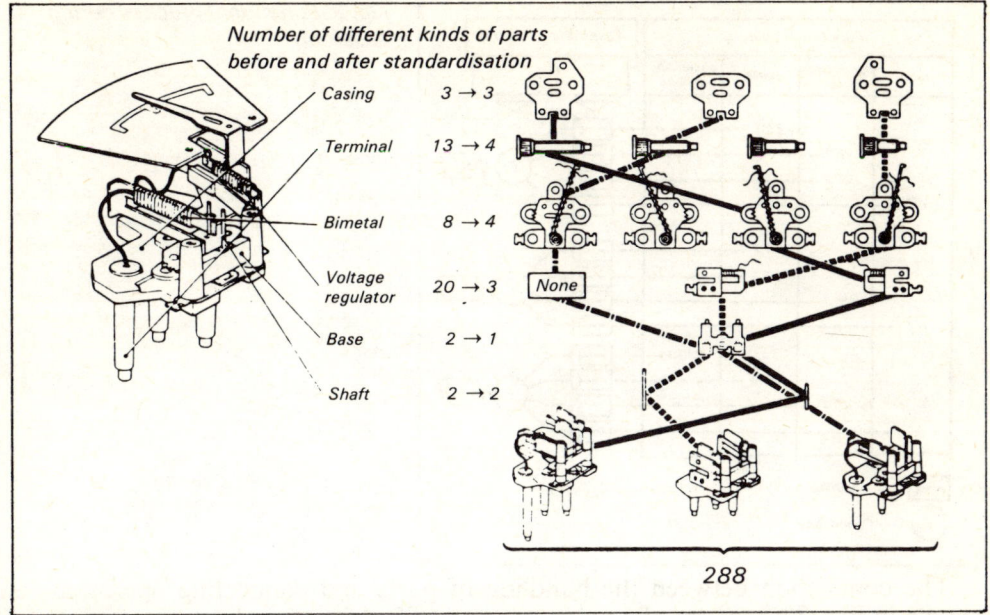

Fig. 2 Standardisation – a possibility for the simplification of assembly

repetition frequency of assembly operations. So above all they touch organisational aspects of assembly automation.

In contrast to this, assembly oriented design serves first of all to reach the easiest way for the technical realisation of assembly processes. It may be assumed that a design that considers automatic realisation of assembly operations will also simplify manual assembly. An example for the systematic structuring and standardising of a product is shown in Fig. 2[6].

Fig. 2 shows the structure of electric indicating instruments as they are used in the automotive industry. By standardising on six structural levels the number of different parts was reduced from 48 to 17. So theoretically 288 different products could be assembled, but only about 110 are really needed. This means, that for an automation of the assembly of these parts the needs for storing and feeding of parts have been simplified. Moreover by means of the standardisation of the parts the organisational expenditure for the parts supply was reduced, the advantages for the manufacturing of the parts are evident.

If, after structuring and standardising the product, it is necessary to change the product's design to achieve an improvement of the conditions for automation, it is at first recommended to look at each single assembly operation. The assembly process consists of a series of assembly functions, such as the connecting, checking and adjusting. These functions help to fulfil the assembly tasks.

During the execution of these assembly functions parts have usually to be handled. So handling of parts is one of the most important additional functions of the assembly process. Often it is hardly possible to distinguish handling and connecting functions, as the connecting operation is often finished with the execution of a handling function. Therefore in many cases assembly oriented design can be characterised as handling oriented design as well.

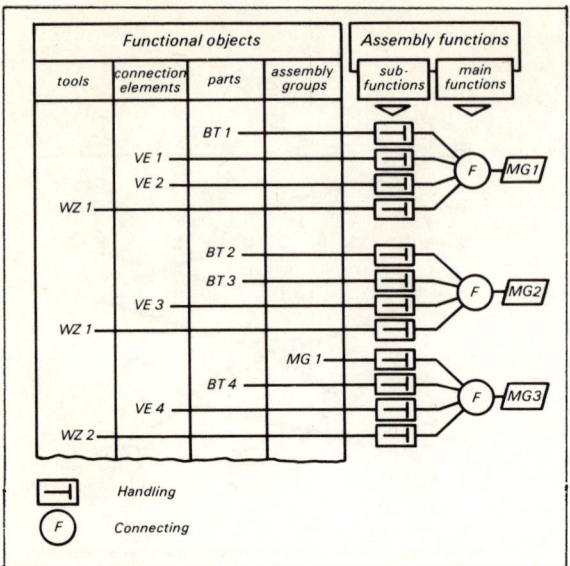

Fig. 3 Functional process during assembly

The connection between the handling of parts and connecting operations is shown in Fig. 3 by means of a very much simplified functional process.

Assembly parts and connecting elements are joined by handling operations. These parts are connected by form, force and/or material locking. The assembly groups have to be handled in the same way, until at the end the total product is assembled.

On the basis of these somewhat theoretical thoughts concerning the components of assembly processes, measures for an assembly oriented design can be developed. There are two principal methods:

○ Assembly operations are to be avoided.
○ Assembly operations are to be simplified.

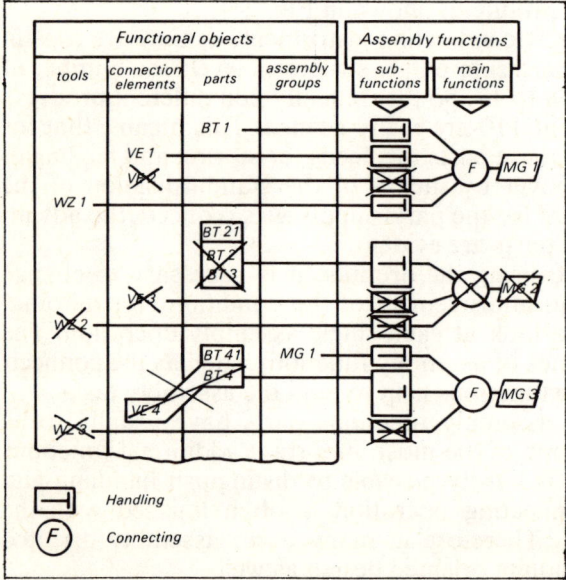

Fig. 4 Possibility for avoiding handling functions

Fig. 5 Integration of single parts in a complex part

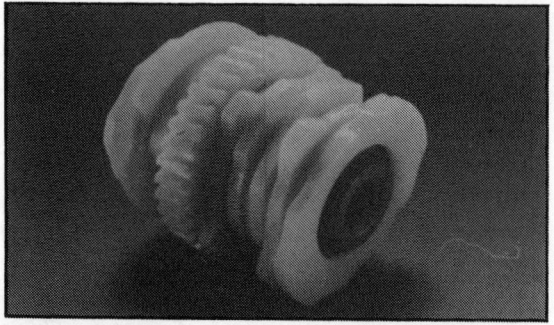

The rationalised reduction of assembly operations refers above all to a reduction of the number of assembly groups, single parts and connection elements that are to be handled. The way in which these measures influence the functional process is demonstrated in Fig. 4.

Avoiding assembly operations, including parts' handling, often can be achieved by checking the number of used connection elements as well as by integrating different parts in one single unit.

This integration of several different parts into more complex single parts is supported by the developments in modern machine tools, that are capable of machining very difficult geometries and materials. For the future it is necessary to use these capabilities in order to support assembly.

An example illustrating the possibility of avoiding assembly operations is shown in Fig. 5. This shows the camshaft of a sewing machine. The unit that was assembled previously from several separated single parts, is now joined in a single complex workpiece [7].

In connection with the integration of single parts it should be mentioned that in doing this the more complex workpiece goemetry leads on the one hand to a reduction of assembly operations. However on the other hand these complex geometries may produce difficulties in handling the parts.

This leads to the second possibility of an assembly oriented design, that is the simplification of assembly operations. As mentioned earlier the simplification of assembly operations means in many cases the simplification of handling operations.

| | | Simplification of handling functions ||
		reduction of the demands on handling by the part	improvement of the suitability for handling
Picking-up	recognition of workpieces	reduction of the demand for orientation	adding of orientation aids
	transfer of forces	reduction of the demand for gripper elements	adding of gripping elements
Moving	change of workpiece position and orientation	elimination of obstacles for movements	adding of feeding aids
	change of quantities	elimination of the tendency for hooking	—
Laying down	reaching of final positions	reduction of the demand for positioning	adding of positioning aids
	releasing of forces	reduction of the demand for gripper elements	adding of suitable gripper elements

Fig. 6 Alternatives for the simplification of handling functions

Looking at the handling functions, these may be regarded as a series of elementary sub-functions — picking-up, moving and laying down — as shown in Fig. 6. According to this more detailed structure of handling functions, there are different possibilities for the simplification of handling operations:

☐ The passive reduction of demands that are made on the capabilities of a handling device by removing obstacles for the execution of handling functions.
☐ The improvement of the handling characteristics of the part by adding special features to the part, that simplify the handling operations.

A commonly known example which creates difficulties for an automated assembly are springs; these usually tend to hook into each other. This habit of hooking can be broken by means of alternate windings and closely rolled wire ends [8]. In the same manner chamfers and guiding slopes are examples to improve the ease of mating workpieces during automated assembly. Other examples are shown in Fig. 7.

The possible measures for a handling oriented design of parts can be stated as aims for the designer and can be handed over to him in form of guide lines — see Tables 1 and 2.

Implementing the measures

The problem for the designer is to reach an assembly oriented design in the most effective way. A checklist, developed by the WZL of the RWTH Aachen, can be

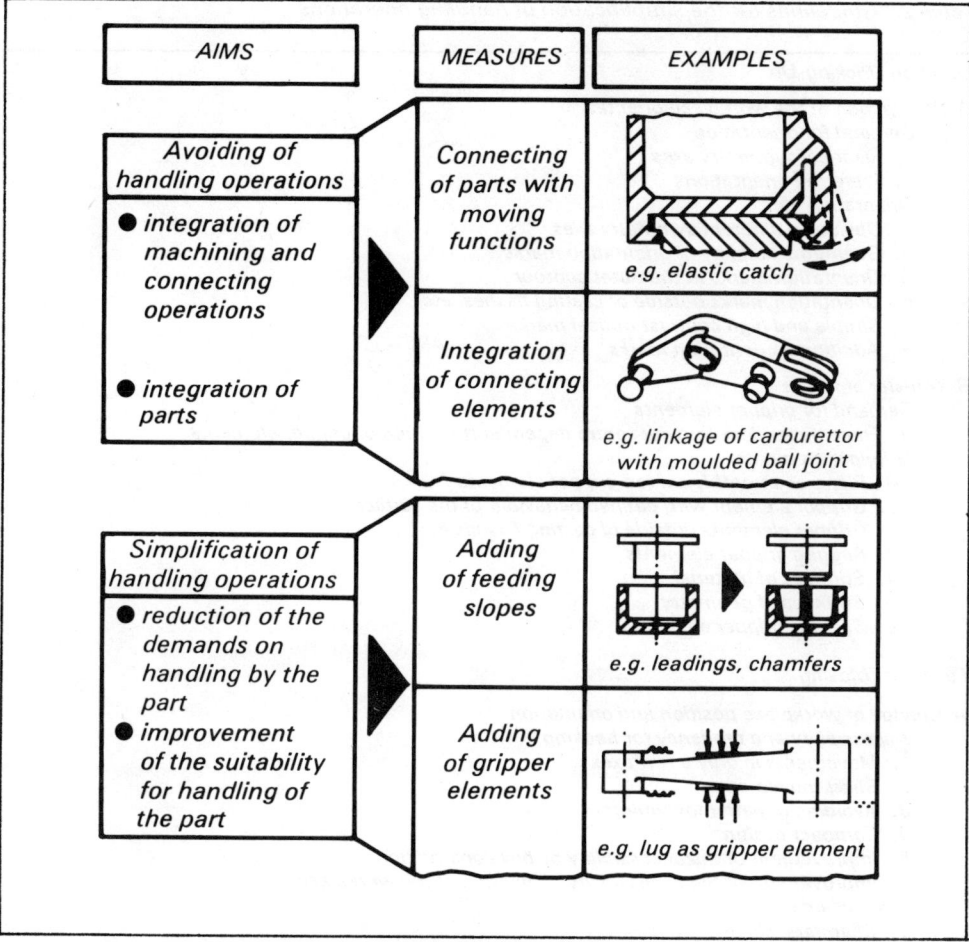

Fig. 7 Examples of handling oriented design

Table 1 Guidelines for the avoiding of handling operations

A. Machining of single parts
   1. Integration of identical or similar parts for parallel machining
   2. Integration of machining operations
   3. Reduction of the number of different machining methods
   4. Reduction of the number of machining operations

B. Checking of single parts
   5. Reduction of the number of controlling operations by modelling machining methods

C. Connecting of parts
   6. Integration of single parts, that have to be rigidly locked
   7. Integration of single parts with moving functions
   8. Integration of assembly and machining operations

D. Handling of connecting elements
   9. Reduction of the number of connecting elements by use of assembly methods without connecting elements
  10. Integration of connecting elements into parts
  11. Integration of connecting element with identical functions
  12. Integration of different connecting functions in one connecting element

Table 2 Guidelines for the simplification of handling operations

---

Function "Picking-Up"

A. Recognition of the workpiece orientation
    Demand for orientation
      1. Identical symmetry axes
      2. Preferred orientations
    Orientation aids
      3. Obviously different symmetry axes
      4. Geometric stability of orientation marks
      5. Orientation marks at the outer contour
      6. Orientation marks outside of casting flashes, etc.
      7. Simple and high contrast optical marks
      8. Additional orientation marks

B. Transfer of forces
    Demand for gripper elements
      9. Force dependent (gravity centre dependent) position of gripper elements
    Grippper elements
      10. Plane or smoothly curved gripper elements
      11. Gripper element with defined behaviour of the surface
      12. Gripper elements outside of casting flashes etc.
      13. Rugged gripper elements
      14. Stiffness of material
      15. Stiffness of geometry
      16. Specific gripper elements

Function "Moving"

A. Change of workpiece position and orientation
    Accessibility and tendency for hooking
      1. Movements in only a few axes
      2. Short movements
      3. Avoiding of path movements
      4. Compact design
      5. Improvement of the accessibility by box construction
      6. Improvement of the accessibility by division of the workpiece
    Feeding aids
      7. Chamfers, slopes, roundings
      8. Leadings
      9. Movements in direction of the gravity
B. Change of workpiece quantities
    Tendency for hooking
      10. Compression of inner profiles and gaps

Function "Laying-Down"

A. Reaching of the final position
    Demand for positioning
      1. Reduction of the number of stop faces
    Positioning aids
      2. Independently working workpiece positioning (notches)
      3. Stop faces, centerings, pins
      4. Stable workpiece positions
B. Releasing of forces
    Demand for gripper elements
      5. Stable geometry of assembly groups
    Gripper elements
      6. Gripper elements with defined behaviour of the surface

ASSEMBLY ORIENTED DESIGN 151

Workpiece characteristics	Dimension of workpiece characteristics		
Form elements	orientation marks at the outer contour	partly orientation by orientation marks at the outer contour	no orientation marks at the outer contour
	simple geometry of the gripper elements	complex geometry of the gripper elements	no gripper elements
	positioning aids	force dependent positioning	no positioning aids
Relative position of the workpieces	one sided accessibility (box construction)	accessibility from several sides (straight-lined)	limited accessibility
Form	compact	complex workpiece	tangling part
Dimensions	middle dimensions	large dimension in one direction	big or extremely small dimension
Position of gravity centre	one stable workpiece position	several stable positions	non-stable workpiece positions
Surface quality	no special demands on transfer of forces	special demands on transfer of forces	surface dirty, rusty, etc.
Weight	middle weight		big or extremely small weights
Strength	rigid	elastic	plastic
Symmetry	multiple symmetric	simple symmetric or obviously non-symmetric	apparently symmetric
Workpiece is:	automatically to be handled	difficult to be handled	hardly to be handled

Fig. 8 Checklist for the suitability of workpieces for automatic handling

a helpful means to check the suitability of single parts, assembly groups and products for automatic handling in a short period (Fig. 8). The checklist shows which characteristics of a workpiece should be redesigned primarily to reach an improved workpiece handling situation.

By means of the checklist, workpiece characteristics that are relevant for automatic handling can be judged. The checklist divides the workpieces into three classes:

O   suitable for automatic handling,
O   difficult to be handled automatically,
O   hardly possible to be handled automatically.

The checklist generally does not give a final judgement about the suitability for handling of the parts, because it does not consider the capabilities of the

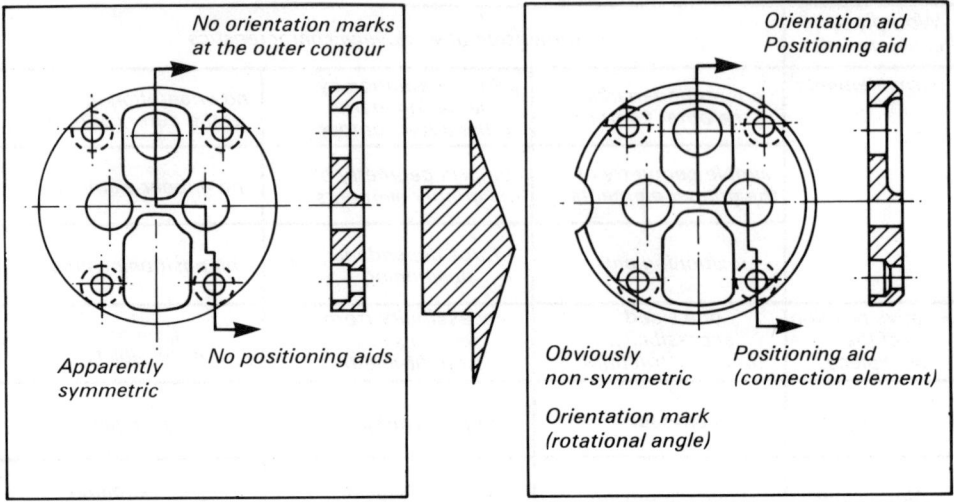

Fig. 9 Measures for a handling oriented design. Example shown is a bearing cover for a lubrication pump

handling devices within a certain factory. It does, however, give essential hints on the difficulties that may occur when automated assembly is planned. By using the above mentioned alternatives for a handling oriented design these difficulties can be eliminated.

Finally, Fig. 9 shows an example for the redesign of a workpiece in the sense of an improved handling by means of the guide lines and the checklist. The illustration shows a bearing cover for a lubrication pump. Firstly, by means of the checklist, weak points in the design of the bearing cover were discovered. Then, according to the guide lines, the bearing cover was redesigned and orientation and positioning aids were added. At the same time the apparently symmetric workpiece was changed into a clearly non-symmetric workpiece. The design created orientation aids on the outer contour of the part, so that the orientation aids could easily be interrogated by simple mechanical sensors.

Conclusion

As shown, assembly oriented design is only one of many alternatives to design a product. It is the task of the designer to divide the product into significant assembly groups and to standardise it as far as possible.

For the automation of assembly processes parts have to be adapted to the capabilities of automated assembly devices, that is, they have to be assembly oriented designed. In particular, handling operations during assembly need to be considered. The designer has the alternatives, either to avoid assembly operations, or to simplify assembly operations. In order to improve the part's suitability for automatic handling, he can make use of guide lines for the handling oriented design of parts, that were developed by the WZL of the RWTH Aachen. By means of a checklist for the parts' suitability for automatic handling weak points of the workpiece's design can be discovered very easily. In the following measures for the redesign of the parts, in the sense of an improved assembly, can be developed.

References

[1] Witte, K. W. Flexibel automatisierte Montage-Notwendigkeit, Voraussetzungen, Lösungen Ind. Anz. 104 (1982) 11, S. 81–85.
[2] Eversheim, W. Organisation in der Produktionstechnik Band 4. Fertigung und Montage Düsseldorf, VDI-Verlag, 1981.
[3] Eversheim, W., Witte, K.W., Peffekoven, K. H. Montage richtig planen Methoden und Hilfsmittel zur rationellen Gestaltung der Montage in Unternehmen mit Einzel- und Serienfertigung Fortschritt-Bericht der VDI-Zeitschriften Reihe 2 Düsseldorf, VDI-Verlag, 1981.
[4] Autorenkollektiv. Flexibel automatisierte Handhabung in Fertigung und Montage Vortrag zum 17. Aachener Werkzeugmaschinen Kolloquium, Aachen, 1981.
[5] Eversheim, W., Müller, W. u.a. Handhabungsgerechte Werkstückgestaltung für den Einsatz automatischer Handhabungseinrichtungen in Fertigung und Montage Forschungsbericht der Deutschen Forschungsgemeinschaft, 1981.
[6] Eversheim, W., Müller, W. Handhabungsgerechte Auslegung der Produkte Ind. Anz. 104 (1982) 11, S. 68–73.
[7] Aoki, K. High speed and flexible automated assembly line – Why has automation successfully advanced in Japan? In, 4th Int. Conf. on Prod. Eng., Tokyo, 1980.
[8] N.N. Technische Kunststoffe – Berechnen, Gestalten, Anwenden Kundeninformation der Firma Hoechst, 1980.
[9] Haeusler, J., Jung., H. Ermittlung handhabungsgerechter Konstruktionen für das manuelle und mechanisierte Zuführen von Schraubenfedern VDI-Berichte (1978) Nr. 323, S. 97–106.

DESIGN FOR AUTOMATION: THE COMPETITIVE EDGE

J. A. Behuniak,
Digital Equipment Corporation, USA

First presented at the 5th International Conference on Assembly Automation,
22–24 May 1984, Paris, France

Abstract

Product design is a key parameter in successful automation. In the most successful automation applications, the product and process have both been designed to facilitate automation.

In a competitive market a product must not only satisfy the customer's need for function, quality and reliability, but also must be produced at a cost that allows the manufacturer a reasonable profit.

The personal computer market is a very competitive area, that is populated with a broad range of excellent product offerings. In this paper the design of several floppy disk drives are analysed for automation potential, cost of the assembly and automation equipment; using design for assembly techniques. Design attributes for automation and automation economics are reviewed, and design rules to facilitate automation are presented.

Introduction

In a competitive market, product design is one of the primary attributes affecting a product's market share and profitability.

The personal computer market has been expanding at a very rapid rate and is expected to continue to expand. There are a large number of manufacturers competing for market share with new product introductions daily. Almost all personal computer product offerings provide the customer with one of the standard and useful software operating systems. As the customer/user expands its system to run useful application programs, some type of mass storage device is usually added to the system. One of the most popular personal computer storage devices is the mini-floppy disk drive.

The major suppliers of floppy disk drives are producing sufficient quantities of the drives to consider automation of their manufacturing. To be profitable and competitive in the area, many manufacturers are designing their products to facilitate automation and reduce cost.

In this paper four floppy disk drives are analysed using design for assembly (DFA) techniques[2] as the basis of comparison. The drives analysed are five $\frac{1}{4}''$ floppy drives with the same storage capacity, equivalent functionality and are reasonably close to having the same quality and reliability. The drives sell for about the same price. Two of the drives are reportedly designed for automatic assembly.

The major difference between the drives is the number of parts and the ease of assembly. Thus for the same product volume the ease with which each is assembled will determine which drive and it's company is the most profitable.

Design for assembly

Design for assembly[2] is a disciplined technique for analysis of a product. The technique is to dissassemble the product one part at a time, noting part and assembly characteristics. Based on these characteristics (difficulty of feeding, handling, insertion and assembly) the drives are evaluated and their manual and automatic assembly costs, manual assembly time and estimated equipment cost are determined.

Characteristics of parts used in the main assembly of each of the drives are shown in Table 1.

Table 1 Part characteristics

Part	Easy to feed (%)	Easy to grasp and manipulate (%)	Easy to orientate (%)
A	59	65	63
B	63	75	69
C	80	90	55
D	80	93	60

The characteristics for feeding, grasping, manipulating and orienting are defined elsewhere[2,4], and were determined for each part using the DFA Software[3]. The characteristics of the parts and drives from an assembly standpoint are shown in Table 2.

Table 2 Assembly characteristics

Part	Straight line from above (%)	View clear (%)	Access clear (%)	Self aligning (%)	Easy to insert (%)	Self locating (%)
A	69	96	94	92	82	61
B	75	98	77	40	79	71
C	77	74	68	26	84	74
D	85	100	95	20	85	65

The DFA analysis for each drive was made using the software[3] and run on a PDP 11/23 minicomputer using a GIGI terminal and colour monitor.

The number of workstations and cost of assembly for the main assembly of the drives is shown in Table 3.

Table 3 Main drive assembly stations and costs

Part	Automatic assembly			Manual assembly		No. of parts	No. of sub-assemblies
	Auto workstation	Manual workstation	Cost (US $)	Manual workstation	Cost (US $)		
A	50	9	3.02	54	3.05	65	3
B	49	10	2.99	58	3.27	61	7
C	28	6	2.00	42	2.35	47	6
D	36	6	2.35	47	2.61	54	4

The number of manual work stations included in the automatic assembly indicates that it was not possible or considered practical to assemble the complete drive automatically.

The estimated capital equipment costs for automatic assembly of the drives is shown in Table 4[1].

Table 4 Capital equipment costs

Part	Automatic assembly (US $)	No. of parts	No. of subassemblies
A	6.3 million	65	3
B	5.9 million	61	7
C	4.5 million	47	6
D	5.2 million	54	4

Assembly costs of the drives ranged from US $2.00 to US $3.02 per drive, a difference of 50% or one million dollars in assembly costs based on the production volume used for the analysis (1 million drives/year). Capital equipment costs range from US $4.5 to US $6.3 million, a difference of US $1.8 million or 40%. These differences are quite significant and are solely a result of product design.

Product design guidelines[2] for assembly automation are summarised as follows:

☐ Minimum number of parts.
☐ Unidirectional insertion of parts.
☐ Stackable insertion and build up of parts from above.
☐ One screw size/minimise use of screw fastening and retaining rings.
☐ No cables on components/or provide handling means.
☐ All electrical connections inserted in same plane as the assembly.
☐ Maximum part weight within the limitations of the handling equipment.
☐ Mating parts located by pins or wells that nest or locate and align suitable for transport prior to fastening.
☐ One handed assembly.
☐ Chamfered mounting holes and shafts for close fitting parts.

One message regarding product design is quite clear: in products with the same functionality, reliability and price, the manufacturing cost of a product designed for assembly can be significantly less than its competitors. Design for assembly practices can and are being used to reduce product cost, improve profitability without decreasing functionality or performance, and is a major contributor to a product's competitiveness.

References

[1] Behuniak, J. A. Product design — The first step in assembly automation. CIRP International Seminar on Manufacturing Systems, Amherst, June 1983.
[2] Boothroyd, G. and Dewhurst, P. Design for assembly. Department of Mechanical Engineering, University of Massachusetts, Amherst, 1983.
[3] Boothroyd, G. and Dewhurst, P. Software design for automatic assembly. Amherst, 1982.
[4] Boothroyd, G., Poli, C. and Murch, L. Automatic Assembly. Marcel Dekker, 1982.

Chapter 4
SENSORS IN ASSEMBLY

If systems are to become truly flexible, sensors of various kinds will have to be used in increasing numbers. Force, touch and vision are some of the techniques now being developed for assembly systems, as the four papers in this chapter illustrate.

ADAPTIVE ROBOT AND SENSORY SYSTEM FOR ATTACHMENT OF ELECTRICAL CONNECTORS TO SOLAR ARRAYS

T. Brooks and R. Cunningham
Jet Propulsion Laboratory, Pasadena, CA, USA

First presented at the Applied Machine Vision Conference, 7–8 April 1982, Cleveland, USA. Reprinted courtesy of the authors and the Society of Manufacturing Engineers, Dearborn, Michigan, USA

Abstract

An advanced robot system for attachment of electrical connectors to solar arrays is described. Using a PUMA robot coupled to a vision system and a force/torque wrist sensor, this system solders a tab to the module buss, lays down a glue bead for sealing, and places a connector housing over the tab. The vision system locates the buss strips and verifies that they are the proper size and configuration, verifies that the tab has been soldered to the buss and determines the soldered tab's location for subsequent steps, and inspects the glue bead for breaks and proper location. The force/torque sensor feels the tab as the connector housing is pushed over it, allowing placement without damaging the fragile tab. System performance and cost are considered in light of the project's goal: low-cost solar array production.

Introduction

The Flat Plate Solar Array Project at the Jet Propulsion Laboratory (JPL) has been actively involved in the development and demonstration of technology to manufacture low-cost solar arrays. Working to reduce a 1978 volume purchase price of US $15.00 per watt to a targeted cost of $0.70 per watt in 1986, the project turned to automation. As a part of this larger effort, the JPL Robotics Group has been developing an advanced robot workcell for automated attachment of electrical power connectors to solar cell modules. This task, considered to be a poor candidate for automation, entails a complicated procedure in which visual and tactile feedback are essential to the successful completion of an assembly. The major difficulty occurs in the last step of module production in which the entire array is laminated to make an environmentally sealed capsule. During this process two paper-thin copper buss strips protruding through slits in

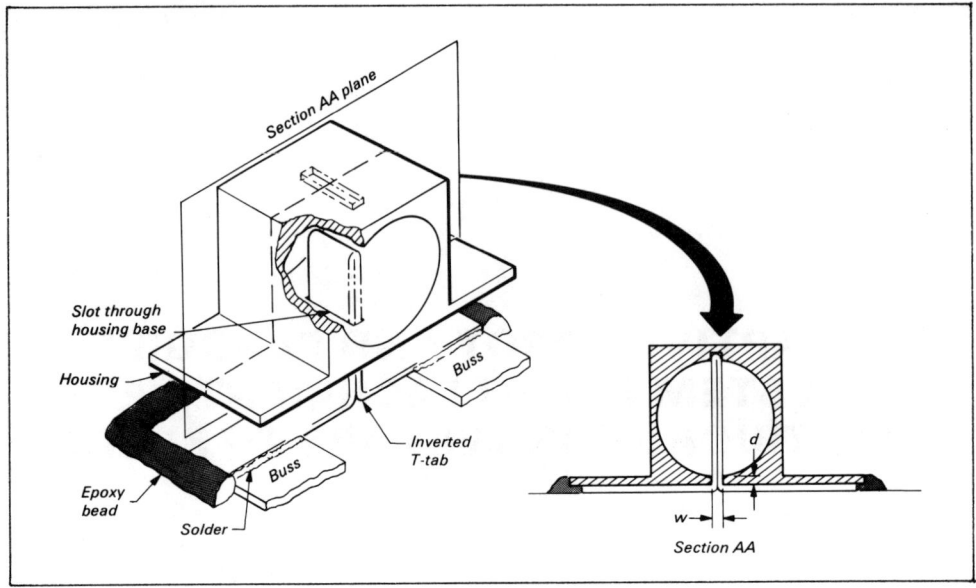

Fig. 1   Basic components of connector assembly

the laminating material, are free to move, resulting in power busses which can be shifted, rotated, missing, embedded, or out-of-tolerance. Due to the degree of adaptability required to assemble a connector under these conditions, manual assembly appeared to be the only alternative. However, with advances in robotic sensing technology which have taken place in the last few years, assembly tasks requiring adaptive interaction are no longer the exclusive domain of man.

This paper describes an adaptive robotic system developed to install electrical power connectors on solar array modules. The connectors are of the AMP IS-230-G type and consist of an inverted copper T tab which is soldered to the module buss, and a plastic housing which is placed over the vertical portion of the T-tab and epoxy sealed to the module. Fig. 1 shows the important characteristics of the connector assembly. Each of the operations necessary to attach a connector to a solar array will now be considered in detail.

Assembly procedure

Fig. 2 shows the complete assembly sequence in order of occurrence. The steps in the figure are described below.

(a) Obtain tab – The PUMA first acquires a T-shaped tab from a gravity fed dispenser. The T-tab is gripped by two 75 watt heating elements controlled by a variac for steady-state temperature adjustment. In a production system the tab dispenser would be instrumented to signal when it is running low on spares or when a part is not present for acquisition. The production system would also include vibratory orienters for feeding the magazine.

(b) Flux tab – The tab is now fluxed by wiping it on a pad soaked with solder flux. It is assumed that the buss strips on the panel are fluxed automatically in transit to the workstation.

(c) Tin tab – The tab is then dipped in a molten solder bath where the robot pauses until the vision system returns with the buss location.

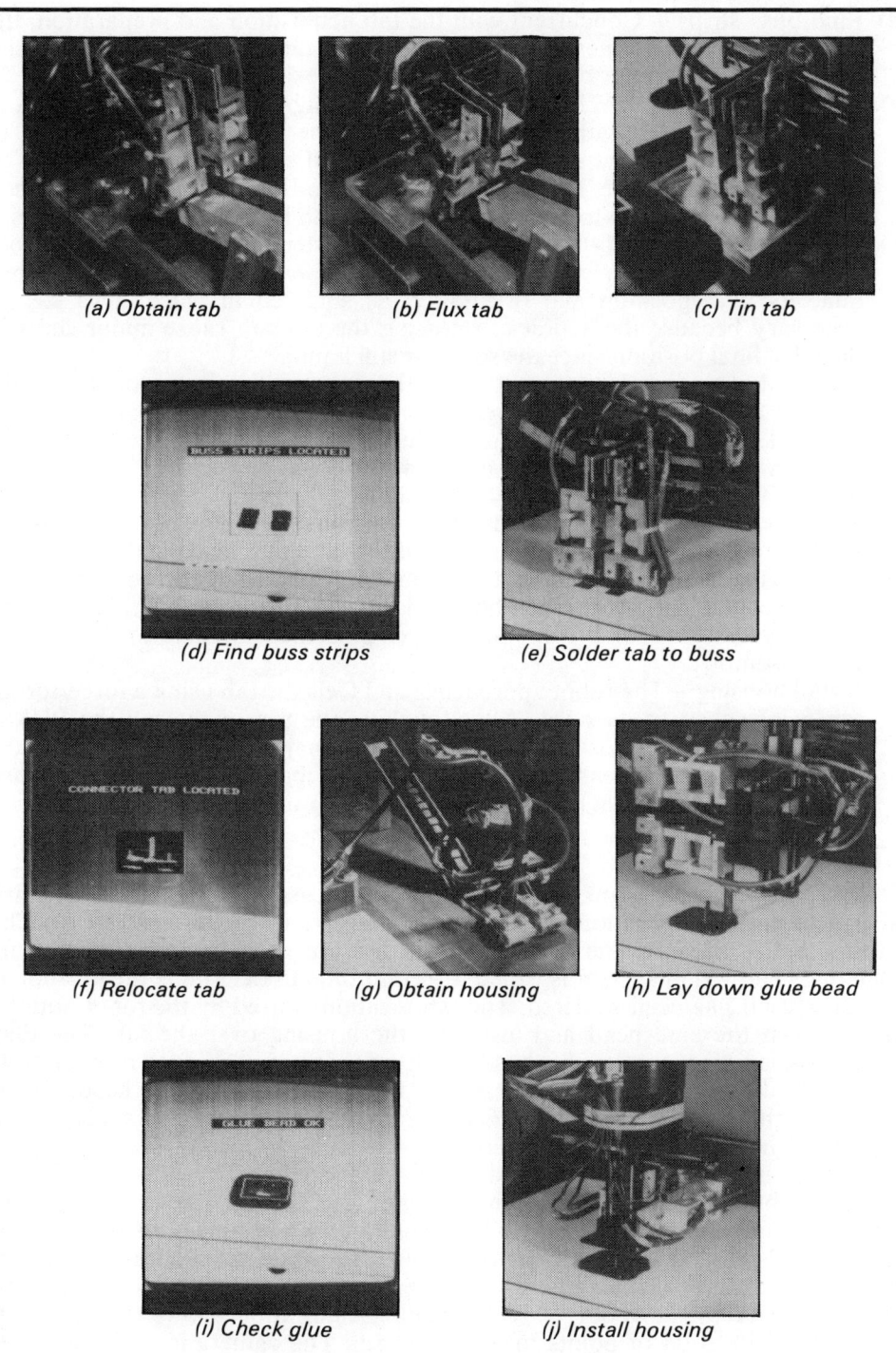

Fig. 2 Connector assembly procedure

(d) Find buss strips – Concurrent with the tab acquisition and preparation, the vision system determines the position and orientation of the buss strips, and ensures that the proper buss strip configuration exists on the module.
(e) Solder tab to buss – When the vision processing is complete the PUMA is signalled to proceed with the soldering task. The PUMA lifts the tab out of the solder bath, gives it a quick jerk to shake off excess solder and moves to the buss location. The T-tab is soldered in place using the visual feedback from the previous step to position it properly with respect to the buss strips.
(f) Relocate tab and verify solder – The vision system now processes a picture of the partial assembly in order to verify the presence of the tab and to determine its final position and orientation as soldered in place. This step is necessary because the action of releasing the tab can cause minor shifts in the tab's final position since the solder is still liquid.
(g) Obtain housing – Concurrent with the tab verification processing, the PUMA moves to another dispenser to acquire the housing which will be installed over the tab. In the final production system the magazine would be fed by a vibratory bowl and its escapement would be instrumented to signal when a housing is not present for acquisition.
(h) Lay down glue bead – Before installing the housing, a glue bead is put down relative to the position and orientation of the tab determined in (f) above.
(i) Check glue – The glue bead is then visually inspected to ensure that there is adequate glue for proper bonding and environmental sealing. This entails checking the glue for breaks and proper placement relative to the tab and buss assembly.
(j) Install housing – The robot approaches and feels the tab using a force/torque sensor attached to its wrist. After touching the top of the tab the PUMA slides the housing across the tab until the tab snaps into a slot in the bottom of the housing. The sensor detects the snap, stops the sideways motion and slides the housing down on to the tab completing the assembly.

Vision

The vision system performs three tasks during the assembly sequence. The first is to determine the location (position and orientation) of the buss strips. This information is used by the robot to place the tab in the proper location for soldering. The vision system is called a second time to determine the location of the tab after it has been soldered. The tab location is used by the robot both for laying down the glue bead and installing the housing over the tab. The third vision task is to inspect the glue bead to ensure that it is positioned properly with respect to the tab and that there are no large gaps in the glue bead. Before discussing the vision algorithms in detail, the vision hardware and some general aspects of the image analysis is described.

The vision system consists of a single black and white TV camera, a digitiser and frame buffer system, and a minicomputer (see Fig. 3). The camera is a GE TN-2000 charge-injection-device (CID) camera which produces an image of 244 lines by 188 pixels (picture elements) per line. The camera is calibrated by a least squares procedure which requires viewing several points whose locations are known relative to the robot [1]. The resulting calibration allows the system to measure the location of points to within 0.05in. The camera is mounted to the side of the assembly area, looking down at approximately a 45° angle and approximately 18in. from the assembly area (see Fig. 4). This arrangement provides adequate views for the three vision tasks with a minimum of

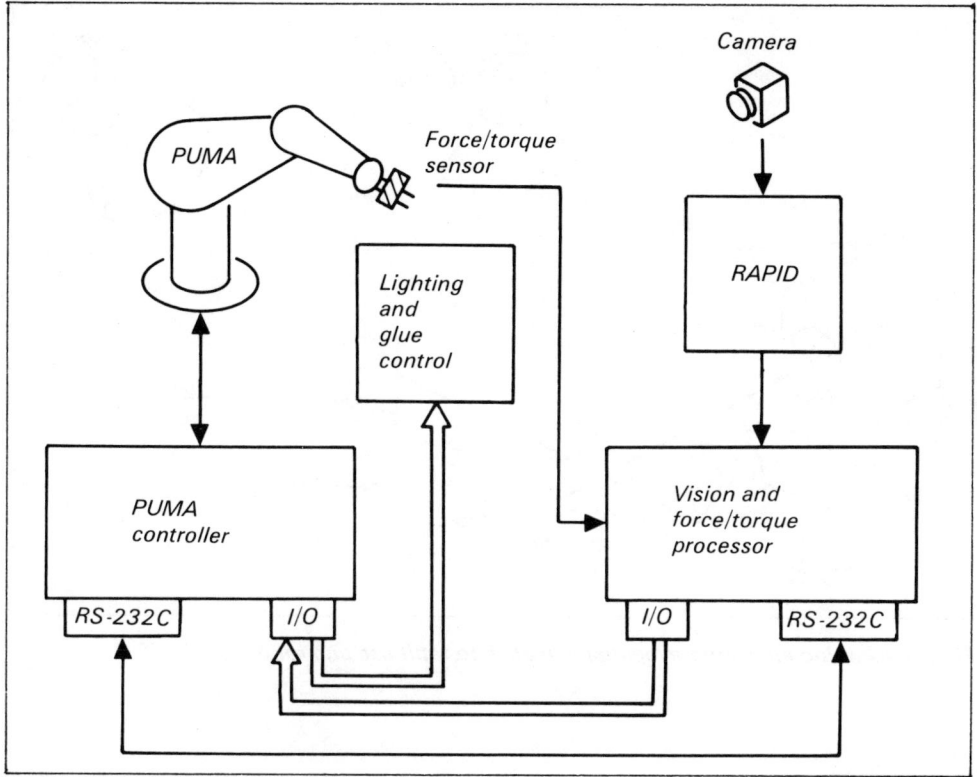

Fig. 3   Block diagram of assembly system

interference between robot and camera. The digitiser and frame buffer system is a device called RAPID (random access picture digitiser) designed and built by the JPL Robotics Project[2]. It is capable of continuously digitising and storing full frame TV images from the GE camera with 8-bits of grey scale while simultaneously allowing concurrent random access to any pixel of the image by the minicomputer. For the current application it is used as a frame grabber, digitising and storing a single frame which is then processed by the computer.

The minicomputer is a General Automation SPC-16/85 with 32K 16-bit words of memory. In addition to being used for computer vision, the SPC-16/85 is interfaced to the force/torque sensor and thus communicates both visual and force feedback to the PUMA robot.

All of the computer vision algorithms operate on binary images obtained by thresholding the grey scale image in RAPID. Thus the objects of interest appear to the computer as light or dark regions on a contrasting background. Suitable thresholds for each task are determined experimentally in trial runs under actual assembly conditions.

The use of binary images is desirable from the standpoint of simplifying subsequent image analysis. Care must be taken, however, in choosing proper illumination of the scene in order to get a faithful representation of objects as

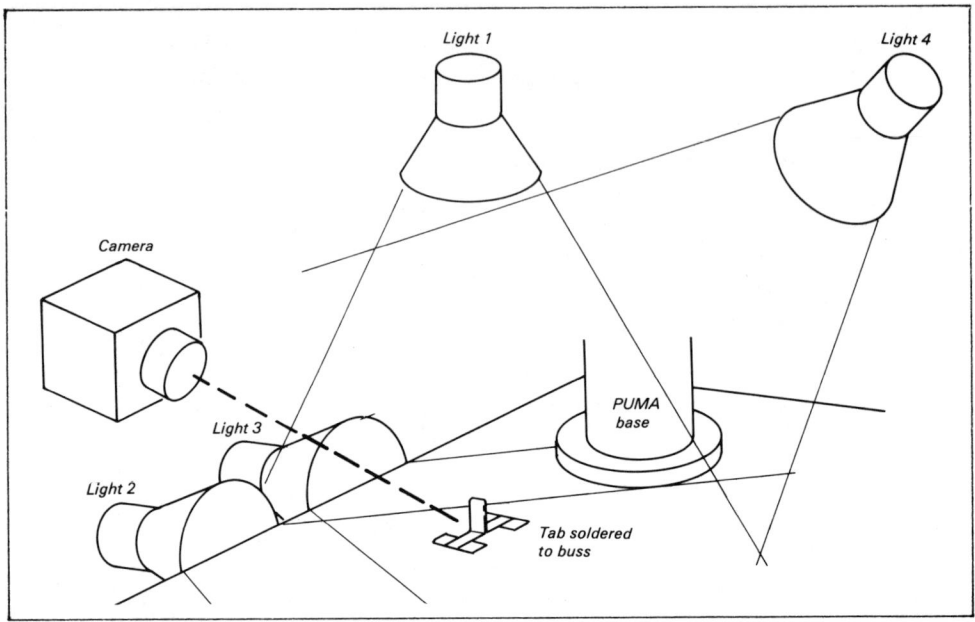

Fig. 4 Lighting and camera locations relative to tab/buss and robot position

light or dark regions when the image is thresholded. The vision system described here requires three different lighting schemes to properly highlight the features of interest. Fig. 4 shows the arrangement of lights used. Light 1, positioned directly above the camera, is used for buss strip location. Under this illumination, the copper buss strips appear almost totally black since the shiny copper surface acts like a mirror, reflecting most of the incident light away from the camera. Lights 2 and 3, positioned on each side of the camera at table level, are used for tab location. Illumination from these lights skims across the solar panel plane so that the only surface reflecting this illumination to the camera is the front edge of the tab which appears to the camera as a white inverted T on a dark background (e.g. see Fig. 2f). Lights 1 and 4 are used for glue inspection. Light 4 is on the opposite side of the table from light 1, located so that it makes approximately a 30° angle with the camera-line-of-sight when aimed at the assembly area. Under either light, the glue bead is nominally a black rectangular strip. To achieve each of the three illumination schemes the lights are switched on and off as needed under computer control. (For convenience, this is done by the PUMA controller since the I/O ports are designed to switch 110V ac.)

Buss strip location
The function of the buss strip location program is (a) to verify that a module is present and ready for tab soldering, and (b) to determine the actual location of the buss strips within a search window. The size of the search window reflects possible variations in process parameters such as how the buss strips are positioned in the module, where the slits in the backing material are made to bring the buss strips out of the module, how much of each buss strip is exposed,

and how the module is positioned relative to the robot at the workstation. Our program uses a fairly large window which covers most of the useful workspace in which the robot can perform the required assembly steps. In a production system the window could be much smaller thus increasing visual processing speed without reducing effectiveness.

To locate the buss strips, the vision system looks for two dark rectangular regions corresponding to the exposed ends of the buss strips. To do this, the image is first thresholded and then a scan is initiated to locate all black regions in the window. The buss stips are modelled on the basis of two statistical features – area and compactness. The area is simply a count of the pixels in a region. Compactness, defined as the ratio of the area to the square of the perimeter (A/P^2), is a measure of shape which is scale invariant. All rectangles with a given aspect ratio (2:1 in the case of the buss strips) have the same compactness value independent of their actual size. The program can reject most false regions due to imaging noise or foreign matter on the module on the basis of these two parameters. In the presence of false regions of nearly identical aspect ratios and areas, the system will suspend assembly until the error is corrected or a new module is placed in the workstation. The system will also suspend assembly if an insufficient number of regions are identified. Figs. 5a,b,c show examples of modules which have been rejected on the basis of too few or too many potential buss strip regions.

Assuming that the buss strips have been located, further tests are required to ensure that their relative locations are correct. For this purpose, the two buss strips are treated as a single unit, hereafter referred to as the buss unit. It is convenient to think of the buss unit as the minimum rectangle containing the buss strips. The orientation of the buss unit is defined as the angle between the x axis of the the PUMA base frame and the line connecting the centroids of the two buss strips and is always in the range 0–180°. The width of the buss unit is the distance between the outer edges of the two buss strips measured along the line connecting the two centroids. The desired width is 1.5in. The position of the buss unit is defined as the midpoint of the segment connecting the edge points described above in the width calculation. Another possible method of calculating the position would be to simply average the two buss strip centroids. If the individual buss strips have different widths however, the computed position will not be centred with respect to the buss unit. The method described here guarantees proper centering and thus proper alignment of the tab and buss unit. In Fig. 2d, grey lines can be seen extending from the buss strip centres to the outer edges, illustrating the orientation line and the edge points used for calculating the width and position. The two grey segments are generated as the program searches out from the buss strip centroids to find the edges. The search terminates when a white pixel is encountered.

At this point, the width and orientation are checked for possible error states. The buss unit is considered unacceptable if the width differs from the desired value of 1.5 in. by more than 0.25 in. If the width is very much greater that 1.5 in., it is impossible to solder the tab to both buss strips. For smaller deviations, the concern is whether or not the housing will cover both buss strips sufficiently to provide environmental sealing and electrical insulation. Fig. 5d shows an example where the width is too large.

Finally, the orientation of the buss unit is checked to verify that it is 90° ±15°. The limiting factor on orientation is not the PUMA which could tolerate a much larger variation, but the performance of the tab location program which degrades significantly outside this range due to illumination problems. In a production setup, the actual buss variation would probably be less that 5°, so this is a very generous tolerance. Fig. 5e shows an example of an orientation error (note that 90° is roughly horizontal in the image).

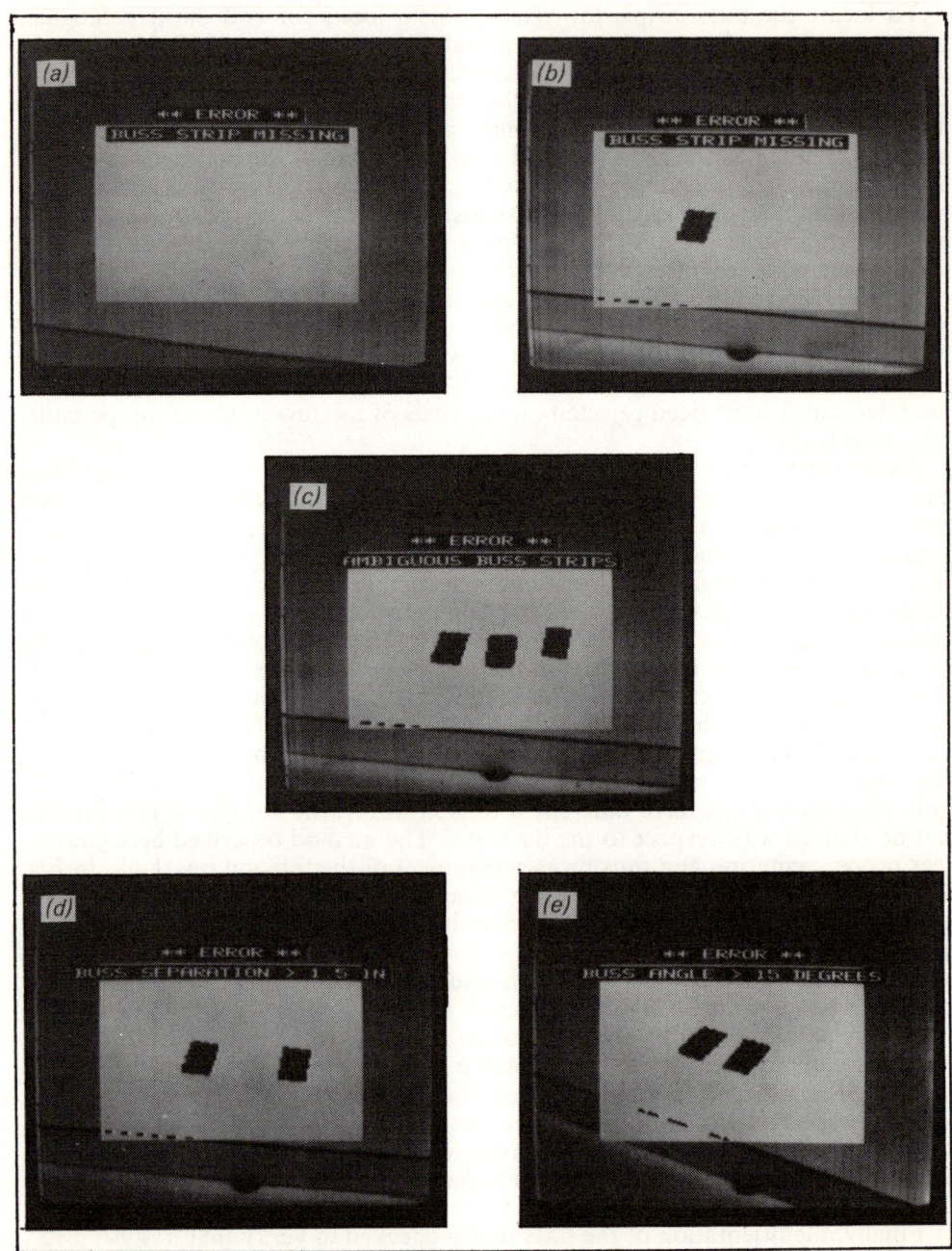

Fig. 5   Buss strip error conditions detected by vision system

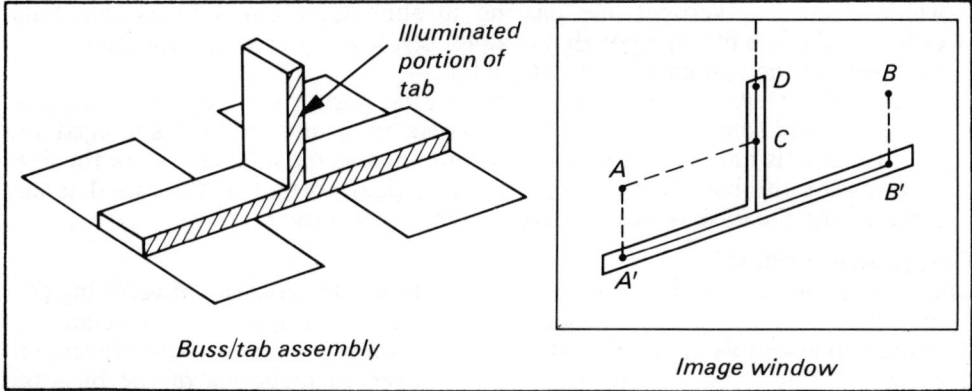

Fig. 6   Tab location scheme

Soldered tab verification and location

The tab location program, to be discussed next, verifies that a tab has been soldered to the buss unit and determines its actual location as soldered to the module. The PUMA uses this measurement in subsequent steps such as laying down the glue bead and installing the housing, rather than assuming that the tab is exactly where it thinks it soldered it. The tab location program operates on a small rectangular image window in which the tab is expected to be. This window is passed on from the buss strip location program. The position of the tab is defined to be the point on the front edge of the tab (with respect to the camera) where the two segments of the inverted T meet. The orientation is defined as the angle between the x axis of the PUMA base frame and the front edge of the base of the tab. Ideally, this is the same as the orientation of the buss unit, but in practice was found to have up to 0.5° misalignment. This discrepancy is easily explained by resolution errors in the buss location passed to the PUMA.

The image of the tab tends to be somewhat noisy due primarily to the presence of excess solder near the edges of the tab. Thus rather than looking explicitly for a T-shaped region, the vision system uses a least squares line fitting approach to locate first the base of the tab, then the vertical portion of the tab. Referring to Fig. 6, the tab location algorithm proceeds as follows:

☐ A downward search is initiated at points A and B to locate two points A' and B' on the base of the tab. This search is several pixels wide to ensure that the base does not go undetected due to a one or two pixel gap caused by some form of noise. The points A' and B' define a line which is an initial estimate of the location of the base. The search is illustrated by two vertical bars near either end of the tab in Fig. 2d.

☐ A least squares line fitting routine is invoked to find the best line which approximates the location of the base. The image is sampled at ten points equally spaced along line A'B'. Each sample consists of five points along a vertical line bisected by A'B'. If a single white pixel is found, this is used in the least squares fit. If two or more are found, the midpoint of the group of white pixels is used. The least squares line fit to these sample points is then assumed to be the base of the tab.

☐ Another search is initiated from point A, this time moving to the right along a line parallel to the base, to locate a point C on the vertical portion of the tab. The top of the tab D is located by searching down from the top of the window along a vertical line passing through C. Using CD as the initial estimate, the line fitting algorithm is again invoked to determine the best fit line representing the vertical portion of the tab.

☐ If the above steps are successful, the position and orientation of the tab are computed and sent to the PUMA, otherwise an error condition is flagged and the assembly is halted. The position is computed as the intersection of the lines representing the base of the tab and the vertical portion of the tab. The line representing the base is used to determine the orientation.

Glue bead inspection

The final vision task involves glue bead inspection. This consists of verifying that a continuous bead of glue has been laid down in the proper location relative to the buss/tab assembly. This procedure operates on an image window expected to contain the glue bead on the basis of the observed locations of the buss/tab assembly.

As mentioned above, the glue inspection program requires two lights for scene illumination. This is due to the shiny surface of the glue which produces bright highlights under virtually any illumination angle. The location of the highlights, however, varies with the angle of illumination. By taking two pictures, one with light 1 on and one with light 4 on (refer to Fig. 4) and thresholding each, we were able to construct a composite image which was black wherever either or both of the original images were black, thus effectively cancelling out the highlights. The composite image is generated by positive AND-ing the two thresholded images. This ANDed image is scanned along a rectangular path that is expected to coincide with the centre of the glue bead. If every point along this path is black, indicating the presence of glue, or if the width of all detected gaps is less than a threshold, the glue bead is considered acceptable and the assembly proceeds. Fig. 2i shows the result of scanning a good glue bead while in Fig. 7 a defective glue bead has been successfully detected. (It is assumed that the glue will flow into the smaller gaps when the housing is pressed against the module. Thus the gap threshold has to be determined experimentally on the basis of the type of glue used and the thickness of the bead.)

Force/torque directed assembly

In the final portion of the connector assembly the robot must place a housing over the tab ensuring that the tip of the tab first slides through a slot at the housing base and then seats firmly into a second slot in the housing roof (see Fig. 1). The tab insertion task can be likened to inserting a pencil through two holes in a paper cup, the second hole is just as difficult to locate as the first and in many cases even more so since the pencil, or tab, is partially constrained by the first slot.

The locations of both housing slots are known accurately since the housing is held firmly in the end effector. The tab location, however, is only as accurate as the camera system used to locate it after soldering. A 188 x 244 pixel camera with a 25mm lens, 18 in. from the tab, will have a visual error in the order of 0.050 in. Since the slot-tab tolerance is approximately 0.020 in., visual feedback is not sufficient for placing the housing on the tab. In addition, even if the slot was perfectly aligned with the tab a bend of only 1.5° would be sufficient to catch or bind on assembly. Hence, some form of assembly compliance is called for.

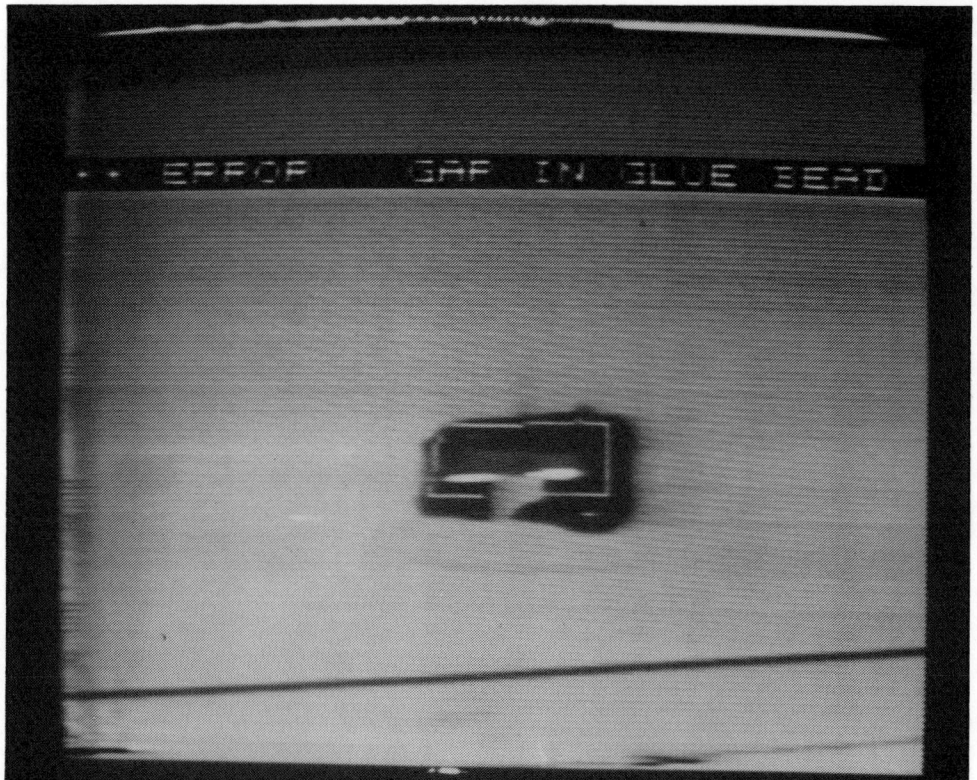

Fig. 7 Vision system detects gap in glue

A remote centre compliance (RCC) device would at first appear to be ideally suited for this task since RCC's are very effective for pin insertion types of tasks. A closer look, however, reveals a fundamental difference between the pin-in-the-hole task and placing a tab in a housing. Specifically, since the width of the slot w (Fig. 1) is greater than its depth d, only single point contact will occur between the tab and slot. As a result, a bent tab as shown in Fig. 8 will create a purely lateral accommodation shift with an RCC device. This is due to the fact that contact occurs on one side of the tab only, and hence double point contact, which is necessary to generate the correcting torques to remove the angular misalignment[3,4], will not be present. Even if the slot depth d were deep enough to generate the correcting torques an RCC could not be used since it is not possible to tilt the housing enough to cause the tab top so snap in, due to the wide base of the housing and short distance that the tab engages the roof of the housing. Hence, the tab must be straight enough to allow insertion in both the bottom and top slots with only minor tilting actions. In other words, if the tab is bent, it is necessary to straighten it to within one degree of vertical before the assembly can be completed. For these reasons it was decided that active force feedback would be required to perform the housing assembly task. Two methods were devised for completing the task through the aid of force feedback. The first was based entirely on force feedback information and the second was a hybrid force feedback plus compliance scheme. Both schemes followed the general assembly

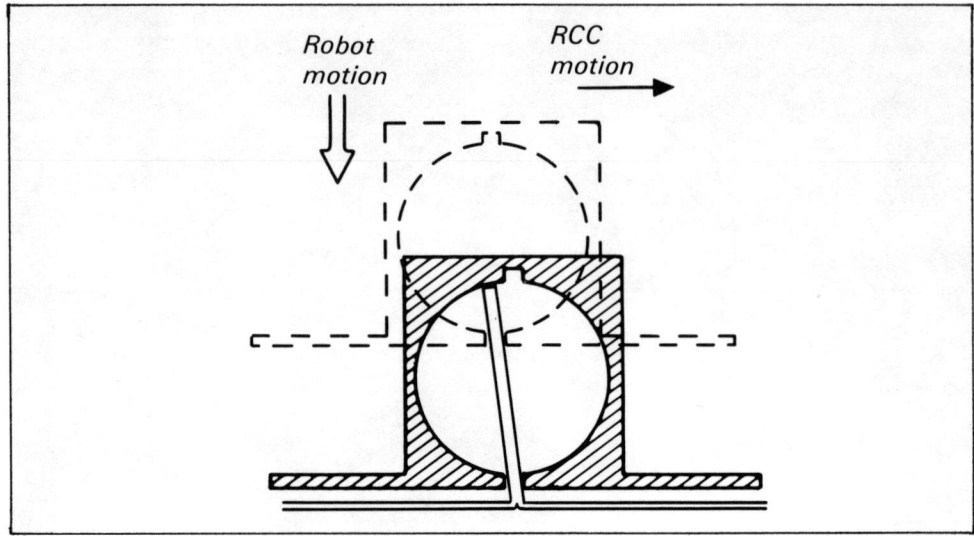

Fig. 8   Assembly using remote centre compliance (RCC)

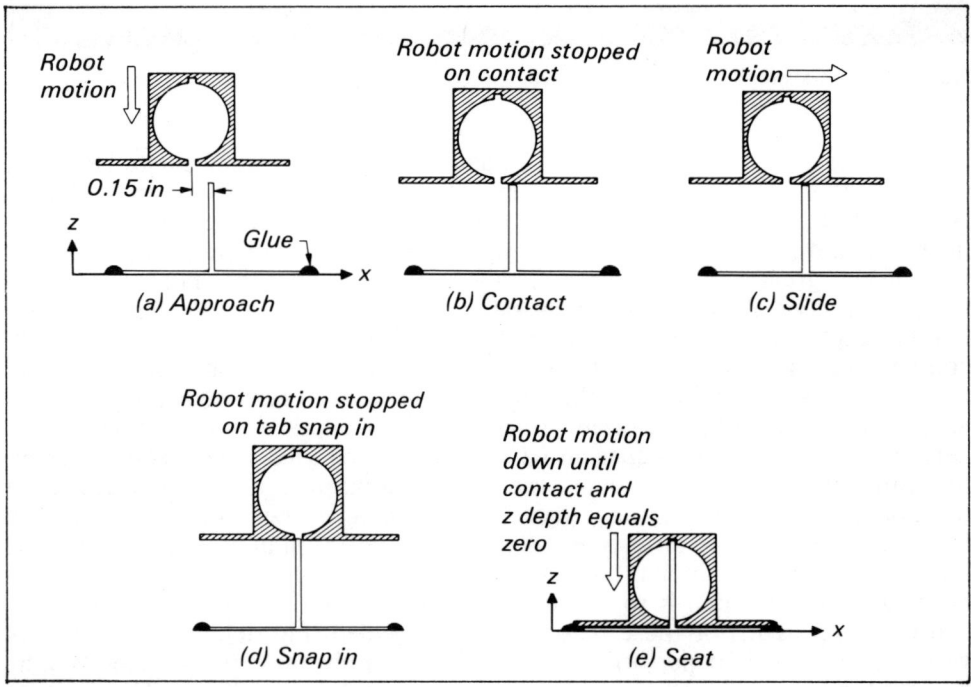

Fig. 9   General assembly procedure

procedure outlined in Fig. 9. This procedure consisted of five assembly subtasks: approach, contact, slide, snap in, and seat. The difference between the pure and hybrid assembly schemes is in the implementation of each of these assembly subtasks.

A conceptual flow diagram for the pure force/torque assembly scheme using an off-the-shelf PUMA robot is shown in Fig. 10. Each of the steps of the force directed assembly shown in Fig. 9 as they apply to the pure force/torque assembly scheme are described below.

○ In the first step the robot moves down on the tab with the housing slot deliberately shifted by 0.15 in. in the x direction of the hand, the hand being directly above the tab with its x axis aligned with the tab base axis.
○ The downward motion of the above step continues until: the housing has moved 0.2 in. beyond the expected tab height of 0.625 in. without finding the tab, in which case the system responds with the error CAN'T FIND TAB; or until contact with the tab occurs and the contact force reaches 6oz (approx. 170g).
○ The robot then slides the housing across the tab top while ensuring that the sliding force never exceeds 16oz (approx. 450g). The 16oz level, derived empirically, was found to be the maximum side force the tab could sustain without permanent elastic deformation. In addition, the z force is continuously checked to ensure that the housing is pressing against the tab. The sliding continues until either the housing has slid 0.3 in. without finding the slot, in which case the system returns with the error CAN'T FIND SLOT, or the slot has been found by the tab top.
○ The system detects the tab snapping into the slot by watching the z force for sudden discontinuities. Once the system finds the tab it pauses and reads the tab/slot location for subsequent use in determining if the tab is bent.
○ Once the tab has been found the system begins moving the housing down over the tab until: the x direction force on the tab increases to greater than 16oz, in which case the system shifts the housing sideways in the direction to relieve the force and then resumes its downward motion; or until the z force increases to greater than 10oz (approx. 280g) signifying that either the tab top has hit the housing roof or that the housing has been properly seated. The system detects whether the housing is seated by checking its height above the panel. If the height is greater than 0.039 in. the housing has not seated and it is assumed that the tab is bent. The system then determines the location of the slot, and hence the bottom of the tab, and with the tap top location found in the above step, calculates the bend of the tab and the necessary shift in the tab top which will straighten the tab.

Unfortunately the housing assembly scheme based entirely on force feedback, although conceptually pure and robust, was not found to be practical when servoing through the PUMA serial port and its I/O module. Theoretically, a PUMA can react to force/torque information at 30Hz rates through its I/O port. Experiments have shown otherwise. In particular, when a react immediate command (REACTI) is used in conjunction with a MOVE command, as shown below, the PUMA will take approximately 0.2 seconds (5Hz) to respond with a BACKOFF movement (i.e. 0.2 seconds from the time contact is made until contact is broken).

174 PROGRAMMABLE ASSEMBLY

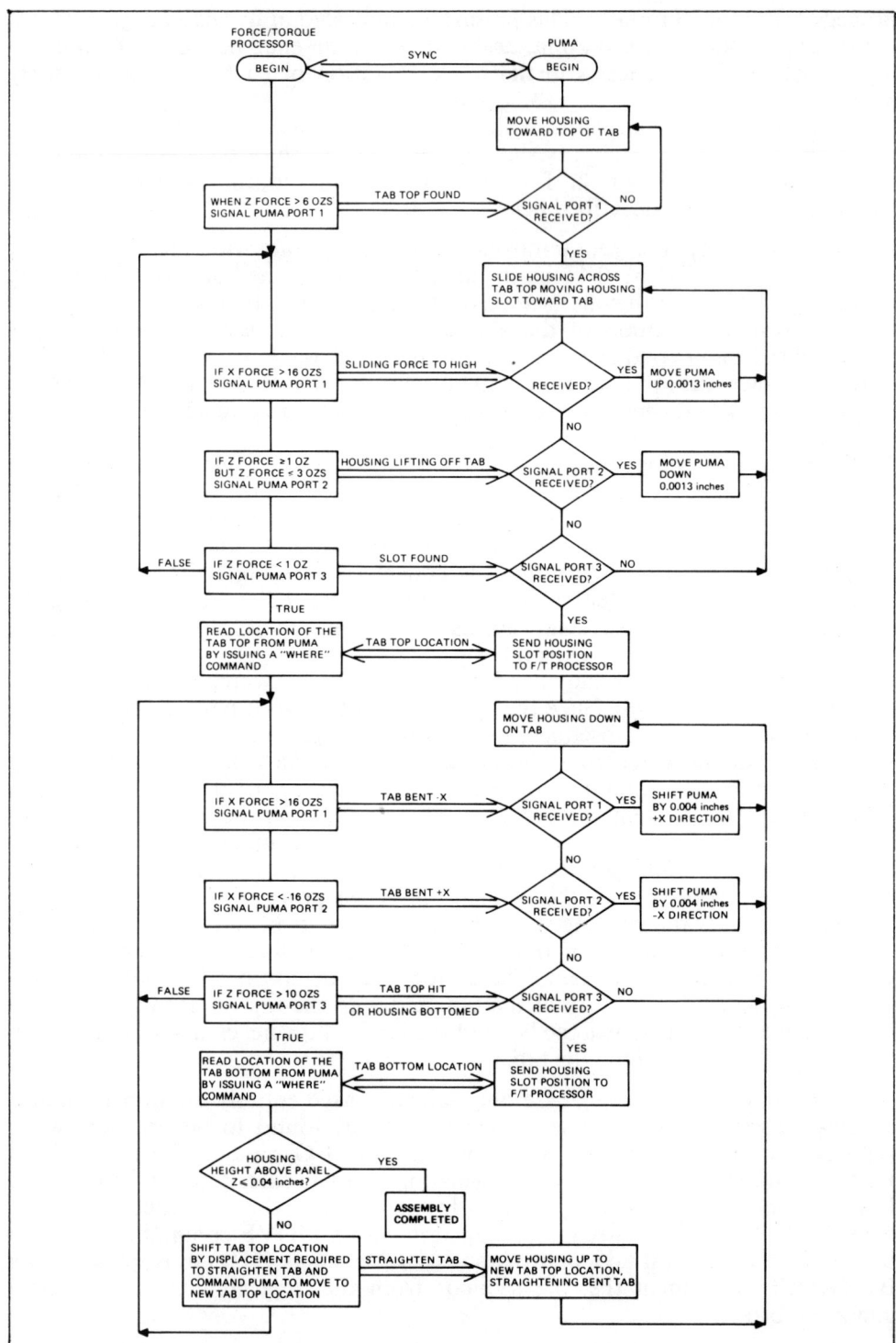

Fig. 10   Pure force/torque sensing assembly scheme

REACTI 1, BACKOFF   During the execution of the next statement if input line 1 goes high, execute subroutine BACKOFF immediately.
MOVES TOWARD   Move toward a microswitch which signals PUMA input line 1 when a collision occurs.

Another force control technique which can be achieved through the VAL controller, is to take small incremental steps and check the I/O ports through an IFSIG command after each step. This technique allows infinite adaptive looping but unfortunately the response time is a function of the step size since the IFSIG command is only executed after the step is finished and not during the step as the REACTI command is. By way of example, a step size of 0.025 in. will result in a 0.5 second (2Hz) reaction time.

It should be kept in mind though that part of the response time is due to the dynamic response of the PUMA arm itself. To determine the portion of the response due to arm actuation lag, a test program which reacts by simply toggling an I/O line was implemented and the response timed. The result was a better than double improvement in response with a worst case time of 0.08 seconds (12.5Hz). Further testing revealed that the PUMA solid state I/O module was responsible for as much as a 35 millisecond delay in switching time. When the I/O module is discarded completely and the signals connected directly to the VAL LSI 11-03 parallel TTL port the reaction time is at worst 45 milliseconds (22Hz) and approximately 30 milliseconds (33Hz) on average.

In light of these empirical results it was decided that implementation of pure force feedback servoing through the PUMA I/O, although possible, would not operate at sufficient speed. Hence, the combined compliant and force feedback scheme shown in Fig. 11 was devised. This scheme effectively removed all of the continuous adaptive force servoing loops and operated strictly on discrete force state thresholding (i.e. when force reaches desired level stop motion and move on to next step). Since each step was implemented with a react immediate command which upon receiving an input signal simply stopped the present motion, the reaction rate was in the order of 20–30Hz.

The hybrid system relies heavily on the compliance in the z direction to maintain the sliding forces within prescribed bounds. The level of compliance necessary to ensure that the sliding force never exceeded 16oz was based on the following assumptions and empirical results:

☐ The coefficient of friction between housing and tab was assumed to be in the order of 0.4, and hence, the maximum downward force which can be exerted on the tab is 40oz if the sideways x force is never to exceed 16oz.
☐ The maximum angle of misalignment of the sliding motion with the plane of the solar panel (see Fig. 9) was assumed to be in the order of +0.5° and hence, for an initial 0.15 in. shift of the slot in Fig. 9 (framed) the maximum height change due to the angular error will be in the order of 0.001 in.
☐ Direct measurements on the PUMA revealed that when incrementing the encoders by one count, stiction and servo gain can result in 0.0000–0.0040 in. jumps in position with an average jump displacement of 0.0013 in. (averaged over 100 increments with the arm in the following configuration: −90°, 180°, 0°, 0°, 0°, 0°).
☐ Imperfections in the sliding surface of the housing base were assumed to be in the order of 0.005 in. maximum.

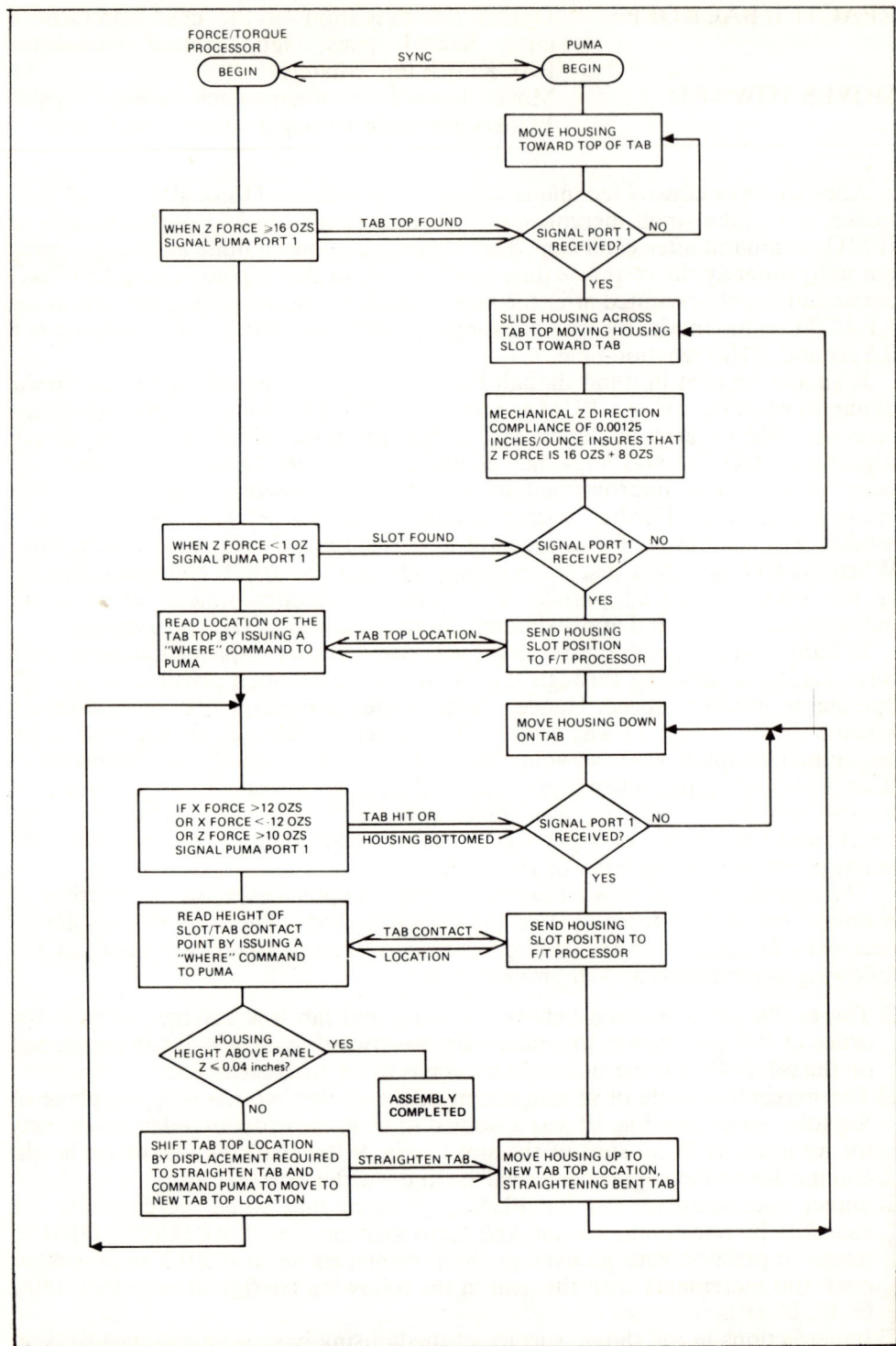

Fig. 11 Hybrid passive compliance and active force/torque assembly scheme

If all of the above conditions were to occur at the same time, the housing base would be pressed into or moved away from the tab by 0.010 in. Now if the x force is never to exceed 16oz, the initial contact force plus the force generated by the above errors should never be greater than 40oz. Also if contact is to be maintained with the tab, the z forces should always be greater than zero. If we select a contact force of 16oz and an 8oz deadband, that is the z force is equal to 16 +8oz, then the required compliance is determined by dividing 0.010 in. by 8oz which yields 0.00125 in./oz. Given that the average PUMA z axis resolution is in the order of +0.0013 in., the force resolution per minimum increment of position is about 1oz with the above compliance. Without compliance, other than the PUMA's normal off-the-shelf value, the force resolution is about 32oz per increment. For the compliant system, the stiction effects mentioned above will result in at worst 3.5oz of force. However, without compliance a 0.004 in. jump in the arm position can mean a jump of 6lbf/on the tab. The result – pure force servoing without added compliance is not practical. Additionally, since the stiction effects are greater than the encoder resolution, it is doubtful that direct current drive force control will improve the situation much. This is not to say that a pure force/torque control scheme would not work, but only that compliance must be added to the system to achieve the desired force resolution even if current servoing is used in place of the PUMA's position servos.

The tab straightening portion of the hybrid routine has not been implemented to date. It is not expected to increase the completion time appreciably, though, since the procedure would be to calculate the offset tab top location required to straighten the tab, move at high speed to the straightening position and then move down cautiously until contact is made again. If the system detected a second bent tab condition it would simply reject the panel rather then become involved in a potentially infinite loop of correction.

Results

The system described here was not intended to be a final production prototype but simply a laboratory demonstration of feasibility. Under those constraints the goals of this project were to demonstrate the feasibility of an advanced robot and sensory system capable of installing 2.3 connectors per minute at a maximum equipment and startup cost of US $110,000. This laboratory demonstration assumed as a baseline a manufacturing process producing 18 modules per hour, 24 hours per day for a total yearly plant output of 20MW of solar panels. The manufacturing process allotted 8,000 hours of plant production time after expected downtime, resulting in a total output of 144,000 modules per year (i.e. 288,000 connector assemblies). We will now consider how the performance and cost goals of this project were derived, how well the feasibility demonstration met those goals and what the results mean.

Rudimentary experiments have shown that human assembly of connectors takes from 16 to 21 seconds with an average of 18.5 seconds. At this rate one full-time assembler is required for the yearly 144,000 module run, assuming 250 available workdays per year (excluding weekends and 11 paid holidays) and a 75% mean time on the job after paid leave, coffee breaks, sicktime, administrative functions, etc. Hence, under the constraints of time actually available for assembly versus the 261 days of paid work, the average paid time per connector is 26 seconds (i.e. 2.3 connectors per minute).

The demonstration system performed one assembly every 38 seconds, 12 seconds short of the human assembly time. A large portion of this time (8 seconds) was spent visually checking the glue bead which, unlike the buss and tab location programs, was not performed in parallel with robot activity (i.e. the robot simply waited during the glue inspection until the vision system was finished). Another area which required a large portion of time (approx. 4 seconds) was the serial communication link between the PUMA and vision processor. In fact to transfer the buss and tab locations from the vision system to the PUMA controller took over 2 seconds. For pragmatic reasons which do not apply to the production system but which account for another 2 seconds of communication time in the demonstration system, actuation of the end effector was also controlled by the vision processor.

There are a number of methods by which the performance time could be reduced, the most obvious being to place end effector actuation directly under the control of the PUMA controller. Another, would be to use hardware image processing instead of software. It is believed that hardware image thresholding and edge detection would result in approximately a 3 second saving in the glue check procedure alone. Savings in the buss and tab location programs would also occur, but since they are already performed in parallel with robot activity no net gain in overall performance is expected.

A second method for speeding up the system entails a modification to the entire glue bead inspection procedure. As mentioned in the 'Vision' section, the shiny surface of the glue presents a problem since it produces bright highlights under virtually any illumination angle. This problem was solved by using two pictures and ANDing the results. However, one possible way to eliminate the need for two images would be to use a polaroid filter since the highlights are likely to be highly polarised. If this worked, there would be significant time savings due to the reduced image processing (6 seconds by eliminating ANDing functions, light switching and thresholding of one image).

Another increase could be obtained by changing the sequence in which the housing is picked up from the magazine. In the present implementation the housing is picked up while the tab is being relocated by the vision system (see 'Assembly procedure' section). Significant time is lost, however, since the tab location processing is considerably quicker than the pickup and make-ready-to-glue actions of the robot. By changing the procedure so that the hand is positioned to lay down the glue bead while the tab is being located and then have the arm move to get a housing while the glue inspection is being performed, another 2 seconds can be shaved off the performance time.

The total cycle time with the above savings would be approximately 28 seconds. With careful attention to feeder and panel location within the workspace and optimisation of robot motion it is believed that the human assembly time of 2.3 connectors per minute could be matched. An even greater saving could be achieved if communication to the PUMA could be performed over a parallel rather than serial link. Unfortunately, the PUMA was not designed for parallel communications and extensive hardware and software modifications would be required.

The above efforts to improve performance time assume that it is necessary or desirable for the robot to assemble two connectors in 52 seconds in order to directly compete with a human worker. It is interesting to note, however, that the plant design calls for 24 hour operation. Hence, while the human worker only puts in 1500 hours of productive work per year, the robot can work a full 8,000 hours of productive time after an assumed downtime of 9%. Hence the robot as implemented in this demonstration could assemble 2.6 times as many connectors

Table 1 Equipment and startup costs (1980, US $)

Robot	45,000
Vision system	35,000
Force/torque sensor	3,000
End effector with built-in soldering tips and glue gun	5,000
Magazines, parts feeders, etc.	12,000
Startup costs	8,000
Total	10,000

per year as its human counterpart. Unfortunately, the robot's capacity cannot be used effectively in this situation since the output of the plant is not limited by the connector assembly, but other upstream processes. This means that the robot will sit idle 62% of the time even if the feasibility demonstration were to be implemented as is (7.4% with suggested improvements). Clearly, if the robot can be put to work on other tasks which utilise the remainder of its available time, the system could be more cost effective. One such candidate task is the connection of finished solar modules to automatic test equipment (ATE). This function would normally require at least one full-time worker and could, depending on the time each module is tested, require attendance around the clock. Additionally, there would be little additional equipment cost since connecting ATE electrical leads to the modules requires the same system capabilities as already incorporated into the robot (i.e. vision and force/torque feedback).

The US $110,000 maximum equipment and startup costs were based on a manufacturing cost model[5,6]. This model relates required revenues to equipment costs, required floor space, cost of material, and labour costs through a convenient equation. The model takes into account startup costs, property and income taxes, minimum 21% return on investment (ROI), allowable tax credits, insurance, work-in-progress, engineering cost estimation errors, and other miscellaneous factors. Assuming that the required floor space and cost of materials are the same for either manual or robotic assembly and assuming the robot performs the work of two unskilled labourers the model predicts that equipment costs greater than US $100,000 with startup costs greater than US $10,000 will not be cost effective. The anticipated equipment and startup costs for the production system are given in Table 1.

Under the assumptions outlined above the cash payback period for the production system which replaces two workers is found to be 2.5 years with an ROI of 37%. Table 2 shows a cash flow analysis for the proposed production system. By way of comparison, if only one unskilled labourer were replaced by the system, the cash payback period would be 4.9 years and the ROI would be only 14%. Since most companies require a minimum ROI of 21%, utilising an advanced robot system would not be warranted under these conditions as was predicted by the economic manufacturing model.

Conclusions

This paper has described an advanced robot assembly system for the attachment of electrical connectors to solar arrays. Designed to demonstrate the feasibility of an advanced robot assembler using off-the-shelf technology, the effort was deliberately restricted to utilising readily available and proven equipment and techniques. The resulting system consisted of a PUMA robot integrated with an

Table 2 Proposed robot assembler cash flow (1980, US $)*

Costs	1	2	3	Year 4	5	6	7
Equipment	−100,000						
Investment credit	10,000						
Start-up costs	−8,000						
tax credit	4,000						
Depreciation (double-rate declining balance with switch to straight-line in year 4)	14,286	10,204	7,289	4,555	4,555	4,555	4,555
Labour cost (wages/overhead/fringes) 2 replaced workers	58,800	63,504	68,584	74,071	79,997	86,396	93,308
tax cost	−29,400	−31,752	−34,292	−37,036	−39,998	−43,198	−46,654
Maintenance (3.75% of cost)	−3,750	−4,050	−4,374	−4,724	−5,102	−5,510	−5,951
tax credit	1,875	2,025	2,187	2,362	2,551	2,755	2,975
Power (8000 hours x 1.5kW x 0.25/kW hr)	−3,000	−3,240	−3,499	−3,779	−4,081	−4,408	−4,761
tax credit	1,500	1,620	1,750	1,890	2,041	2,204	2,380
Savings on human assembly stations	3,000						
tax cost	−1,500						
Totals	−52,189	38,311	37,645	37,339	39,962	42,794	45,853

*50% tax rate, 8% inflation added per year, cash payback period 2.5 years, return on investment 37%.

SRI-type vision system, a Scheinman wrist sensor, and simple on/off microswitch touch sensors on the end effector. The visual, force, and tactile sensing used in the various steps of the assembly serve two purposes. First, the sensory feedback allows real-time determination of component locations so that assembly can be performed successfully in spite of uncertainties due to fixturing or previous assembly steps. Second, the system is able to detect error conditions which would require remedial actions before the assembly process could continue. The results show that implementation of a practical advanced assembly system can be achieved with off-the-shelf technology. However, the following conclusions were also derived from this project:

○ When an advanced robot system is not effective for a job which under utilises its capabilities it can be made cost effective if the system can be used for other tasks during the time in which the investment would otherwise lie idle. This is true with any robot, but particularly so with an advanced robot since it can readily adapt to new tasks.
○ Although available robotic technology is quite sophisticated, it is not developed sufficiently yet to support advanced sensory systems without considerable user effort.
○ Advanced robot systems are more difficult to economically justify than simple pick-and-place robots due to their higher initial costs and slower performance resulting from the limitations of robot controllers and inefficient robot communication links. Future reductions in costs and improvements in communications and controllers are essential.
○ Advanced robots represent long-term investments which require longer payback periods traditionally frowned upon by American managers.

Acknowledgement

The research described was carried out by the Jet Propulsion Laboratory, California Institute of Technology, and was sponsored by the United States Department of Energy through an agreement with the National Aeronautics and Space Administration.

References

[1] Yakimovsky, Y. and Cunningham, R. A system for extracting three-dimensional measurements from a stereo pair of TV cameras. Computer Graphics and Image Processing, Vol. 7, 1978, pp. 195–210.
[2] Eskenazi, R. and Cunningham, R. A random access picture digitizer, display, and memory system. Proc. 5th International Joint Conference on Artificial intelligence, Cambridge, MA, August 1977.
[3] Drake, S. H., Spencer, R. M. and Simunovic, S. N. Using compliance in assembly – An engineering approach to float. SME paper, MS79–873, 1979.
[4] Drake, S. H. Using Compliance in Lieu of Sensory Feedback for Automatic Assembly. Sc.D. Thesis, MIT, September 1977.
[5] Aster, R. W. LSA Project Progress Report 14, DOE/JPL 1012–42, 355–357, December 1979.
[6] Chamberland, R. G. A Normative Price for a Manufacturing Product: The Samics Methodology, Vol. 1, Executive Summary, JPL Doc. 5101–93, 15 January 1979.

A RESEARCH PROGRAMME IN SENSOR GUIDED ASSEMBLY

A. Pugh, P. M. Taylor, J. J. Hill, G. E. Taylor, D. G. Whitehead,
A. M. Ali, I. Mitchell, D. C. Burgess, K. K. W. Selke, D. R. Kemp and
C. Stubbings,
University of Hull, UK

First presented at the 6th British Robot Association Annual Conference/Automan '83,
16–19 May 1983, Birmingham, UK

Abstract

This paper describes the activities of one of the SERC/Industry funded research groups within the SERC robotics initiative. The experimental facility incorporating PUMA robots is detailed together with the philosophy of robot assembly using vision sensors. Three industrial assemblies have been identified from two companies and these are used to guide the direction of research. The specific assembly problems are derived from high-power semiconductor products as well as fabric handling in the garment manufacturing industry.

Introduction

Following the announcement of the SERC robotics initiative in 1979[1] the Department of Electronic Engineering at the University of Hull was successful in negotiating collaborative research programmes with industrial partners. The first contract involves the GEC Research Laboratories at Chelmsford with specific assembly problems drawn from GEC product companies. Two projects are currently under investigation with Marconi Electronic Devices Limited in Lincoln and a third project has been identified with Hotpoint plc at Peterborough. The second contract is in the garment industry and is being undertaken with Corah plc of Leicester and Marks and Spencer plc of London.

The role which the Robotics Research Unit at Hull occupies in the SERC programme is in the general field of sensor guided assembly using specific assembly problems to provide the direction which the research should take. The approach implies the use of solutions which promise a high degree of success and the fundamental ingredient is simplicity in the techniques chosen. The challenge for the group is to model the assembly area but concentrating on the difficult sensory aspects of each assembly process. In choosing the assembly problems, attention is given to products manufactured in small batches where there is variability either within the range of assemblies or in the components used. Typically, these assemblies have no possibility of assistance from conventional automated handling and must rely on the technology of flexible manufacturing using sensors to retrieve and insert components while monitoring the total operation. This is the only possibility for automated assembly to be associated with batch manufacture where the quantities of assemblies manufactured cannot support the economics of 'hard' automation.

The Hull group makes extensive use of the Unimation PUMA robot which is used in a precision role. An important feature of the assembly area and robot tooling is the high quality of mechanical fabrication which is seen as essential to offer convincing demonstrations to industry. At the time of writing, the research programme has been operational for a little under two years and is approximately at the mid-point of the contract period.

Work area and tooling

The PUMA robots are installed within large work tables of sturdy construction so essential for precision work. It was established at the outset that separate assembly projects would need to be constructed and evaluated simultaneously and it would not be possible to dedicate the robot work area to a specific problem on a permanent basis. The concept of work pallets was introduced which are located in dowels under the robot manipulator. Each researcher assembles an experiment on a demountable pallet which can be placed in the work area in a matter of seconds. This innovation has proved to be extremely effective in making maximum use of a robot facility and allows several projects to be implemented simultaneously. This pallet system can be commended to other research groups and a typical assembly pallet is illustrated in Fig. 1. The

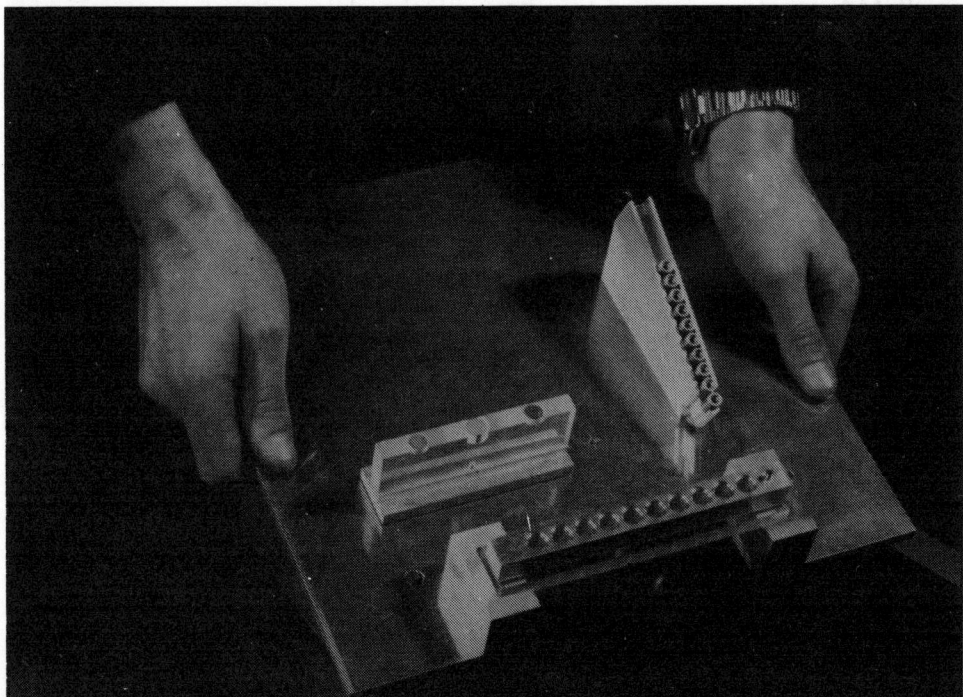

Fig. 1   Pallet containing magazines and jig for an assembly experiment

magazines and fixtures are attached to the aluminium base of the pallet using double-sided adhesive tape which has proved effective for experimental purposes. The advantage of this unconventional fixing is that it makes the work area tolerant to collisions which are unavoidable in robot assembly work. In all other respects, feeders, jigs and fixtures are fabricated to a high degree of precision.

Apart from reprogramming, it is only the robot tooling which needs to be changed with the work pallet. It is quite common during intensive periods of work to change the function of a PUMA twice or three times in one day with negligible 'down time' between each experiment.

Sensory gripping
Proportional gripper
It was recognised at an early stage that a 'high technology' gripper would be needed to support commands initiated by a variety of sensors. A proportional electrically actuated gripper illustrated in Fig. 2 uses a rare-earth frameless motor to apply force through the gripper jaws via a rack and pinion gear with each rack mounted on a precision linear bearing. The position of the motor shaft is measured either by potentiometer or resolver.

The modular design of the gripper housing accommodates several motor sizes. On the prototype, which weighs 1kg, the maximum motor torque is 20 oz-in., giving a gripping force in the region of 10 Newtons. Simulation of the gripper mechanism, with suitable compensation, reveals a closed-loop bandwidth of 15Hz and the jaws will reach 95% of their final position in approximately 60ms. The travel is presently set at 32mm but the design is able to increase this

Fig. 2   Electrically actuated proportional gripper (PUMA compatible)

movement as necessary. The motor is designed to operate 'stalled' and the gripper housing is provided with cooling air actuated by a temperature sensor. However, excessive temperature rise has not proved to be a significant problem.

Concepts in vision sensors

Experience in visual feedback extends back to the Nottingham SIRCH robot[2] which was an integrated hand-eye robot incorporating gripper selection within a rotating turret. The advantages of combining sensors with the gripper were well proven in the SIRCH experiments and the same concept has been retained in the current projects. Over the past decade, most interactive vision research has been implemented using remote (fixed) cameras over the work area and this approach has severe limitations in applications of robots to assembly, not least of which is that the robot arm obscures the work area. When the camera is mounted on the gripper, resolution can be reduced and coordinate transformation between vision and robot axis is very much simplified. In fact the ideal approach is to combine both remote or work-area monitor cameras with a gripper mounted camera.

There are difficult problems in constructing a camera which is suitable for gripper mounting. Clearly the camera cannot dominate the gripper structure and it must not add significantly to the weight of the end effector. We were fortunate to discover the Periphicon 511 Image Digitiser dynamic RAM camera at an early stage in our research. The camera in its commercial housing is not suitable and we have re-engineered the product in a form much more suitable for robot applications. The camera, illustrated in Fig. 3, measures a mere 33 x 28 x 15mm which also incorporates a 9mm f1.9 four element lens. The 1K dynamic RAM provides a resolution of 32 x 32 which is adequate to 'fine-tune' the position of the PUMA prior to the act of assembly[3].

Other dynamic RAM alternatives are currently under investigation with the objective of increasing resolution to include the capability of component recognition and inspection within the gripper facility. At present our concept here is to 'show' components to a fixed camera to accommodate the need for inspection or measurement of component position after gripping.

Some assembly problems require precise measurement of range or distance and for these applications we are evaluating the Optocator laser probe. In its commercial form it is too large to be carried on the robot gripper and experiments are presently confined to a fixed probe.

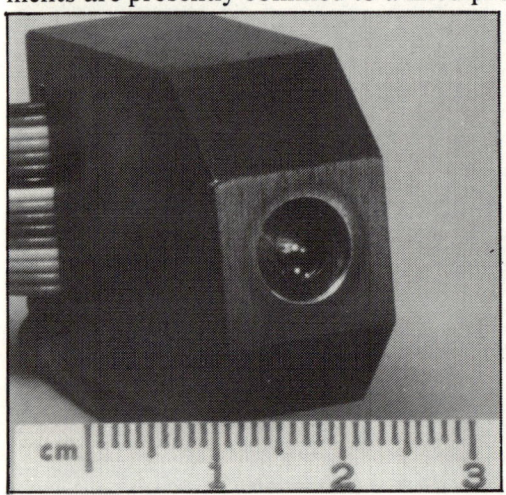

Fig. 3 Dynamic RAM area array camera

There is no doubt that there is a serious crisis in the lack of availability of vision sensors for robot applications. Commercial pressures demand sensors for the television industry and the excellent CCD arrays in cameras currently available are too large and too expensive for robot assembly applications. In common with other research groups, we are using CCD cameras of this kind in the hope that solid-state cameras will eventually become competitive in price. Our guiding philosophy is simplicity with preference given to cheap, rugged sensors which will look 'comfortable' in a shop-floor situation.

Multiprocessor sensor interface
A complex sensory robot assembly facility incorporates a number of sensors operating simultaneously with each providing data to support the assembly process. In most cases, the computational support for each sensor is quite small but the problem arises when several sensors need to be serviced at the same time. Further, a communicating path to the robot controller is best implemented through a single 'port'. Our solution to this requirement is a multiprocessor architecture which consigns all data processing and special I/O requirements for a given sensor (or gripper) to an array of 'slave' processors. A 'master' processor allocates tasks to these slaves as and when required within any specific task.

Sensory tasks are initiated by the master processor and are executed by the appropriate slave subsystem using software and hardware within each slave and dedicated to the particular sensor requirements. Dialogue between master and slaves is executed by means of a simple 'semaphore' message passed via a standard bus which is used universally within the research group. A task 'tree' or 'graph' can be evolved which optimises the use of concurrent processes thereby reducing operational cycle times and increasing workstation efficiency. This task tree is loaded into the master processor which then performs the assembly operation by stepping through the tree initiating sub-tasks as and when required.

It is implicit in this approach that all sensors (and grippers) become 'smart' systems which can execute autonomous tasks as they are instructed by the master processor. The master has *a priori* knowledge about the attributes of each smart system and is aware of the protocol required in order to invoke a particular sensory task. It is expected that the next generation of the multiprocessor system will allow all task code, protocol specification and initialising information to be down-line loaded as output from a program written in a high level language for assembly.

Communication with the PUMA controller has necessitated modifications to the software to provide facilities for communication through the controlling language VAL. A parallel port has been reassigned which allows control of the PUMA to be transferred to a supervisory program running on a DEC LSI 11/02. In addition, a software package called VALADD is incorporated with the VAL software in the PUMA controller.

The prototype multiprocessor system incorporating slave processors for a dynamic RAM vision sensor, laser probe, instrumented mechanical probe and proportional gripper is illustrated in Fig. 4.

Experiments in the assembly of high-power semi-conductor devices
Assemblies involving diode pellets
Diode pellets (*p-n* junction diode wafers) are fabricated by etching from a silicon wafer either in circular or hexagonal format. The pellets are often poorly

188 PROGRAMMABLE ASSEMBLY

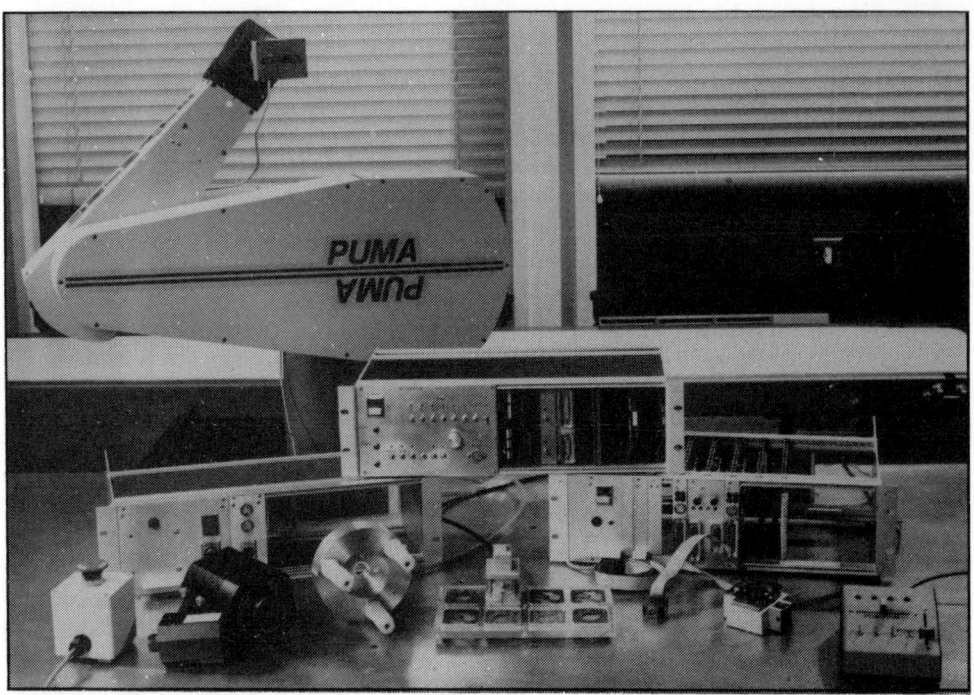

Fig. 4   Multiprocessor hardware and a variety of sensors and grippers

Fig. 5   Diode pellets

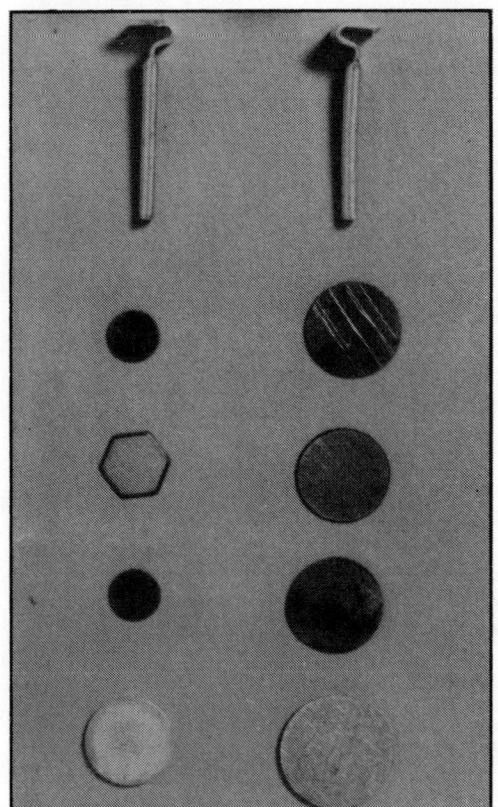

Fig. 6   Diode components

formed and contain a large fraction of distorted shapes. A typical selection is illustrated in Fig. 5. This raises the problem of parts inspection and presentation which cannot be solved by traditional hard automation methods because of batch size. The intention here is to use vision in the role of retrieval followed by inspection to insert the pellet into graphite jig as one of a number of components. Incidentally, the pellet also needs to be checked for polarity.

Each assembly consists of the five components illustrated in Fig. 6 for two different diode specifications. Starting at the bottom of Fig. 6, the first disc is a molybdenum base followed by a solder preform, the diode pellet, a second solder preform and the terminating 'C' crimp. A graphite weight holds the assembly in place under gravity during the time when the jig is heated in a furnace.

Both solder preforms and the molybdenum base are fed from magazines and inspection of these is felt to be unnecessary. The 'C' crimps are fed from a bowl feeder and picked off the end of the track by a vacuum sucker fitted to the gripper. The shank of the 'C' crimp cannot be assumed to be perpendicular to the ohmic contact and vision is used to determine the amount of correction needed to position the 'C' crimp shank in the hole through the graphite weight. In this case the camera forms part of a fixed visual inspection station (see Fig. 7). This project has been described in greater detail elsewhere[4].

Assembly of thyristor power block

A second project under this heading is the assembly of a thyristor power block which comprises two thyristor pellets mounted on a copper heatsink and con-

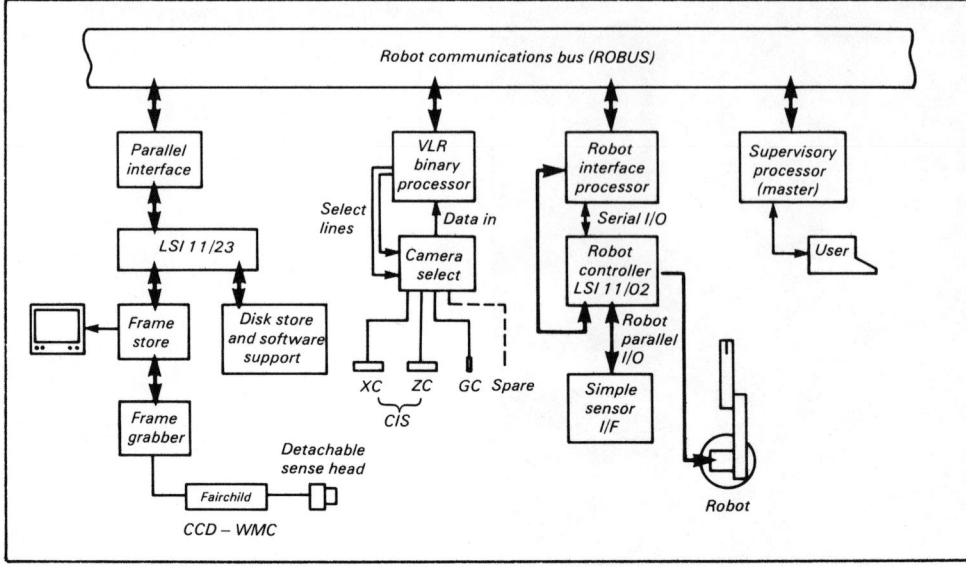

Fig. 7 Schematic diagram showing vision system architecture. CIS, camera inspection station; XC, ZC, axis cameras; WMC, workspace monitor camera; GC, gripper camera

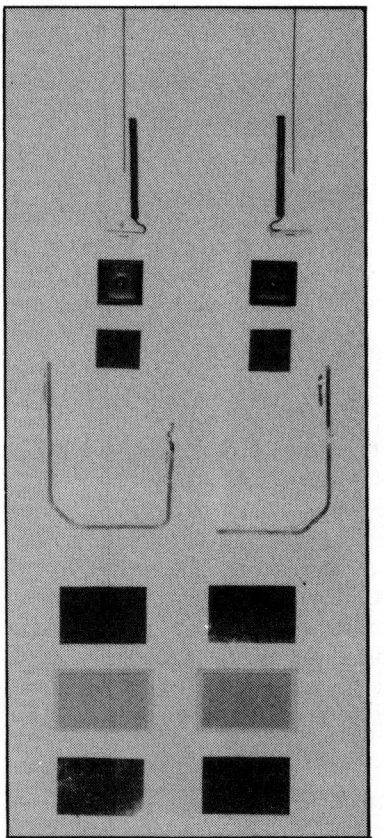

Fig. 8 Thyristor components

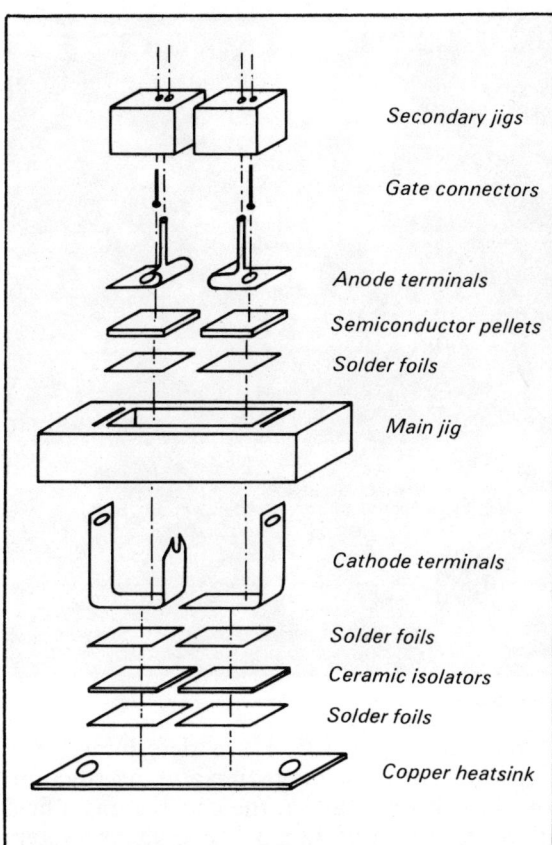

Fig. 9   Power block assembly

tained inside a plastic injection moulding. The components involved in the power block are illustrated in Fig. 8.

The challenge of this assembly is its complexity and requirement to handle difficult components – notably a fine wire. The assembly sequence is illustrated in Fig. 9 which shows how the components are related one with the other and how they are retained in graphite jigs. The jigs are 'inverted' to execute the assembly and restored to the correct position before thermal treatment. The secondary jigs act as weights to keep components together as the solder melts.

The experimental area is illustrated in Fig. 10 which simulates rather than replicates the production area. The main and secondary jigs are delivered on chutes and under gravity and are assembled 'blind' on a fixture in the work area. The next part of the assembly operation requires the insertion of the anode terminals or 'C' crimps using vision to correct for any misalignment as discussed previously. These terminals are inserted in the larger of two holes in each secondary jig so that the small gate connector hole is concentric with the perforation in the 'anode terminal'.

The next operation is the most delicate and takes the gate connector wire (0.5mm diameter) from its magazine for insertion into its retaining hole in the jig (0.7mm diameter). The position of the jig is not known to within about ±1mm because a space tolerance must be included to insert the cathode terminals later. Further, we are working at the limits of the PUMA accuracy which cannot be relied upon to do this operation blind. The hole in the jig is located visually using

Fig. 10   Assembly area for thyristor power block

the gripper camera and minor corrections for position are determined and retained for future use. The wire is retrieved from the magazine and inserted into the hole at its predetermined location. For this operation, the camera has a field of view approximately 3 x 3mm with a resolution of 32 x 32. The vision system is able to locate the hole to an accuracy of about 0.2mm which is sufficient for inserting the wire without the need to chamfer the hole. The hole is illuminated from beneath with an infrared light-emitting diode. Extensive tests have been conducted on this particular operation with a collection of secondary jigs manufactured with the gate connector hole drilled in eight different positions on a pitch-circle diameter equal to the hole diameter of 0.7mm. This is to simulate the kind of manufacturing tolerance which might be expected in practice. The success rate after several hundred operations promises to be better than 99%. The gate wire at the point of insertion is illustrated in Fig. 11.

At the time of writing the remainder of the assembly has not been attempted but will be simpler than the delicate operations already successfully accomplished. This particular assembly has demonstrated its suitability for use as an assembly 'benchmark' using real component parts.

Textile handling and assembly

At the outset of this project the fundamental problem was the ability to handle fabric panels from a cut stack with all the attendant difficulties which are implied. The evolution of our prototype sensory gripper for this purpose has been an interesting experience but this account concentrates on the device after development which has proved remarkably successful although a little slow in its present form.

SENSOR GUIDED ASSEMBLY 193

Fig. 11   Gate wire positioning after visual sensing of hole position. (Wire is 0.5mm in diameter and hole is 0.7mm in diameter.)

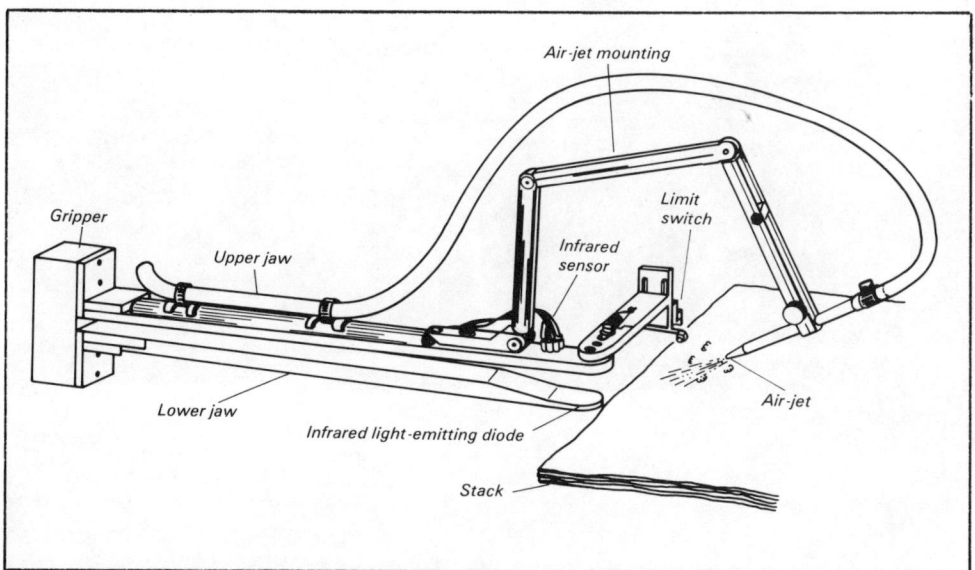

Fig. 12   Schematic drawing of sensory gripper for textile handling

The gripper has two parallel jaws 450mm long and 20mm wide. The lower jaw of the gripper contains infrared light-emitting diodes which supply the triggering for cross-fire sensors mounted in the upper jaw. On the upper jaw is an adjustable mounting to direct an air jet at any desired angle on to the cloth, and at the tip of the upper jaw is a mounting for a limit switch. The gripper is illustrated in schematic form in Fig. 12.

The operating cycle begins with the gripper at the start position at a point 5mm above and 5mm over the edge of the cut stack. The start point is updated after each cycle has been completed. The air jet is activated which causes the top layer of the material to vibrate and the tip of the lower jaw comes into contact with the vibrating cloth. The robot searches downwards so that on continued incremental movement the fabric 'pops' above the lower jaw. The sensor at the tip of the jaw detects the cloth and the air jet is turned off. The gripper is moved horizontally towards the stack inserting the lower jaw between the layers. When the gripper is extended fully into the stack the jaws are closed and the fabric panel is 'peeled' away from the top of the stack and positioned in the work area. The development and operation of the gripper has been published previously[5]. The gripper during operation is illustrated in Fig. 13.

The remaining part of this project required the orientation and placement of an embroidered motif on the fabric panel using vision guidance. In this case, a fixed vidicon camera has been used and a resolution of 64 x 64 has proved satisfactory for recognition and orientation. The motif is fastened to the fabric by thermally 'fusing' a nylon thread built into each motif. This project is continuing but results on the vision system used have been described recently[6].

Future work

The work on the current range of products has been producing results for a short period of time but sufficient experience has been gained to shape plans for the

Fig. 13 Textile gripper in operation

future. The robot as a stand-alone programmable assembly tool has deficiencies of accuracy, resolution and repeatability. The need for a sensory assembly table is evident in common with the findings of other research groups. This will be able to position the assembly jig or fixture beneath the robot gripper with greater precision and incorporate tactile transducers to monitor the assembly process. While no tactile transducers have been realised by the group at this stage, we remain very interested in the integration of tactile sensors with our work on vision.

Recently we have directed some effort towards evaluating RAPT as a language for off-time programming assembly operations. The thyristor assembly is being used as a benchmark and ultimately we hope to demonstrate and compare this approach to direct on-line programming using the VAL language.

An important aspect of our future plans is to identify similar assembly problems in industry in an attempt to specify a general purpose sensory robot for such applications. A survey of vision research projects reveals that solutions to sensory problems produce results which are often specific to one particular problem or a group of problems. Before sensory methods can become a standard option for commercial products, some acceptable common denominator must be found so that the sensors and subsequent data processing can experience a wide application. A collection of papers on robot vision has been published recently[7].

Concluding remarks

The UK launched its initiative in robotics to stimulate research in the university sector, in line with international interest in the subject. The first results of the funding under this initiative are now beginning to emerge from industry supported research groups in the universities. The work described in this paper summarises progress in one such group which is responding to the challenge. A secondary product of this development is to provide the knowledge based within the universities to train postgraduate researchers and undergraduate students alike to the ultimate benefit of British Industry. The involvement of industry as a fundamental requirement has proved enormously successful to this initiative.

Acknowledgements

The authors acknowledge with gratitude the support and guidance of their industrial partners: The Marconi Research Centre, Chelmsford; MEDL, Lincoln; Hotpoint plc, Peterborough; Corah plc, Leicester; and Marks and Spencer plc, London. These companies have previously given their time to partake in discussions and offer advice, as well supplying component parts for experimental purposes. The projects described in this paper have been supported through funding from the Science Engineering Research Council under Contract Nos. GR/B/1769.3, GR/B/4586.3, GR/B/5940.2 and GR/B/8090.1. The construction and installation of experimental hardware was undertaken by J. Hodgson, A. Shiels and K. A. Welsh.

References

[1] Davey, P. G. Robot research in the SERC/Industry national programme. In, Proceedings of the 6th British Robot Association Annual Conference/Automan '83, Birmingham, UK, Nov 1983, pp. 193–204. IFS (Publications) Ltd., 1983.

[2] Heginbotham, W. B., Pugh, A., Kitchin, P. W., Page, C. J. and Gatehouse, D. W. The Nottingham 'SIRCH' assembly robot. In, 1st Conference on Industrial Robot Technology, Nottingham, March 1973, pp. 129–42. IFS (Publications) Ltd., 1973.

[3] Taylor, P. M., Selke, K. K. W. and Taylor, G. E. Closed-loop control of an industrial robot using visual feedback from a sensory gripper. In, 12th ISIR (l'Association Francaise de Robotique Industrielle), Paris, June 1982, pp. 79–86.

[4] Burgess, D. C., Hill, J. J. and Pugh, A. Vision processing for robot inspection and assembly. SPIE Proceedings on Robots and Industrial Inspection, Vol. 360, 1982.

[5] Kemp, D. R., Taylor, G. E., Taylor, P. M. and Pugh, A. A sensory gripper for handling textiles. In, 13th ISIR. Robot Institute of America, Chicago, April 1983.

[6] Taylor, G. E., Kemp., D. R., Taylor, P. M. and Pugh, A. Vision applied to the orientation of embroidered motifs in the textile industry. In, Proceedings of the 2nd Conference on Robot Vision and Sensory Controls, Stuttgart, November 1982. IFS (Publications) Ltd., 1982.

[7] Pugh, A. Robot Vision. IFS (Publications) Ltd. and Springer-Verlag, 1983.

A MODULAR PROGRAMMABLE ASSEMBLY STATION

R. C. Smith and D. Nitzan, SRI International, USA

First presented at the 13th International Symposium on Industrial Robots and Robots 7, 17–21 April 1983, Chicago, USA. Reprinted courtesy of the authors and the Society of Manufacturing Engineers, Dearborn, USA

Abstract

A hierarchical programmable assembly system, consisting of assembly stations and a transfer mechanism, was designed conceptually and is being developed. A station consists of functional modules, each including a major device, such as a robot arm, on a sensor as well as auxiliary devices or auxiliary sensors. Each module is controlled by a computer, which stores reflex, bootstrap, and program routines that are required to perform the module functions. The module computers are interconnected with each other and the station computer by means of a communication network using a single coaxial-cable bus. Based on this design, a programmable assembly station has been implemented. The station modules incorporate two PUMA robots with force and visual sensors, an SRI vision module, a table with general-purpose fixtures, and an auto-place manipulator on a servoed rotary table. Employing a 1MHz communication network, this station was used to demonstrate programmable assembly of computer-terminal components.

Introduction

The automation of discrete-parts manufacturing, spurred in its practical development by both economic and social considerations, is being implemented worldwide on an ever-increasing scale. Advanced automation expands the productivity of labour, combats inflation, and makes it possible to compete effectively in world markets; it improves the everyday working conditions of the labour force and raises the standard of the population as a whole. Automation is especially important for batch manufacturing, which is extremely labour-intensive and accounts for the largest portion of the total cost of discrete-product manufacturing[1]. Unlike the application of hard automation to mass production, where the expense of acquiring special-purpose equipment can be justified by the resulting high volume of production, the automation of batch manufacturing must be programmable. Programmable industrial automation[2] is characterised by three salient features:

☐ *Flexibility* – the capability of a machine system to perform different actions for a variety of tasks.
☐ *Ease of training* – the facility with which a person can efficiently program a machine system to execute a desired task.
☐ *Artificial intelligence* – the ability of a machine system to perceive new conditions (whether anticipated or not), decide what actions must be performed under those conditions, and plan these actions accordingly.

Although suitable primarily for increasing the productivity of batch manufacturing, programmable automation may also be applicable to mass production for the following purposes:

○ Reduction of the setup time for manufacturing short-lived products in a competitive world market.
○ Lowering the cost of production equipment by using components, such as robots, sensors, and computers, that are commercially available and recyclable.
○ Producing a sufficiently large sample of new products to enable their technical and market performance to be tested before investing in a costly hard automation system that would mass produce them.

The factory of tomorrow will be characterised by integration of manual, hard automation, and programmable automation activities in proportions designed to minimise the total cost of manufacturing and servicing. The ultimate goal of programmable industrial automation is an automated factory based on the advanced technology of programmable computer-aided manufacturing (CAM). Eight CAM functions are distinguished: product design, part fabrication (including metal cutting by numerically controlled machines), part storage and transportation, logistics, materials handling, assembly, inspection, and process planning. At present only the first four functions may be found in a factory employing a flexible manufacturing system (FMS). The other four are still objects of investigation in research centres worldwide.

Among these latter four research topics, programmable assembly is the most challenging for two reasons. First, programmable assembly is important; it may replace manual assembly, which constitutes the largest portion (approximately 22%) of the total labour cost for all durable goods[3]. Secondly, programmable assembly is complex; it includes materials handling and in-process inspection, as well as programmable part presentation, trajectory planning, collision avoidance, arm control guided by multisensory feedback, part mating, and multi-manipulator cooperation.

In this paper we describe recent research on programmable assembly conducted by the Robotics Department of SRI International. The paper covers two major topics:

☐ The characteristics of an assembly system we are developing, including hierarchical organisation, modularity, distributed processing, and robustness.
☐ Implementation of an assembly station with the above characteristics and its use in demonstration of a programmable assembly task.

Assembly system characteristics

The characteristics of the assembly system we are developing are as follows:

○ *Hierarchical organisation* – to simplify system software development by par-

titioning the overall control problem into manageable subsets that can be dealt with separately and simultaneously[4].
○ *Modularity* – for rapid reconfigurability of system components, and easier modelling of those components because their interconnections and capabilities are well defined.
○ *Distributed processing* – to take advantage of the opportunities for parallel computation by means of multiple processors at the points of sensing, control and command.
○ *Robustness* – to minimise the need for manual intervention in recovering from various fault conditions.

Each of these characteristics, as well as some instances of their implementation, are discussed below.

Hierarchical organisation

An assembly *system* computer will control the activities of a number of assembly *stations* and a *transfer mechanism* carrying subassemblies between them. An initial design criterion is that every station must be available for experimentation, training, calibration, setup and debugging. For such purposes, each station will be operated independently of the rest of the system. In contrast, when the entire system performs a task, some coordination of the individual stations and the transfer mechanism will be necessary. Station-level integration will be accomplished by the system computer.

We next divide each assembly station into its major functional components, called *modules* (e.g. manipulators, vision modules, part presenters, and support tables). A station could be composed of these modules in different ways and configurations, thus providing flexibility for batch assembly of a variety of products.

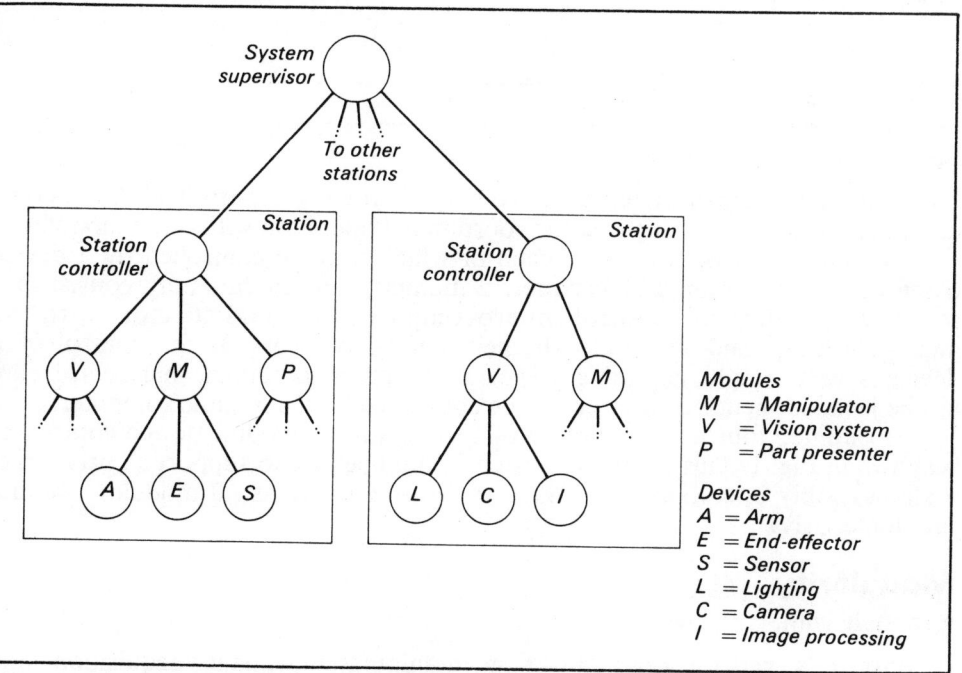

Fig. 1   Hierarchy of control for the assembly system

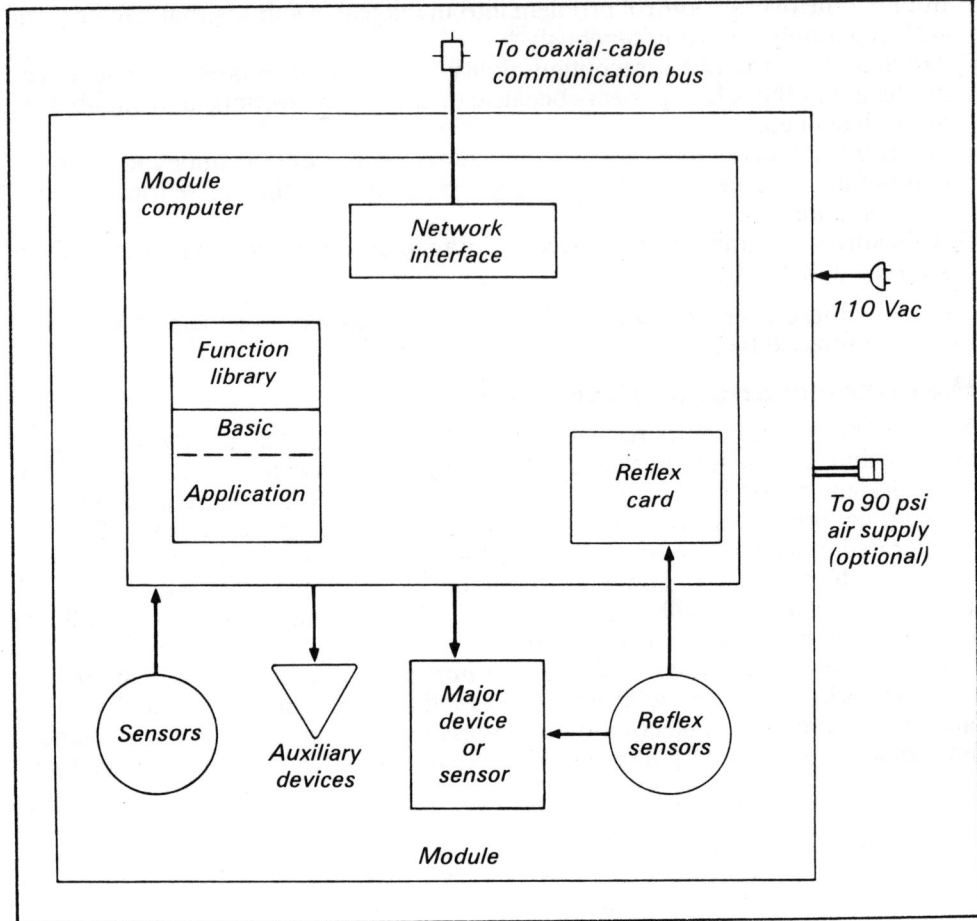

Fig. 2   Basic module organisation

Each module is encapsulated to make it self-contained, controllable by its own computer, and easily configured for coordinated operation with other modules.

A module consists of *devices,* each of which may be controlled by a device computer or processor. For instance, a manipulation module may consist of a robot arm with its control microcomputer, an end-effector with its microprocessor, and sensors with their microprocessors. If the control of a device is very simple, assigning a special computer to control that device may not be justified; hence the device will be controlled by the module computer.

The outlined four-level hierarchy of system-station-module-device computers is shown in Fig. 1. This computer system should be able to support a fairly large-scale assembly operation (e.g. a factory with as many as 50 modules working simultaneously).

Modularity

A module will:

☐ Perform a generic operation (e.g. manipulation, image acquisition and processing, or part feeding) and control any devices used in that operation.

- Use its computer to provide an external interface to these generic operations at a high level.
- Use auxiliary sensors to verify expected local conditions; e.g. a manipulator will have a sensor to verify that it is 'holding an object'.
- Execute reflex actions, under predefined conditions, that can be detected by the local sensors.
- Operate independently of other modules.

Fig. 2 shows the basic components of a general module with the above listed attributes. These components are discussed in the subsequent sections of this article.

The module computer contains a processor, network interface cards, memory and input/output interface cards for the analogue and digital signals from or to the auxiliary sensors and devices of a module. Some specific module types are described below, along with examples of the basic functions they provide. In general, these functions entail intermodular communication of high-level information, rather than large amounts of raw data.

Manipulator module

We have developed two manipulator modules, each consisting of a Unimation PUMA 560 robot and an end-effector. The end-effector of one arm consists of an electrically actuated two-fingered hand, a remote-centre-compliance (RCC) device, and a remote-head video camera. The end-effector on our second PUMA 560 consists of a six-axis force/torque sensor mounted on the wrist, and a pneumatic two-fingered hand. The module computer and the PUMA controller for each manipulator are mounted under the arm's supporting stand. Fig. 3 shows the hardware control and sensing functions associated with the manipulator module, including planned extensions for proximity and touch sensors on the end-effector. The module computer, in this case, will provide a means for controlling the hand, reading sensors in the hand and wrist, and moving the arm (indirectly, by communicating with the PUMA controller).

A few examples of the manipulator module functions are:

Where(Result)
   Returns the location (position and orientation) of the manipulator's end-effector.
MoveTo(Location)
   Move to the specified location.
Grasp(GraspLocation)
   Approach the given grasp location, open the hand, move to that location, close the fingers, and depart along the approach vector.
StopOnForceZ (Frame, StopForce, Overshoot)
   Move along the z axis of a specified coordinate frame, stopping if a contact force along the z axis exceeds Stopforce, or if the manipulator travels farther than Overshoot along the z axis.

Binary vision module

As shown in Fig. 4, the binary vision module includes an SRI vision module[5] with up to four 128 × 128-element solid-state cameras attached, a module computer, and illumination and camera control (to be implemented in the future).

In addition to invoking SRI vision-module functions, the module computer could be used, as needed, to:

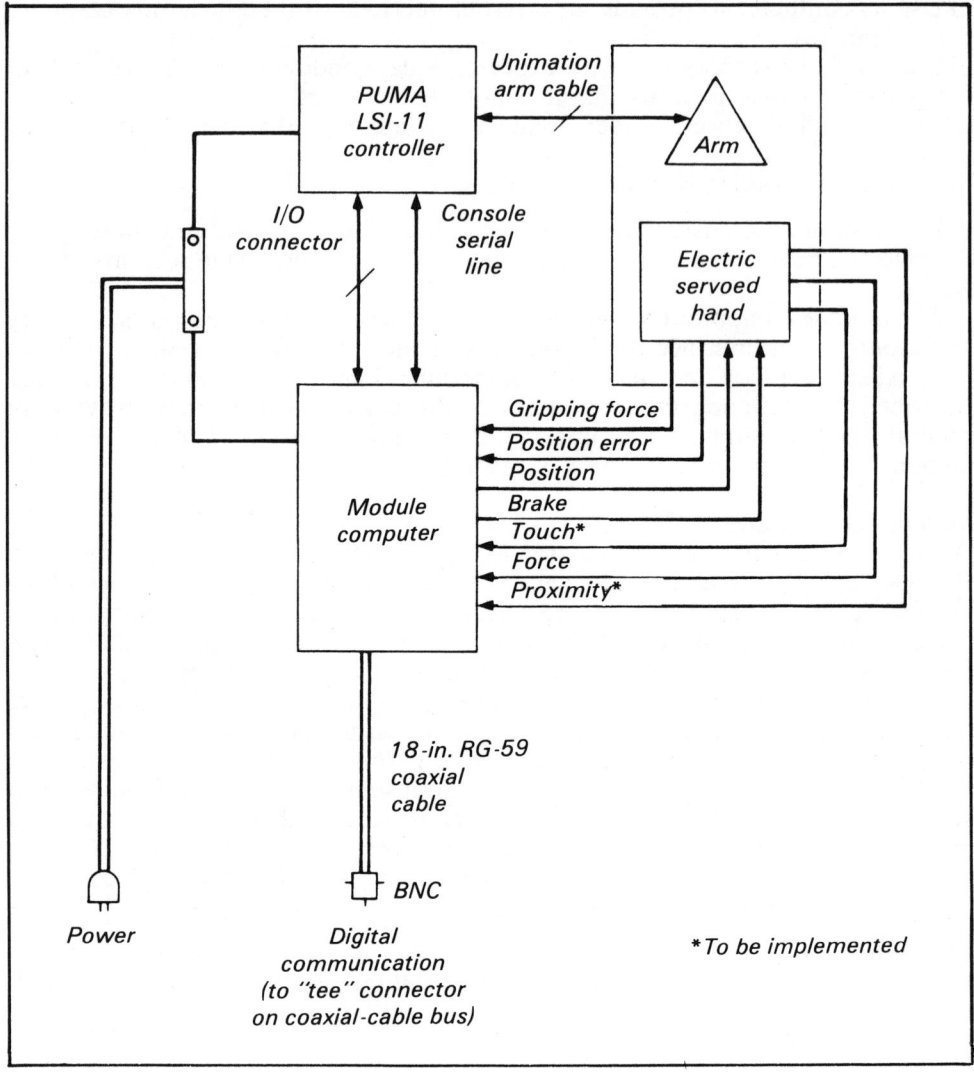

Fig. 3   Schematic of the manipulator module

○ Store the application-dependent routines developed around the basic functions of prototype training, recognition, and feature extraction.
○ Control such auxiliary devices as zoom lenses or lens turrets, and implement control of focus, directional lighting (front/back), and lighting intensity.
○ Measure illumination uniformity with work surface sensors, and track intensity changes. This measurement would ensure that the lighting conditions during object recognition and object training are identical.

The electronic components of the SRI vision module and the module computer are stored inside a stand like the one on which each manipulator is mounted. The top of the stand forms a work surface and can hold the aforementioned intensity meters and/or a light table for backlighting operations. Cameras are attached to long cables and can be mounted anywhere within the

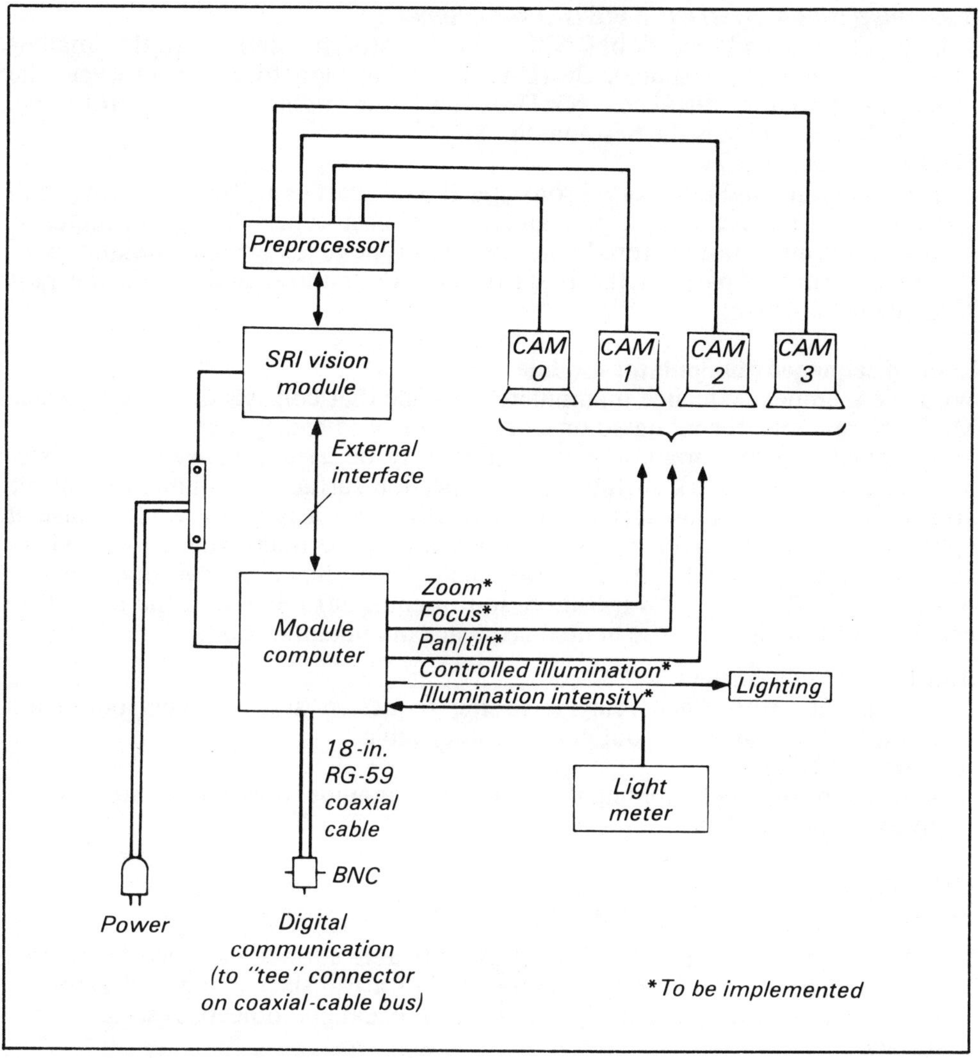

Fig. 4 Schematic of binary vision module

assembly station. One camera is mounted on the hand of one of the manipulators.

Some examples of binary-vision-module functions are:

Picture(BlobCnt)
   Take a picture, perform connectivity analysis of the image, and return the number of connected regions (blobs), BlobCnt, that meet a certain blob criterion, such as minimum size and roundness.

GetFeature(BlobN,FeatN,Result)
   Return the value of a blob feature (indicated by the index FeatN) of a selected blob (indexed according to decreasing size by blobN). Features are, for example, blob area, perimeter length and moments.

Recognise(BlobN,BestDist,NxtDist,ProtoName)
  Compare a given blob, BlobN, with a set of prototypes and return the smallest distance (in feature space), BestDist, from the blob to any prototype, the second smallest distance, NxtDist, and the name of the prototype, ProtoName, which is the best match.

Find(PartType,Result)
  Take a picture and look for a prototypical blob, PartType, by comparing each blob in the picture with a pretrained set of prototypes. If a good match is found, use information stored with the prototype to determine a unique set of 3-D axes in the part on the basis of its 2-D features, and return the part location.

Limited-sequence manipulator module
We have a limited-sequence manipulator module that consists of an Auto-Place Model 50 manipulator mounted on a servo-controlled rotary table.

A limited-sequence manipulator module may be used effectively to transfer parts to and from a work surface or an inspection surface. The limited-sequence arm is less costly than servoed manipulators, is usually faster (because it operates between fixed mechanical stops), and is normally rugged enough to carry a greater payload. The limited-sequence arm, however, can operate only on parts in a fixed location (position and orientation) relative to the arm. Two examples of functions for the limited-sequence manipulator module are:

AutoCmnd(RelayStates)
  Set up the Auto-Place relay(s) to the requested state(s). This command actuates the pneumatics, but not the rotary table.

APTMove(Theta)
  Turn the rotary table holding the Auto-Place manipulator to absolute position given by theta radians.

Other modules
We have been developing or planning other modules, such as:

☐ *X-Y theta table* – an X-Y table whose movable surface can also be rotated about the z axis. The table is equipped with a translucent top and a row of fluorescent lamps underneath, so we can backlight objects resting on the surface.

☐ *Part presenters* – programmable part presenters previously developed at SRI will be incorporated into the assembly system and extensions of their capabilities explored. One example is the SRI 'Eye Bowl' – a standard bowl feeder that utilises vision rather than mechanical blades for part sorting and feeding[6].

☐ *Programmable jig* – we wish to explore the concept of a multipurpose, computer-controlled jig that is capable of accepting and rigidly holding parts of many different shapes, which it would then present, upon request, to a manipulator or vision system in a specified orientation. This might take the form of a rugged hand and a three degree-of-freedom wrist.

Distributed processing

A local-area network is a natural organisation of a system in which processing is distributed among numerous computers, which are often separated from one another by, say, one metre or more. Hardware supporting various network topologies is available commercially. We use a system in which a coaxial cable forms a communication *bus* connecting all the communicating computers. The bus-network organisation:

○ Allows direct communication between any two computers connected to the coaxial-cable bus.
○ Promotes modularity of system components by requiring only a standard network interface for systemwide communication.
○ Facilitates reconfigurability of components by permitting them to be connected to the network at any point on the coaxial-cable bus.
○ Permits the sharing of expensive system resources, such as printers, graphics devices and file-storage units.

Each computer connected to the communication bus contains a network interface with a unique name (a number) assigned to it. Names are used to identify both the source and the destination of a message. When a message is transmitted on the bus, every network interface compares its name with the message destination and receives the message only if there is a match. A special type of broadcast message can be addressed so that all the computers (except the sender) will receive it. This type of message is useful when a module needs help from the 'system', but does not know the name of the unit that can furnish such help.

A communication software package has been written to provide flexible communication capabilities through the network interface. The selected protocol and the characteristics of the network interface are described in detail elsewhere[7]. Briefly, the package supports a 'random-access' protocol whereby any computer may send a message to any other computer or computers if the communication bus is idle. The message is usually one of the following:

□ A command to a module to perform one of its functions with the parameters given in the message. The command may be to supply information, to request information, or to perform a specified activity.
□ A reply to a command, containing any results or requested information. The reply also serves to confirm completion of a commanded activity, which may otherwise not have returned any results.

The communication package performs the following functions:

○ Message creation, retrieval, buffering, and deletion.
○ Message receipt, acknowledgement, and transmission.
○ Automatic retransmission of unacknowledged messages.
○ Detection of any special broadcast messages for later action.

A module will never have more than one buffered message to transmit at any time. However, it can receive and buffer numerous messages and either attend the oldest one or search in its buffer for an expected message of a certain type, from a certain source, or both.

Robustness

We have recently implemented three levels of processing in each module computer:

□ *Reflex level* to detect hardware or software faults and set the module hardware devices to predetermined states.
□ *Bootstrap level* to set the module-computer program at a predetermined state

in response to reflex activation, and to notify the rest of the system about this event.
☐ *Program level* to implement the main functions of the module. A sensor-monitor routine will be implemented at this level in the future to detect conditions beyond the capabilities of the reflex level.

Reflex and bootstrap levels
At the lowest processing level the modules (particularly those incorporating manipulators) need self-protective mechanisms that act as reflexes – hardware responses to a set of external or internal fault conditions. Such conditions include loss of operating power, loss of program control, and human intrusion into the assembly area. If necessary, special sensors will be assigned to detect these conditions. Once enabled, a reflex device will watch for the fault condition it guards against and be triggered if that condition arises. We have designed a reflex card to implement this function and fulfil the following responsibilities:

○ To provide a mechanism that forces the module's devices or sensors into 'safe' default when a reflex is triggered.
○ To protect the module from power loss.
○ To protect the module from loss of program control.
○ To provide sensor-triggered hardware reflexes.
○ To provide programmable-condition reflexes.

In addition, activation of a reflex may optionally force the program control in the module computer to transfer to a simple bootstrap program resident in a nonvolatile memory on the reflex card. The module subsequently executes a simple program at the bootstrap level that will:

☐ Initialise the network interface for communication.
☐ Use the network to broadcast a message notifying other module computers of this module's current state.
☐ Provide capabilities for loading the program of this module through the network interface when commanded by another computer.
☐ Supply tools for remote diagnosis of this module through the network interface.

Fig. 5 depicts the relationship between the reflex and bootstrap control levels. More information about reflexes and the bootstrap may be found elsewhere[6]. The above-listed responsibilities at the reflex level are described below.

Default state conditioning. Each device or sensor connected to the module computer should be set to a safe default state whenever a reflex is triggered. It is common practice to provide a computer bus line that carries an INITIALISE signal to all the computer interfaces. The reflex card triggers generation of this signal when a reflex condition occurs. The INITIALISE signal, for example, may halt any moving device if the signal is applied directly as an override or a shutdown signal to that device.

Power loss protection. Like many computers, the module computer has the capability of detecting imminent power loss through a sensing circuit in its power supply. A signal indicating this event is supplied to the reflex card from the computer power supply. The reflex card triggers an INITIALISE signal on the bus, thus initiating a command to halt moving devices, for instance, prior to the power loss in the module computer. In a 'power-up' sequence, the INITIALISE signal is generated again, and program control is transferred to the bootstrap code.

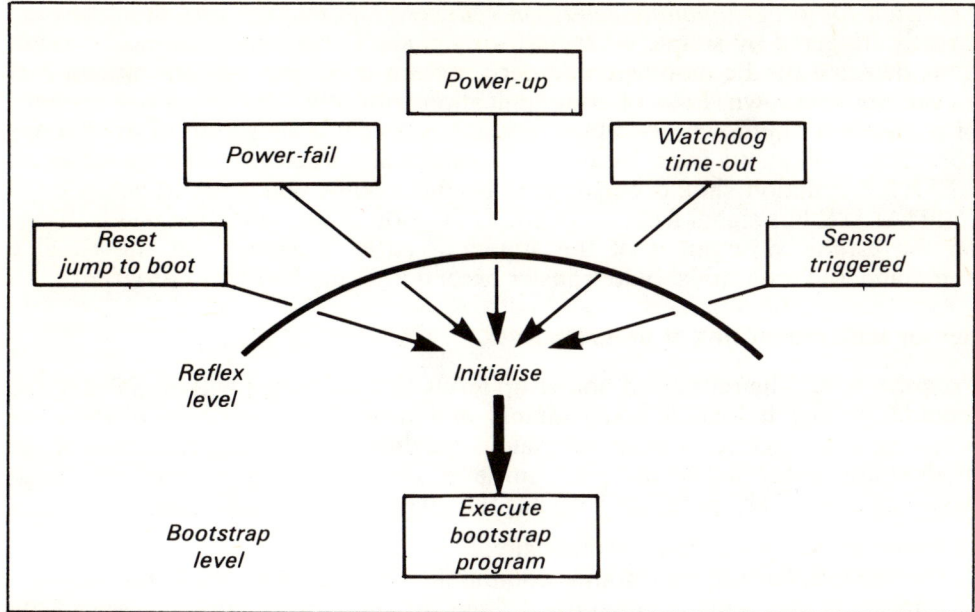

Fig. 5   Reflex level

Protection from loss of program control. The module computer's program may not always be running correctly; e.g. it may become deadlocked, halted, or contain errors. For these contingencies, an independent timer, called a 'watchdog timer,' is included on the reflex card. The watchdog timer must be reset periodically by a properly executing program in the module computer. If it is not reset, the watchdog timer reaches a 'time-out' state. This state is a reflex condition that, like any other, will cause generation of an INITIALISE signal via the reflex card and transfer the module computer's program control to the bootstrap code.

Provision for sensor-triggered reflexes. Certain events detected by sensors connected to the module computer may require an emergency response from the module. The reflex card provides such a response capability by accepting binary signals from these sensors. The binary signals could, for example, indicate the state of contact/noncontact or proximity/nonproximity of objects with respect to a manipulator's end-effector. One reflexive response might be to halt a moving manipulator anytime an intruder is detected in the workspace. As another example, a proximity sensor on a manipulator's hand may be used to trigger a 'stop-arm' reflex to prevent collision with unexpected obstacles; however, sometimes this reflex must be disabled to permit the hand to reach a target object. The computer program may disable any reflex circuit on the reflex card by transmitting a special code word. This encryption reduces the possibility that the reflexes may be disabled accidentally. Reflex devices will always be placed in the 'reflex disabled' state following the INITIALISE signal, so that the module will react to the fault condition once, rather than repeatedly. The triggering condition should be determined by reading the reflex status and then be removed before the reflex is enabled again. When the module program begins, it enables those reflex sensors that should be active at that time.

Provision for programmable-condition reflexes. The reflexes just described are directly triggered by simple binary sensor signals. Certain more complex conditions detected by the module-computer program may also warrant initiation of an orderly shutdown. Loss of communication with other devices is one instance of a potential shutdown condition; another example is detection of anomalous conditions computed from local sensor values and internal program states. A RESET instruction should be supplied by the module computer to activate the INITIALISE bus signal from software. Utilisation of this provision will be based on the estimated urgency of the situation. After it carries out the RESET command, the program should transfer execution to the bootstrap program.

Sensor state monitoring at program level

Program level. The reflex and bootstrap levels are concerned with initialising the module, getting it loaded and running, and providing a uniform method for detecting and reacting to local anomalous conditions. The main functions of the module are performed at the program level, i.e. controlling the module's main device or sensor. The program level has the following responsibilities:

○ Reset the watchdog timer periodically.
○ Provide full, flexible intermodule communication via the network interface.
○ Implement the generic functions defined for the module type to control the module's devices or sensors.
○ Provide periodic sensor-state monitoring.

Sensor state monitoring. A background process for monitoring the numerous sensors associated with a module is under development; it is presented as a principal component of the program level of each module. The monitoring process is intended to read the values of local sensors periodically and to compare these values with their expected range, which is given in a table. When an actual sensor value is found outside the expected range, a programmable-condition reflex may be executed or, less drastically, a message reporting the anomaly may be broadcast. Range entries in the sensor table will be made in one of two ways:

☐ Explicitly, through a MONITOR command given to the module, indicating which sensor to use and the unexpected range values of that sensor in the next interval.
☐ Implicitly, by execution of a generic function that imposes known constraints upon a sensor's values.

A MONITOR command can be used, for example, to instruct a manipulator module to monitor the forces and torques acting upon its end-effector and to assure that they remain within a specified range for a given application. Monitoring will then proceed independently until the MONITOR command is withdrawn.

In some cases the effect of a generic function on a set of sensors is known *a priori* and sensor monitoring can be initiated implicitly. For example, let us consider two functions for a manipulator hand: GRASP and RELEASE. The value of a binary touch sensor on the hand's fingers after execution of the GRASP command is expected to be ON. The GRASP routine itself will verify this condition and enter the tolerance range for the sensor value (unnecessary in this instance) in the sensor range table. Execution of the RELEASE routine will generate an entry in the table corresponding to the OFF value for the contact sensors (if, indeed, they were off). The sensor value corresponding to the ON or OFF state can be checked repeatedly by a sensor monitor routine. Thus, an external application program directing the manipulator module need not verify continuously that an object in the hand is still there, but will instead be notified immediately if the object is dropped.

Other operations similarly impose anticipated constraints on associated sensors. Should a discrepancy occur between the sensor table range and the actual value of a sensor, the safest approach will be to execute a 'programmable-condition' shutdown reflex and enter the bootstrap level. Since the reflex does not destroy the resident program, state information can be retrieved from the module by means of the diagnostic routines available in the module's bootstrap program. Such information may be used in future work to determine why a module failed and to direct the recovery of that module or the entire station accordingly.

Assembly station

Station configuration

Fig. 6 shows the current configuration of our assembly station. It consists of a station controller, a binary vision module with three video cameras, and two manipulator modules whose black and white arms are called arm 1 and arm 2, respectively. The end-effector of arm 1 consists of a plastic remote-centre-compliance device, the front end of a video camera (the back end is mounted on the manipulator arm), and a two-fingered hand. The end-effector of arm 2 consists of a six-axis wrist force/torque sensor and a two-fingered hand. Not included in this configuration are the limited-sequence-manipulator module and the other modules previously described; these modules may be incorporated into the assembly station as needed during the performance of other assembly tasks.

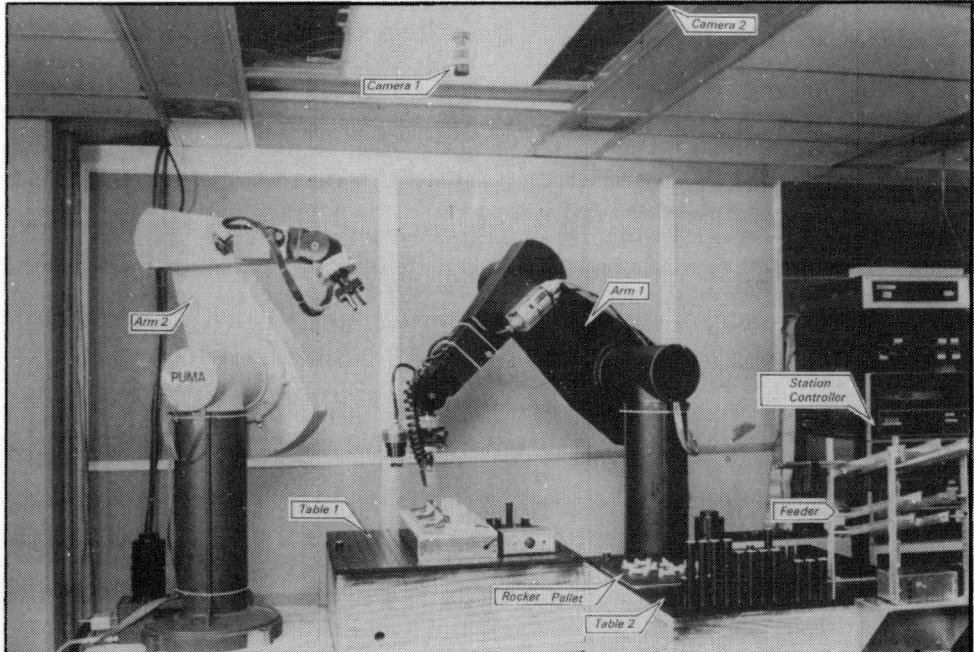

Fig. 6   Assembly station

This station configuration includes two support surfaces. The first, located between the manipulators, is the assembly area (the binary vision module, excluding its cameras, is mounted beneath it). The second surface, located near arm 1, supports general-purpose part feeders and pallets within reach of that arm. Both surfaces have fixed cameras mounted above them, and both have a grid of holes, which are used to provide mechanical support components and to aid in calibrating the coordinate frames of the manipulators, the support surfaces, and the cameras relative to one another.

The station controller, a small computer, is used to control the sequential and parallel operations of the modules by means of commands on the communication network. The station controller also controls a printer, a speech output device, and two disks. Messages from the modules to the operator are printed or spoken; files stored on the disks are sent to the module computers upon request. In addition, the station controller has menus of module commands that can be executed interactively by the user, as well as a cross-network debugger that allows him to examine, alter, and set breakpoints in the programs of the module computers.

Calibration of coordinate frames

Communication of object locations among modules is facilitated if there is a common coordinate system, or *reference frame,* in which to define these locations. For instance, a camera may be used to determine the location of a part; that information, expressed in terms of the coordinates of the reference frame, will enable any manipulator within reach to access the part. Each manipulator module will need to know only its relationship with the reference frame rather than all its pairwise relationships with other coordinate frames in the assembly station. We, therefore, calibrate all the station modules to a station reference frame.

In Fig. 7 we define the base coordinate frames of two manipulators, R1 for arm 1 and R2 for arm 2, the coordinate frames of their respective end-effectors, E1 and E2, and two coordinate frames, T1 and T2, on table 1 and table 2, respectively. Camera 1 overlooks table 1 and camera 2 overlooks table 2. Frame T1 is chosen to be the reference frame and the other frames must be related to it, as indicated by the dashed arrows, by means of transforms – 4 x 4 homogeneous coordinate-transformation matrices[8].

The following general notation is used. The position vector $(x, y, z, 1)$ of a point P in an arbitrary homogeneous coordinate frame F is denoted by P(F). The same point may be described by P(F1) or P(F2), where F1 and F2 are two different frames. We denote the transform from frame F1 to frame F2 by [F1/F2], where P(F1) = [F1/F2] * P(F2). Using this notation, note that P(F2) = [F2/F1] * P(F1), where [F2/F1] is the inverse of [F1/F2]. We depict [F1/F2] by an arrow pointing from the origin of frame F1 to that of frame F2 (see examples in Fig. 7); note that the inverse of [F1/F2] would be depicted by an arrow in the opposite direction.

Using the above notation, we obtain P(R1) = [R1/T1] * P(T1). The elements of transform [R1/T1] can be determined by mounting a pencil-like tool on the wrist of robot 1 and leading that robot so that its tool tip touches three points in frame T1 – its origin, a point on its $+x$ axis, and a point on its $+y$ axis. Reading the robot positions in frame R1, the coordinates corresponding to the origin yield the translation vector of transform [R1/T1], while those of the other two points yield its rotation matrix. A similar sequence using arm 2 and the same

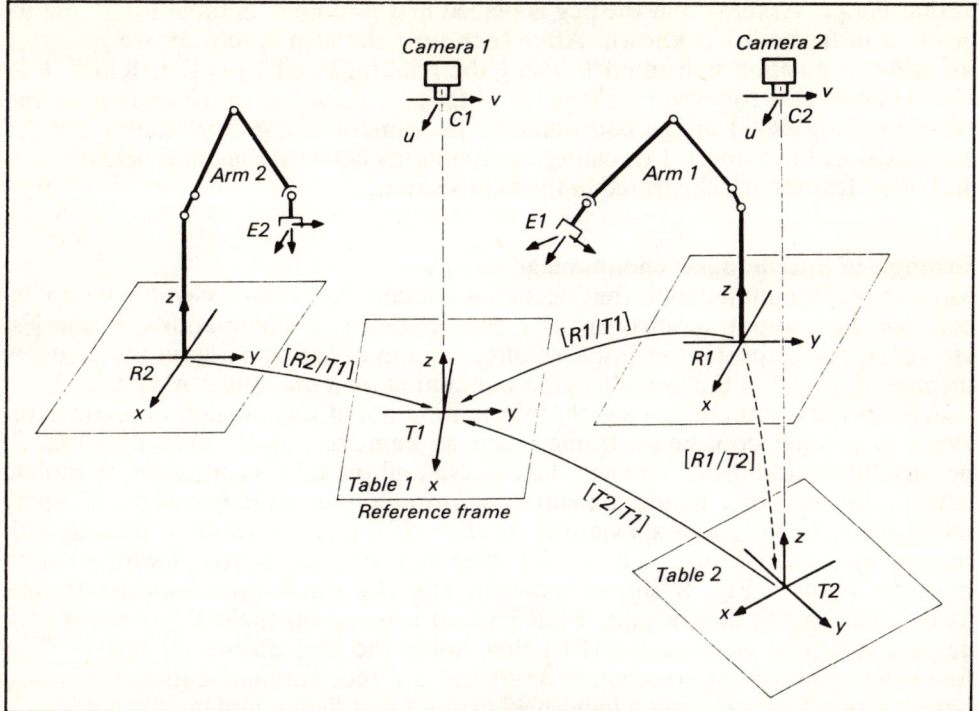

Fig. 7 Station coordinate frames and transforms

calibration points in T1 may be performed to derive [R2/T1]. At this stage, the position of a part given in frame T1 can be transformed into the frames of both arms.

Next, points in the 2-D image plane of each camera are related to a 3-D space. Consider first camera 1, which is calibrated directly to the table 1 grid it views. Pegs of varying lengths are inserted into the grid holes, which are easily identifiable integer coordinates in the grid framework. The white tops of the pegs can be seen as bright spots by the camera. A set of known peg-top (x, y, z) positions in frame T1 and the corresponding set of (u, v) image coordinates in a 2-D frame C1 are utilised, using a least-mean-squares fitting method, to produce the camera calibration matrix[9]. The camera calibration matrix can be used to compute the image coordinates (u, v) of a given point (x, y, z) in the calibration frame. Generally, however, we do the reverse – the x and y coordinates of a point in the calibration frame are obtained as a function of the camera calibration matrix, the image point (u, v), and known z coordinate of that point. For camera 1, the calibration frame is equivalent to the reference frame (frame T1).

Calibration of camera 2 is similar to that of camera 1, except that the relation between the calibration frame (frame T2) and the reference frame (frame T1) must be derived. The relation [T2/T1] is determined by making the tool tip of arm 1 touch points on the origin, the $+x$ axis, and the $+y$ axis of frame T2 to derive [R1/T2], and computing [T2/T1] = [T2/R1] * [R1/T1].

The hand-held camera is calibrated similarly to the other cameras, except that only one peg is used and the camera is moved to view from different locations. The camera is rigidly attached to the tool-mounting flange on the wrist of arm 1; hence, points seen by the camera will be initially referenced to a frame (FL) fixed in that flange. Assume that the peg is placed in a grid hole on table 1 and that its position in frame T1 is known. After each time the arm is moved, we compute the peg-top position in frame FL, using the relation Peg(FL) = [FL/R1] * [R1/T1] * Peg (T1). From the resulting list of (x, y, z) positions in frame FL and the corresponding (u, v) image coordinates, the camera calibration matrix can be computed as for camera 1 or camera 2. Relations between the hand-held camera and other frames are illustrated in the next section.

Example of intermodular communication

Most of the communication that occurs in the context of our assembly station is between the station controller and the modules it commands or queries. However, one important example of direct communication between the modules themselves involves the use of a camera mounted on a manipulator's hand.

A stationary camera can supply information about the location of an object it views in a fixed coordinate frame, such as camera 1 over table 1 in Fig. 7, because the relationship between the camera and the table is constant. A mobile camera, on the other hand, can supply information about the position of a part relative only to the camera's viewing location. If the part location is desired with respect to a fixed frame, such as T1, then the location of the viewing camera must be known. Fig. 8 shows schematically the transforms between frames associated with finding a part, PARTXI, in a pallet on table 1 by means of a camera attached to a flange (FL) that holds the end-effector of arm 1. The following sequence of commands illustrates a direct communication between a binary vision module using a hand-held camera and the manipulator module:

MoveTo(ARM1, AbovePallet)

The station controller commands arm 1 to move to Location AbovePallet above a pallet with a desired part. Since AbovePallet is described with respect to T1, arm 1 converts AbovePallet to a location in its own frame, using the relation [R1/AbovePallet] = [R1/T1] * [T1/AbovePallet], and moves to that location. After the move, frames E1 and AbovePallet will coincide.

SetCamera(VM1, HandCamera)

The station controller commands vision module VM1 to select the hand-held camera for input.

Find(VM1, PARTX, Result)

The station controller commands VM1 to take a picture with the selected camera, compute the location of a part of type PARTX in reference coordinates, and return this result.

The task of VM1 is thus to compute transform [T1/PARTX1]. Since [T1/PARTX1] = [T1/FL] * [FL/PARTX1], VM1 will first ask and obtain from arm 1 the value of [T1/FL]; it will then find PARTX1 and compute its transform, [T1/PARTX1], as follows:

WhereFL(ARM1, Result)

The vision module asks arm 1 for the location in T1 of its tool-mounting flange, FL, to which the camera is attached. The camera has been previously calibrated so that points it views will be referenced to a coordinate system fixed to the flange.

Reply(VM1, Result)

Reading its current value of [R1/E1], arm 1 computes [T1/FL], using the expression [T1/FL] = [T1/R1] * [R1/E1] * [E1/FL], and gives its value to the vision module. The latter then takes a picture with the hand-held camera, recognises an instance (PARTX1) of PARTX, if present, and determines [FL/PARTX1] on the basis of 2-D image features and additional prototype information.

Reply(StationController,Result)
The vision module tells the station controller where PARTX1 is in T1, using the relation [T1/PARTX1] = [T1/FL] * [FL/PARTX1].

Assembly demonstration

A few demonstrations of the assembly station have been performed, the most recent one involving the assembly of part of a DEC LA-34 printer carriage[10]. That assembly contains four part types (see Fig. 9):

○ A friction shaft with four plastic rocker arms, each containing a hole.
○ Four plastic rockers, each of which snaps into the hole of one of the rocker arms.
○ Two small rollers, each of which snaps into the front end of two adjacent rockers.
○ Two large rollers, each of which snaps into the back end of two adjacent rockers.

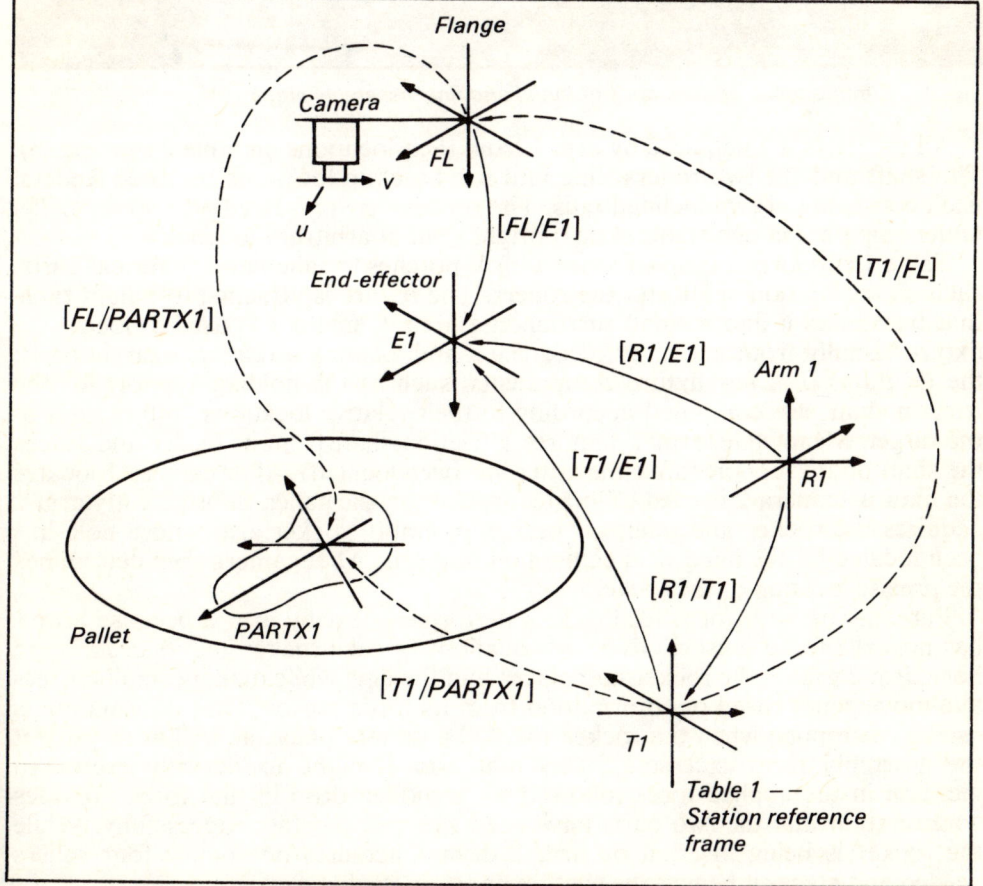

Fig. 8   Using the hand-held camera

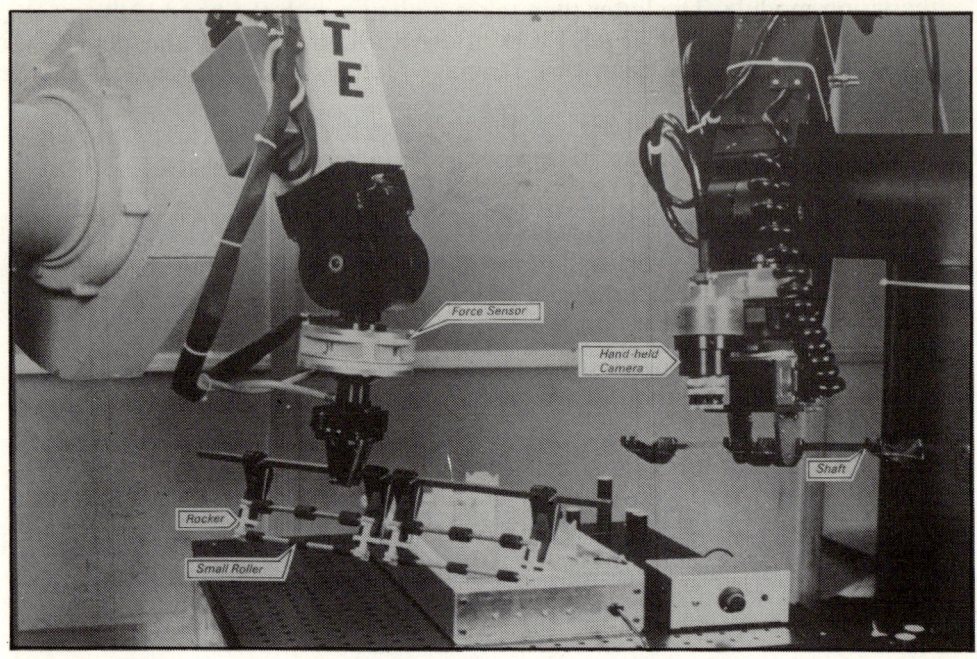

Fig. 9   Simultaneous performance of initial and final assembly steps

All the parts are acquired by arm 1 from their locations on table 2 (see Fig. 6). The shaft and the two rollers slide into fixed pickup locations on three feeders, each consisting of two inclined rails. The rockers are presented on a sticky pallet under camera 2 in one stable state (upright), but at arbitrary locations.

Table 1 supports a simple fixture with V notches for aligning cylindrical parts, such as the friction shaft and the rollers. The fixture is attached to a light table that backlights it and a small surrounding area. Camera 1 is used to locate the fixture visually from above by recognising and locating a reference target on it; the locations of a few fixture components, such as the holding support for the friction shaft, are computed according to their relative locations with respect to the target. Meanwhile, arm 1 acquires a friction shaft from its feeder and places the shaft on its support (after the latter has been located). After camera 1 locates the fixture, camera 2 is used to locate a rocker on the pallet. Subsequently arm 1 acquires the rocker and places it beside an empty rocker arm whose hole has been located by the hand-held camera on that arm. That camera then determines the precise location of the rocker.

Part mating with force feedback is performed next by arm 2 because arm 1 has no wrist force sensor. Given the locations of each rocker and its destination hole, arm 2 grasps the rocker and places it in the hole while making small corrective movements based on information from its force sensor. That information is used to determine when the rocker should be set into place, as well as to protect the assembly from excessive forces that arm 2 might accidentally exert. An increase in the applied force followed by a sudden drop in that force provides confirmation that the two parts have been snapped together successfully. While the rocker is being inserted by arm 2, arm 1 acquires one of the four rollers needed and places it behind the shaft support on the fixture.

The above cycle is repeated until all four rockers have been inserted in their

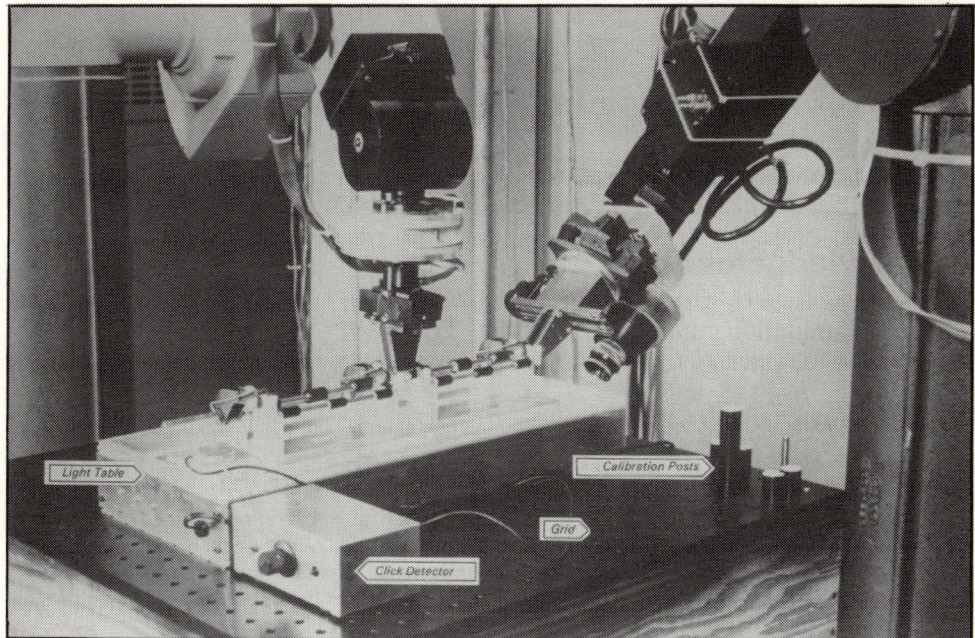

Fig. 10 Pressing rockers on to rollers

respective rocker arms and all four rollers have been placed in their single fixture. Subsequently, arm 1 turns the friction shaft subassembly over onto the fixtured rollers and holds the shaft in place, as shown in Fig. 10. Arm 2 then presses every rocker down until it is snapped on to the corresponding front and back rollers; force sensing is used again to verify this operation. Finally, arm 1 releases the friction shaft and arm 2 removes the completed assembly from the assembly area.

Conclusions

The assembly station and assembly demonstration described in this paper have exhibited the following capabilities:

- ☐ No special-purpose fixtures – some parts are presented to the manipulator on a pallet, others are on a slide. The assembly is assisted by a simple fixture with 'V' notches for centering cylindrical workpieces.
- ☐ Utilisation of medium-level module commands, such as MOVETO, FIND and GRASP.
- ☐ Binary vision to locate and identify parts on a pallet for acquisition. Three cameras are used during the assembly – two fixed in the ceiling and one attached to the end-effector of one of the manipulators.
- ☐ Force feedback to actively control compliance and insertion forces in snap-together, part-mating operations.
- ☐ A stationwide calibration scheme for two manipulators and multiple cameras, enabling part locations to be given with respect to a common reference frame.
- ☐ Concurrent operation of two manipulators in a central assembly area, while the binary-vision system recognises and locates parts elsewhere.
- ☐ Synchronisation of manipulator motions to avoid collisions in the central assembly area.

☐ Simple sensors to verify operation, including the use of finger separation to ascertain the presence of a part in the hand, and a click detector (employing a microphone mechanically coupled to the assembly area) to verify that semi-rigid parts have snapped together successfully, or to detect a drop of a part.

These capabilities are important for the development of not only programmable assembly but also of programmable automation in general.

Acknowledgements

This work was performed as part of a programme supported by the National Science Foundation under Grant No. DAR-8023130, and by thirty US Industrial companies affiliated with that programme.

References

[1] Cook, N. H. Computer-managed part manufacture. Scientific American, Vol. 232, pp. 86–93, February 1975.
[2] Nitzan, D. and Rosen, C. A. Programmable industrial automation. IEEE Trans. Computers, Vol. C–25, pp. 1259–1270, December 1976.
[3] Nevins, J. et al. Exploratory research in industrial modular assembly. NSF Grants GI–39432X and ATA74–18173A01, Reports 1–3. Draper Laboratories, Cambridge, Massachusetts (1 June 1973 to 31 August 1975).
[4] Albus, J. J. S. Brains, Behaviour and Robotics. BYTE Books (Subsidiary of McGraw-Hill), 1981.
[5] Gleason, G. J. and Agin, G. J. A modular vision system for sensor-controlled manipulation and inspection. In, Proceedings of the 9th International Symposium and Exposition on Industrial Robots, Washington, DC, March 1979.
[6] Nitzan, D. et al. Machine intelligence applied to industrial automation. Eleventh Report. NSF Grant DAR80–23130. SRI International, January 1982.
[7] Smith, R. C. Design of a modular programmable assembly system. Technical Note. SRI International, December 1982.
[8] Paul, R. P. Robot Manipulators. MIT Press, 1981.
[9] Bolles, R. C., Kremers, J. H. and Cain, R. A. A simple sensor to gather three-dimensional data. AI Centre Technical Note 249. SRI International, July 1981.
[10] Nitzan, D. et al. Machine intelligence applied to industrial automation, Twelfth Report. NSF Grant DAR80–23130. SRI International, January 1983.

PRECISE ASSEMBLY BY LOW PRECISION MACHINES

A. Romiti, G. Belforte and N. D'Alfio,
Politecnico di Torino, Italy

First published (in Italian) in *Notiziaro Tecnico Amma* (No. 10, 1982)

Abstract

With both dedicated and flexible automatic assembly devices, success, at this time, is only achieved by replacing human skill with precision. Thus high precision machines and workpieces with strictly controlled tolerances are generally required.

This paper considers 'peg-in-hole' insertion by low precision machines. Active and passive assembly devices are discussed, as is the SAGA (self-adaptive guided assembler) system which has been developed in order to allow large positioning errors.

Introduction

The majority of assembly operations for mechanical workpieces at this time are still carried out manually. Such operations require skilled work of a particular nature exploiting human capabilities; in particular, adaptivity related to solving different problems. This is due to the need for inter-related motion and human sensory perception that are subconsciously monitored and interpreted by the human nervous system.

For instance, an operator does not find particular difficulty in positioning pieces together. Firstly, his hand is guided by visual observation to position the pieces near the correct place for assembly, and secondly, during contact between the pieces, the tactile sensations received by the hand allow the operator to correct the relative position and orientation of the pieces till mating is complete.

For repetitive assembly, the operation is performed automatically without much conscious use of the human intellect. Nevertheless, the process is made up by complex interaction and feedback utilising elements of the extremely complicated human cybernetic system.

A classic example of the difficulties involved in the insertion of a component into a close fitting cavity is provided by the example of a drawer that must slide into a piece of furniture. If the initial position of the drawer and/or the direction of the insertion force is incorrect, wedging occurs readily. In order to overcome the wedging action appropriate manoeuvres are needed, i.e. an intermittent side-to-side application of force and perhaps a sharp blow from time to time. This means that it is necessary to formulate and apply an assembly strategy that can change according to the initial conditions and to the mode of application of the forces during the operation. Such strategies do not necessarily repeat because they are dependent on the historical pattern of the insertion.

Automatic assembly devices may be dedicated or flexible. Dedicated devices are specialised single function machines for large batches. Flexible devices are meant to deal with smaller batches involving the manipulation of different components. In both cases however, at this time success is only achieved by replacing human skill by precision, thus high precision machines and workpieces with strictly controlled tolerances are generally required. For the case of a peg in a hole insertion, the centering and orienting precision must be sufficient to avoid the jamming drawer problem. Assembly robots can be used by generating a sequence according to the characteristic features of the workpiece. The use of visual sensors for assembly is still in the laboratory stage and the main barrier to the application of artificial vision is cost and operating speed combined with satisfactory reliability.

Active and passive assembly

Active assembly devices have sensors providing relative positioning error measurement and control elements to provide correction. Passive assembly devices rely on contact forces to provide correction without feedback. Such devices have grippers mounted on elastic supports so that they use such compliance to allow small deflections to take place, thus cancelling the error. In the case of active devices the gripping system may have high stiffness because small displacements are sufficient to measure the forces, i.e. through deflection of ligaments or interpreting the output from strain gauges. Alternatively, the contact forces can be measured by potentiometers actuated by the displacement of helical springs providing support for the gripper[2].

Different sensors have been used by research workers, i.e. sensitive finger tips made from silicon rubber having pressure dependent electric resistance characteristics.[3] At the Karl Marx Stadt University of East Berlin the peg-in-hole insertion has been controlled by four fluidic sensors. Their jets operate on the peg surface. The relative position of the peg and hole is revealed by differential pressures between diametrically opposite pairs of sensors. Though active assembly systems have theoretically great adaptivity, they also have some drawbacks. They require positioning feedback with many controlled axes, high precision and low friction actuators, and sensors able to detect very small forces or deflections.

The systems are quite complex and the insertion times can be long, but can assemble sharp edged components into a sharp edged cavity. Passive assembly systems are simpler and they use only contact forces to achieve the final relative setting after the first contact between the pieces. Such systems, however, need 'lead in' to control the initial contact and to bring the components into line.

The self adaptivity of the pieces is conditioned by the elastic support of the gripper; the joint must have at least four degrees of freedom. Furthermore, the first contact of the tip of the inserted piece must occur at the wall of the cavity in the opposite piece, or inside a chamfered zone. A chamfer may permit quite large initial errors but a properly designed elastic connection between gripper and piece is essential for reliable assembly.

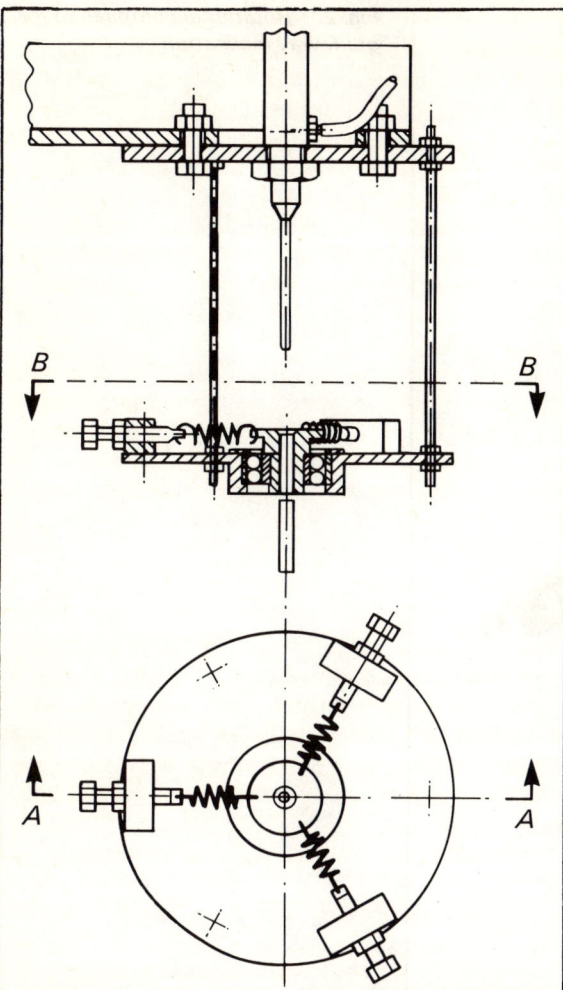

Fig. 1 Guideway mounting for peg assembly

A few devices have been studied in order to avoid such conditions. The best example of this is the Remote Centre Compliance System developed at the Charles Stark Draper Laboratories.[1] In this instance the piece in the gripper is suspended by a carefully designed elastic support, designed so that its tip centre approximates to the instantaneous centre of rotation of the part (plane of engagement). The elastic support is made in such a way that, if a force acts in the plane of the tip centre, the piece translates in the direction of the force accordingly. If a moment acts around the same point, the piece also rotates accordingly.

The system is not very sensitive to the real position of the single-piece tip centre, but is based on the approximate assumption that a single-point contact force passing through the tip centre, and a two-point contact induces a simple moment around the same centre. The limits of the approximation are related to the piece shape; the approximation is good for slender bodies, not so good for pieces with large tip faces. Furthermore, the gripper is actuated through the elastic support, so that only small insertion forces are tolerable in order to avoid large deflections.

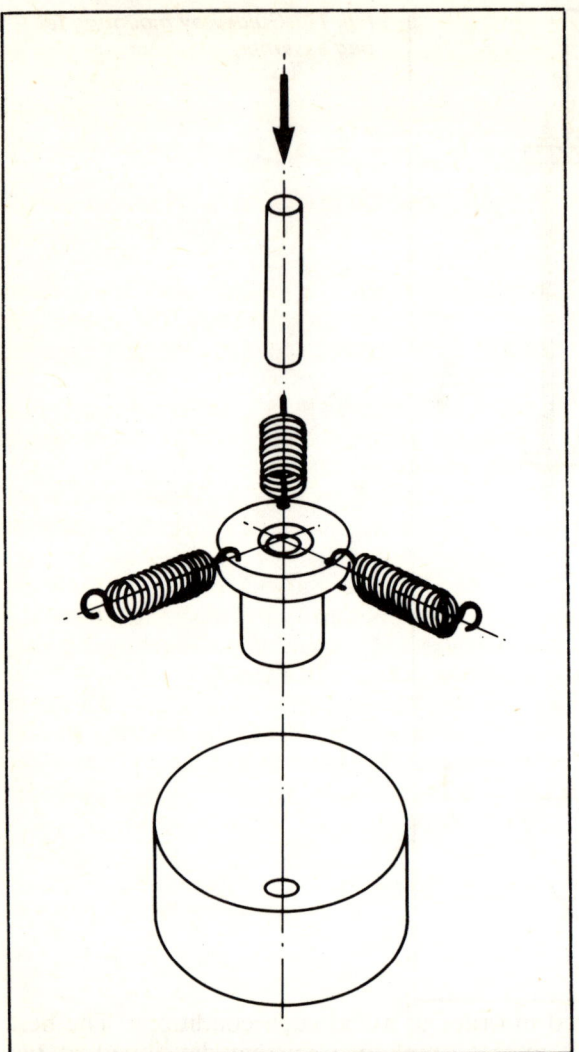

Fig. 2   Rotational compliance for guideway inclination

Another passive assembly device has been built at the Yamanashi University in Japan (assembly robot Scara)[4]. The total achitecture of the assembly robot is arranged to have a low resistance to movement in a direction tangent to the plane of insertion and a high resistance to rotation about an axis normal to the plane of insertion. Thus in general the relative angular insertion errors are limited and the centering error correction is obtained by low contact forces.

The SAGA system

A Self-Adaptive Guided Assembler (SAGA) device has been developed in order to allow quite large positioning errors. Such errors take into account combinations of workpiece tolerances, and workpiece fixture and positioning machine quality levels.

The system is itself quite inexpensive. However, the system can overcome some problems of insertion by kinematically disconnecting the insertion elements from the gripping elements. Such a feature has been obtained by an elastically supported guideway. The insertion actuator acts on the guided piece; the insertion force has no fixed direction, subsequent motion must however be parallel to the guideway axis.

As in the Scara robot, different compliances have been chosen for different axes of motion. Such compliances are referred to the guideway suspension and have no influence on the insertion system. As a result, the insertion forces never give rise to appreciable moments acting on the guideway. Rotation of the guideway is due to moments caused by contact forces, that have a straightening effect.

The first device that was built according to such a system was designed to insert pegs into holes; the pegs were introduced directly into the elastically supported guideway, and were pushed at their upper end by the tip of a cylinder rod.

The system was not very flexible because a change in peg diameter required a change of the guideway. However, that is an inexpensive and easily interchangeable component[5, 6]. Fig. 1 schematically shows its features.

Peg feeding is performed by one of a few different ways. A guideway is made of low friction material and is connected to a platform by a spherical joint (ball bearing). The bearing is kept in a fixed position by three symmetrical springs attached to the platform (Fig. 2). The platform is connected to the frame by three symmetrical flexural bars that provide the guideway translational stiffness.

The insertion actuator is a pneumatic cylinder with a spherical rod tip that presses on top of the peg. Such contact allows easy relative sliding.

Tests have shown that the sliding friction has negligible effects on sliding; in any case friction can be reduced to a very small value by fitting a rotating ball on the rod tip. The peg feeding is performed in two ways. Fig. 3 shows the pegs loaded horizontally on a slide; feeding is controlled by a ratchet actuated by a pneumatic membrane motor.

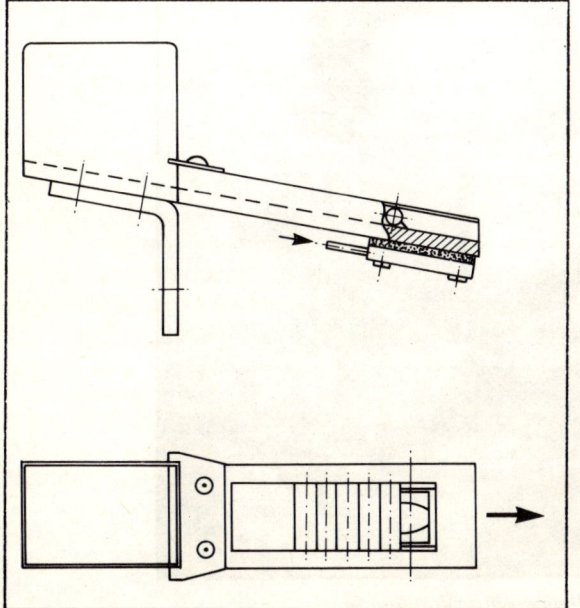

Fig. 3   Slider for peg feeding

222   PROGRAMMABLE ASSEMBLY

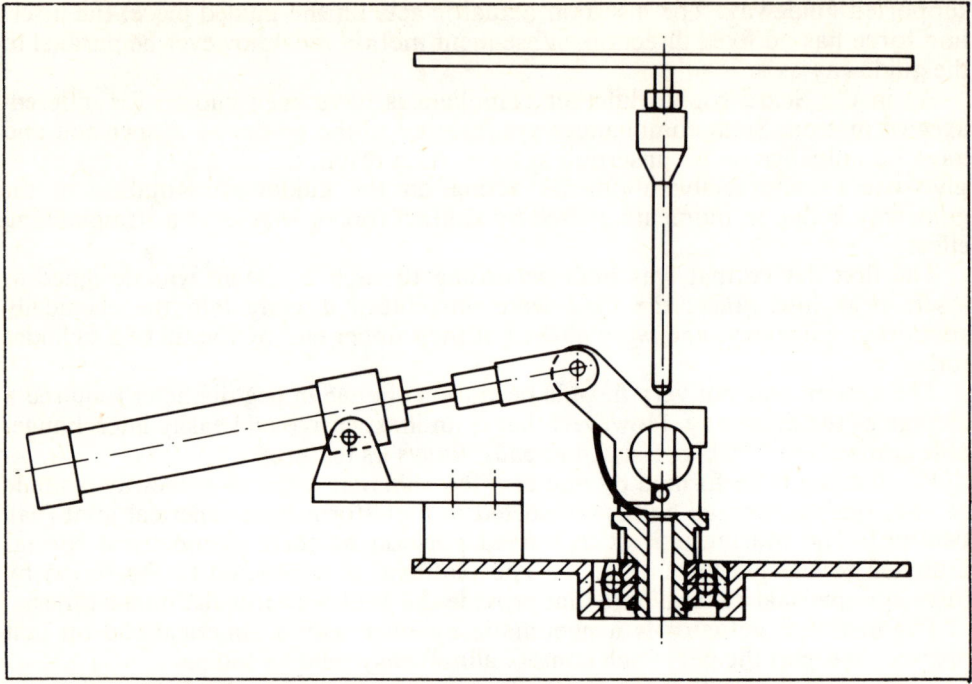

Fig. 4   Manipulator for peg straightening and positioning on the guideway

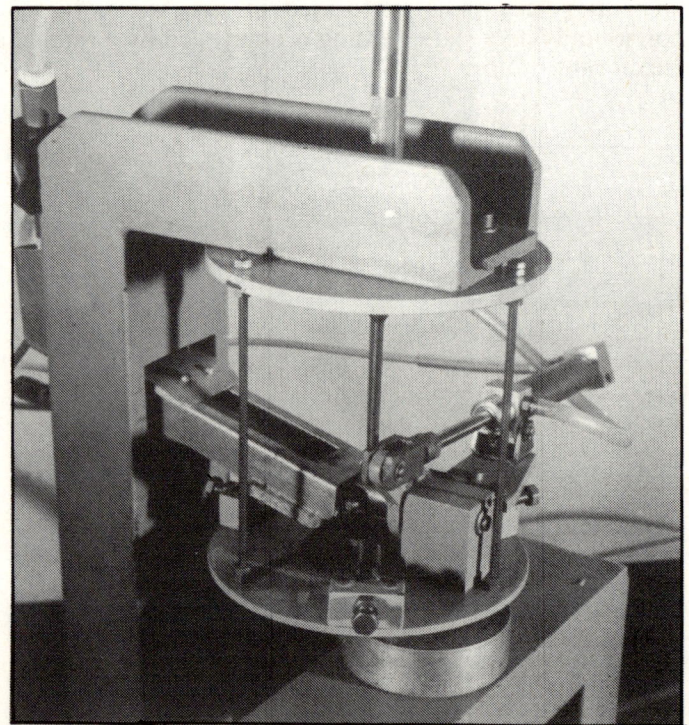

Fig. 5   Peg insertion device with slide feeder

PRECISE ASSEMBLY BY LOW PRECISION MACHINES 223

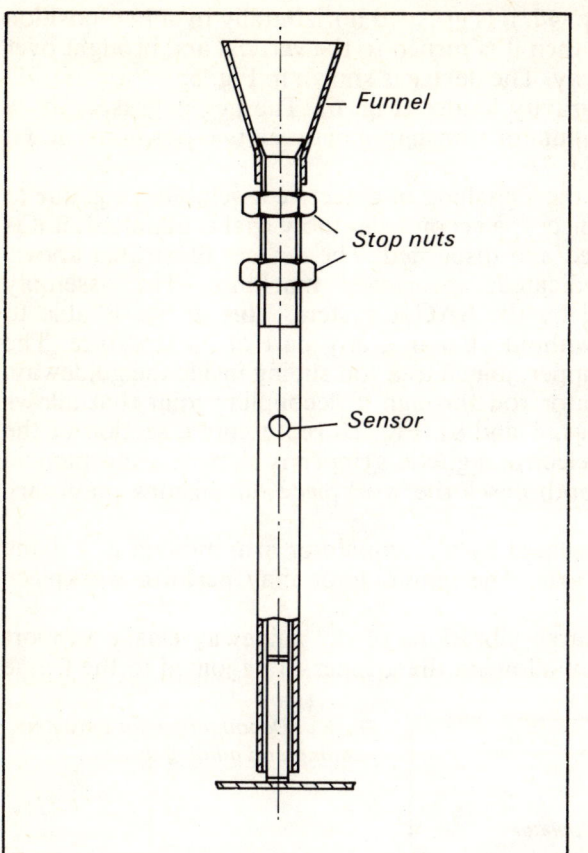

Fig. 6   Peg feeding pipe

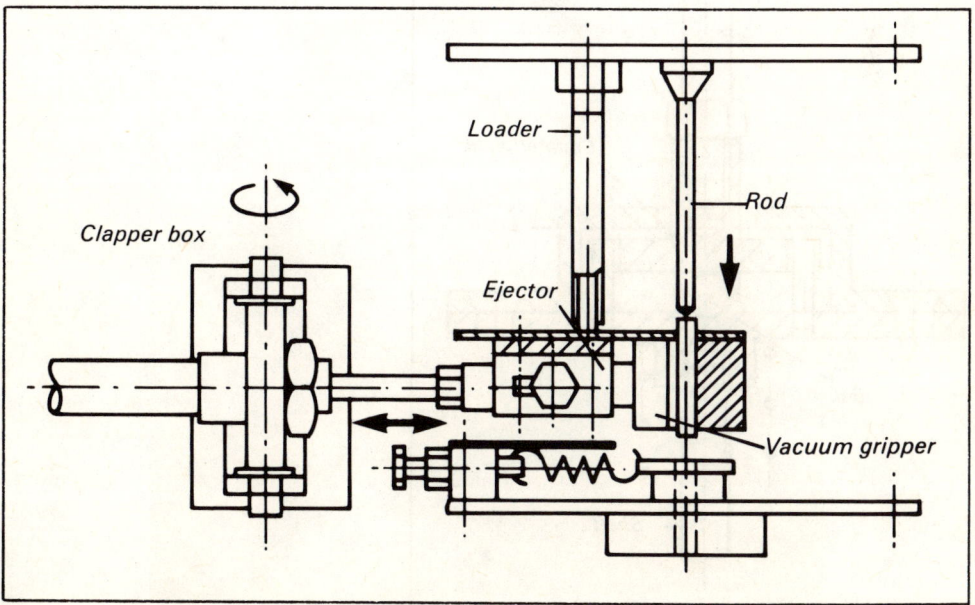

Fig. 7   Manipulator with vacuum gripper for peg positioning on the guideway

When the peg is freed by the pawl, it is grasped horizontally by a two-position pneumatic manipulator (Fig. 4), then it is turned to the vertical and brought over the axis of the chamfered guideway. The device is shown in Fig. 5.

Another feeder is a vertical gravity loader (Fig. 6). The peg is grasped by a vacuum gripper joined to a manipulator translating between two positions, and it is brought over the guideway (Fig. 7).

A few ancillary devices allow the signalling of defective assemblies, e.g. due to geometrical defects of the workpieces; a second assembly trial is admitted; if it is unsuccessful, the peg is extracted and discarded. The devices illustrated above, are particularly suited for dedicated, specialised machines. The assembly capabilities have been extended by the SAGA system. This device is able to assemble different workpieces without changing any part of its structure. The feeder has been replaced by a gripper, joined to a rod sliding inside the guideway. This rod is pushed by the actuator rod through a decoupling joint that allows small relative displacements (Figs. 8 and 9). Fig. 10 represents a section of the whole device. In this figure an electromagnetic gripper is shown; a mechanical gripper may also be used. In both cases the workpiece dimensions may vary over quite a large range.

The whole device may be presented by a manipulator arm moving in a plane perpendicular to the guideway axis. The manipulator may perform workpiece gripping, transport and insertion.

During manipulator motion large vibrations of the guideway elastic support could occur; they are avoided by allowing the gripper to be joined to the frame

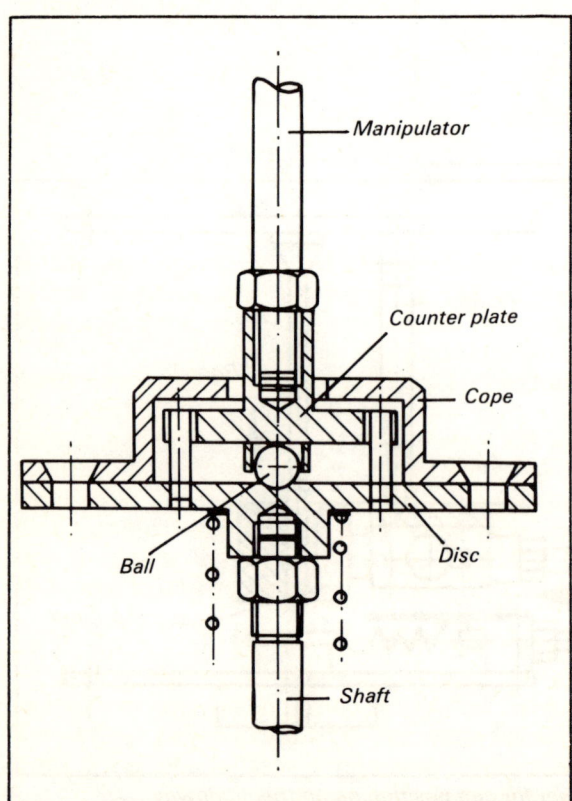

Fig. 8 Decoupling joint between pushing and guiding devices

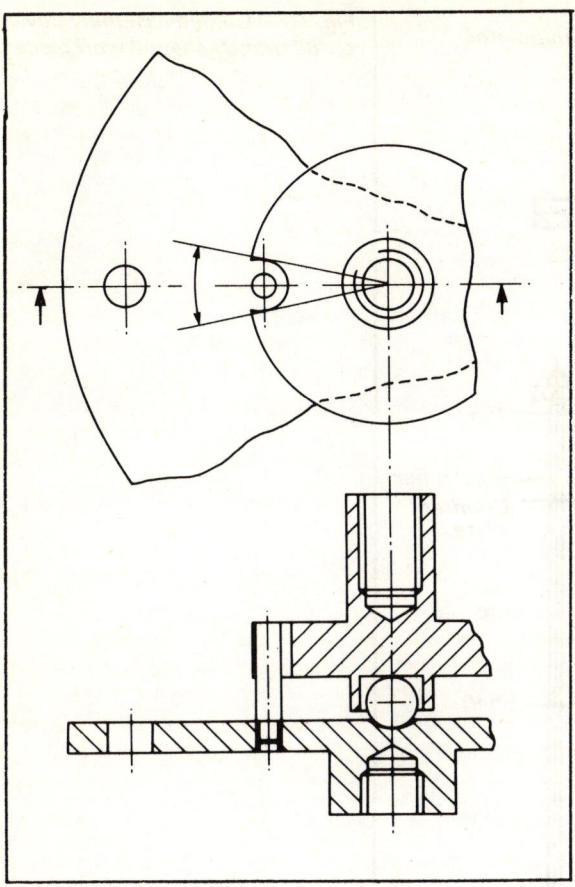

Fig. 9 Decoupling joint sections

during motion. The part of the joint between actuating and driven rods that is connected to the driven rod is shaped like a disc with vertical holes. Before starting a manipulator arm displacement, the rods are lifted until the disc holes are engaged with chamfered pins mounted vertically on the frame.

Fig. 11 shows the system control logic. Fig. 12 shows a picture of an experimental assembly device of the type that was described above. The tests have been performed with workpieces of the type shown in Fig. 13. The pieces were axisymmetrical, with a 10mm hole depth, and a 45° chamfer with 1mm width.

Some pieces had 20mm diameter and 0.05mm clearance. They were completely inserted even with initial eccentricity up to a limit of 0.8–0.9mm. Other pieces had 30mm diameter and 0.02mm clearance, and full insertion was possible up to eccentricity values of 0.5–0.6mm. Angular errors of about 1° could be tolerated.

By increasing the insertion force, it is possible to perform some fitted assemblies.

Laboratory tests have been carried out using components and workpieces made by coarse machining. It appears that this very inexpensive system will permit extending the range of use of transfer assembly machines, and it will allow them, when properly developed, to perform precise assembly tasks.

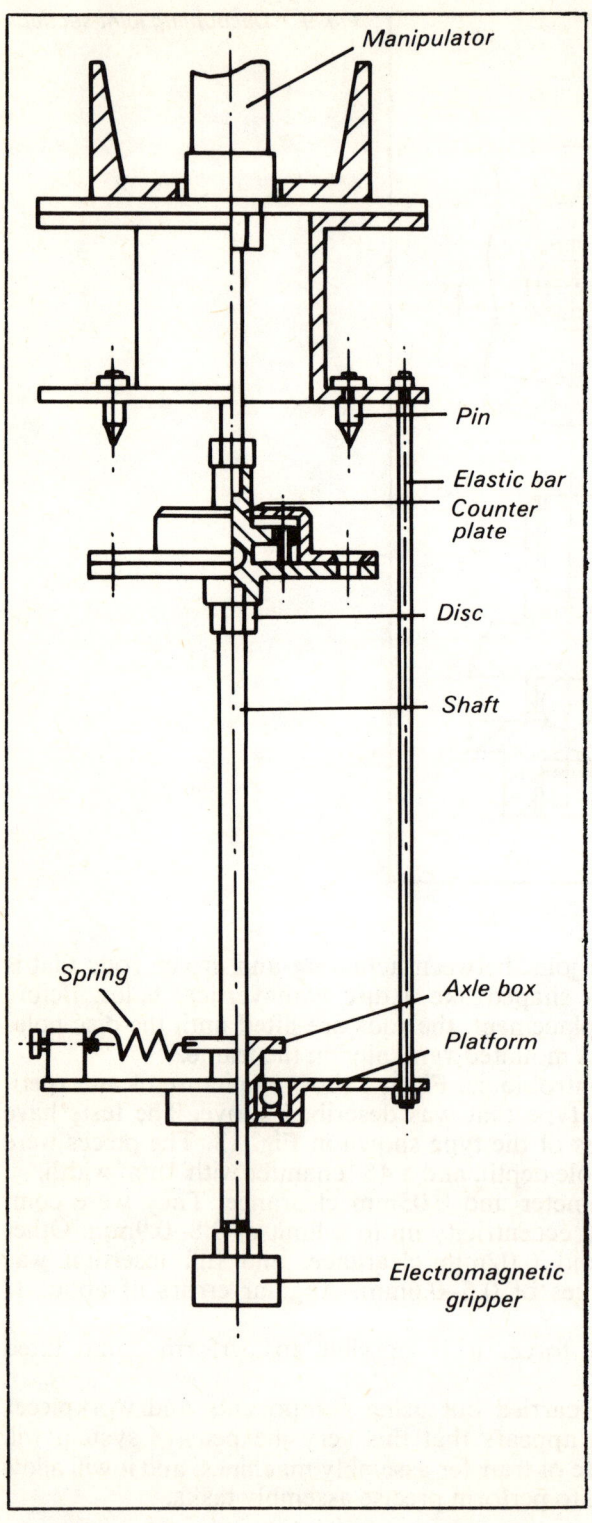

Fig. 10 Guided assembly devices for differently shaped workpieces

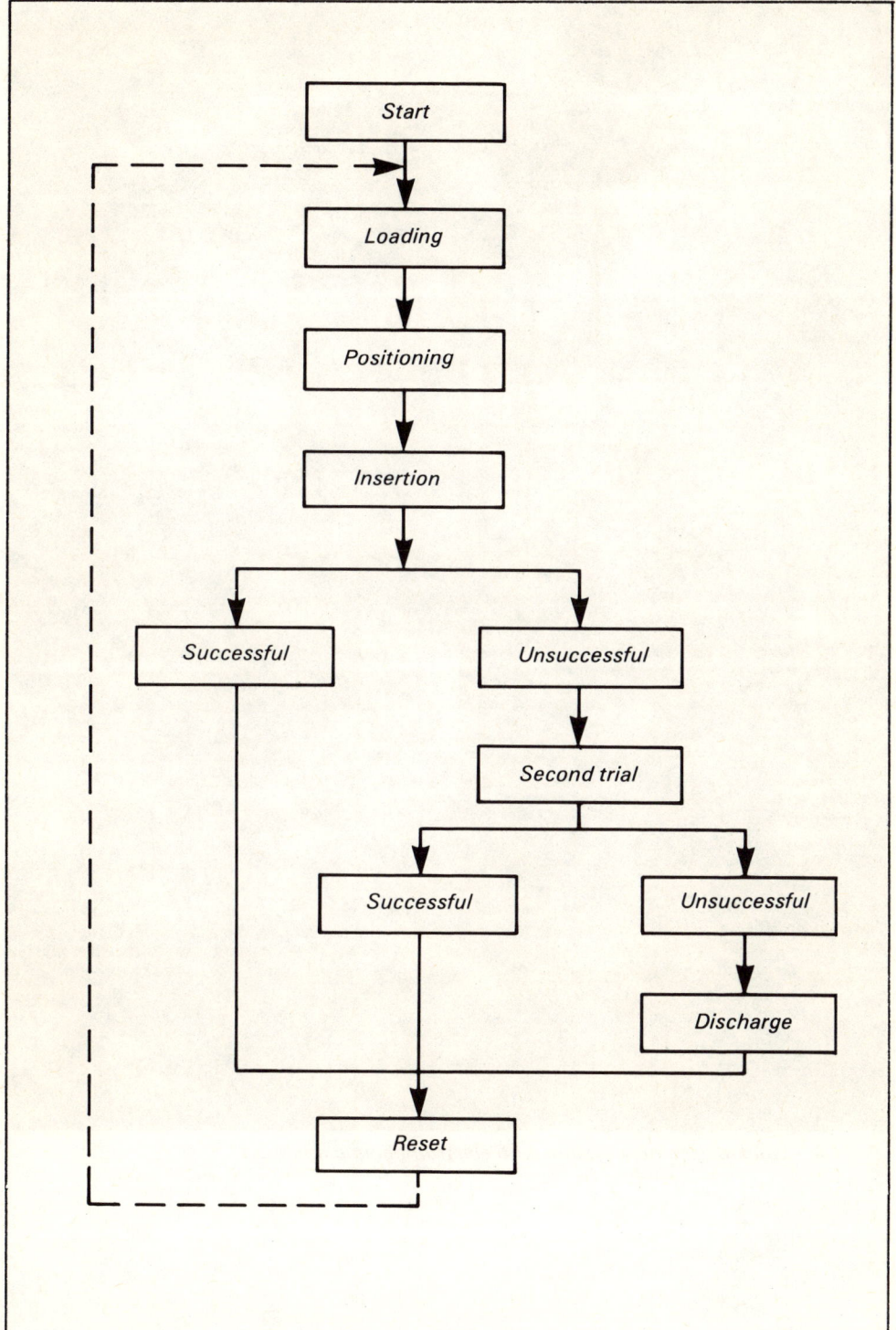

Fig. 11   Assembly system logics

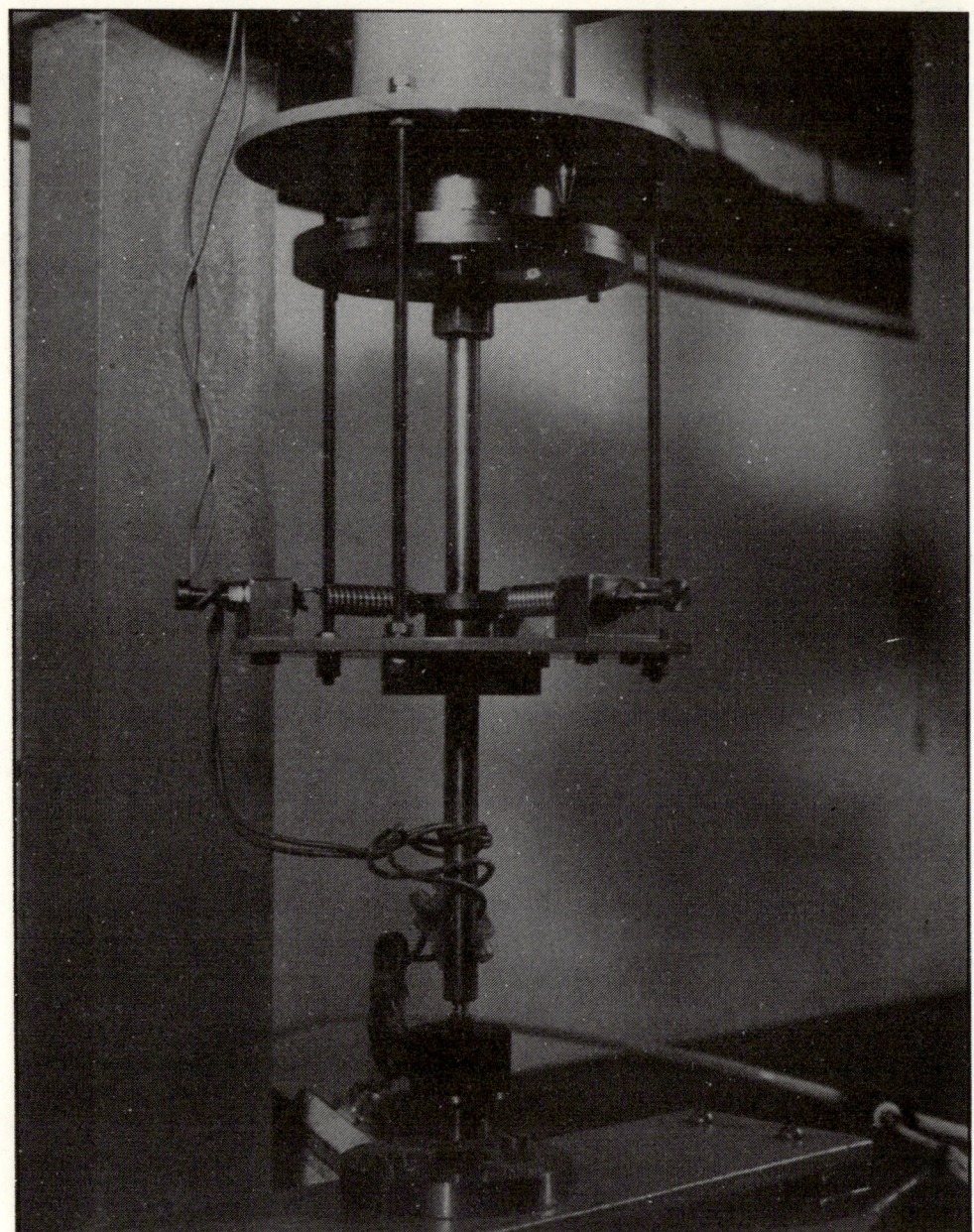

Fig. 12   Guided assembly system with electromagnetic gripper

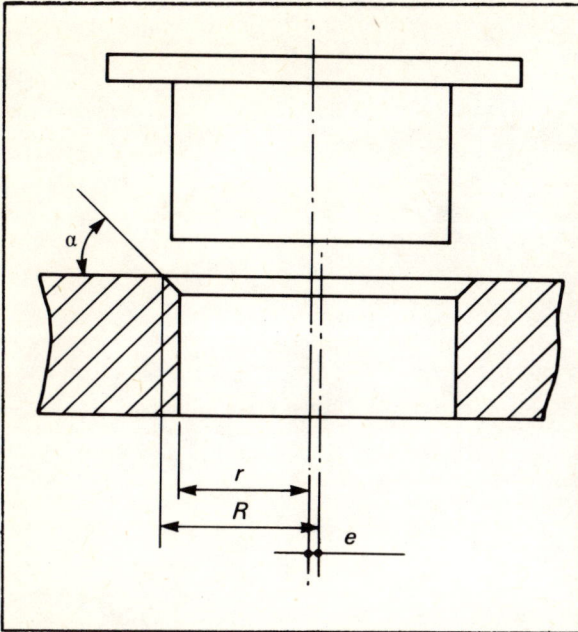

Fig. 13 Assembled workpieces

References

[1] Nevins J. L., Whitney D. E. Research issues for automatic assembly. In, IFAC Symposium on Manufacturing Technology, Tokyo, 1977.
[2] Hanafusa H., Amada H. A robot hand with elastic fingers and its application to assembly process. In, IFAC Symposium on Manufacturing Technology, Tokyo, 1977.
[3] Purtrick, J. A. A force transducer employing conductive silicone rubber. In, Proceedings of the 1st International Conference on Robot Vision and Sensory Controls, Stratford-upon-Avon 1981. IFS (Publications) Ltd., 1981.
[4] Makino H., Furuya N. Selective Compliance Assembly Robot Arm. In, Proceeding of the 1st International Conference on Assembly Automation, Brighton 1980. IFS (Publications) Ltd., 1980.
[5] Romiti A., Belforte G. Manipolatori sensorizzati. Bollettino Tecnico AMMA, No. 10, 1981.
[6] Romiti A., Belforte G., D'Alfio N., Quagliotti F. A passive assembly device for the speedy assembly of pegs in holes. Assembly Automation, 1981.
[7] Romiti A., Belforte G., D'Alfio N. A self-adaptive guided assembler (SAGA). Robots VI Conference, SME, Detroit, 1982.

Chapter 5

PARTS HANDLING AND FEEDING

The 'achilles heel' of flexible assembly is the feeding and handling of the individual parts. The papers in this chapter highlight some of the techniques now under development that will allow varieties of parts to be presented to the system.

COMPUTER-CONTROLLED MAGAZINING SYSTEM

M. Schweizer and I. Schmidt, Fraunhofer Institut für Produktionstechnik und Automatisierung (IPA), West Germany

First presented at the 3rd International Conference on Assembly Automation and 14th IPA Conference, 25–27 May 1982, Boeblingen, West Germany

Abstract

At IPA, Stuttgart a completely computer-controlled magazining system has been developed. The purpose of the system is to feed workpieces to magazines, to unload the workpieces from magazines, and to store the unloaded workpieces in boxes.

The system consists of an industrial robot with six axes and a further eight axes. Tactile sensors are used to control the gripping and joining forces. A TV camera and a reflective light barrier to recognise the workpieces are also included. The workpieces are orientated by a flexible gripper. The system is set up under computer control about every hour to one of more than one hundred workpiece types.

Introduction

The use of magazines in manufacturing engineering is of interest for various reasons, enabling:

☐ careful transport of the workpieces,
☐ minimising the handling effort during loading and unloading production units,
☐ simultaneous processing of several workpieces.

The task to load and unload magazines has been automated in the past but only in large-scale production. This is because there have not been the technical possibilities for flexible automation. Using industrial robots together with sensors gives new opportunities in this field.

Description of the task

For coating the surface of plastic parts, several magazining operations have to be carried out (Fig. 1). Depending on the type of production the untreated workpieces have to be fed to a magazine before the coating process; the workpieces are transported on this magazine during production. After coating,

234 PROGRAMMABLE ASSEMBLY

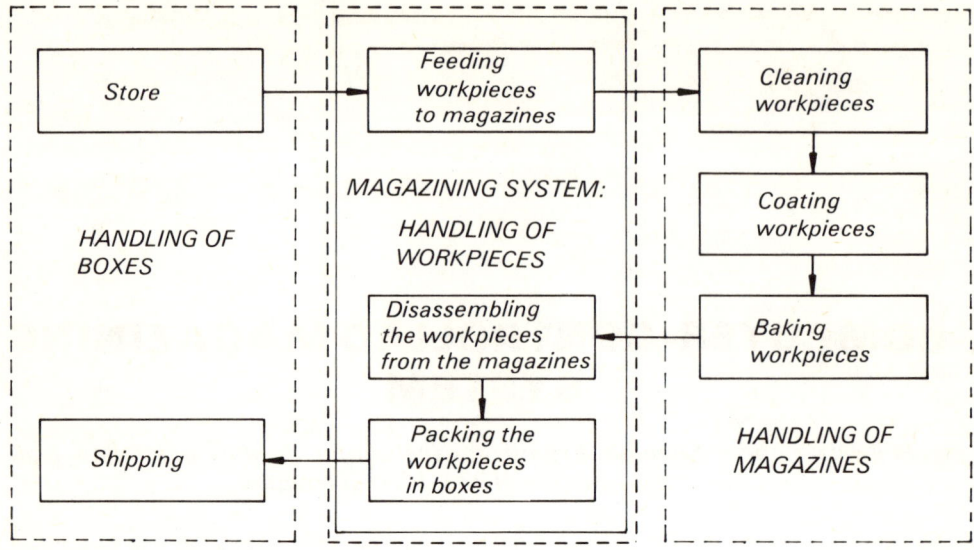

Fig. 1 Production cycle for the surface coating of plastic parts

Fig. 2 Selection of workpieces that have to be handled

the workpieces have to be taken off the magazine and stored in palletised boxes. Once a pallet is completely loaded it is immediately replaced.

The magazining system includes all tasks as depicted in the centre of Fig. 1. The following operation modes can occur:

○ Feeding magazines with uncoated workpieces.
○ Unloading coated workpieces from the magazine and storing the workpieces in boxes.
○ Combination of the above operation modes. First, the finished parts are removed from the magazine, which is then loaded with uncoated workpieces.

Workpieces

About 150 different types of plastic workpiece have to be handled, a selection of which is shown in Fig.2. About 70% of all workpiece types are cylindrical with a central blind hole, their weights ranging from 1 to about 50g. The diameters range from 7.5 to 100mm, with about 40mm being the average. The average workpiece length is about 26mm, the range being from 9 to 51mm.

Magazines

The magazines are partly standardised and consist of modules assembled for each workpiece type in a special combination. The number of workpieces per magazine is different for each workpiece type. The magazines are essentially the same for each workpiece type; however they have unfavourable tolerances which can build up to several millimetres. Furthermore, the fact that the workpieces are not distributed symmetrically on the magazine makes handling more difficult. Also there is a handle in the middle of the magazine and as a result, during transport several workpiece fixtures can be broken off. In addition, the three-dimensional distribution of the fixtures on the magazine means that the joining directions are very different. The design of the magazine largely depends on the type of production and therefore the design can only be varied within a small range. The joining forces range from 0.15 to 25N, depending on the workpiece and the magazine type.

Important considerations

Customer production. Due to the customer's production little can be said of the future of the workpiece spectrum and its handlability. Analysis which identifies workpieces to be produced in large quantities that are automated economically is therefore possible only for a shorter time. This timescale, which is imporant for planning, depending on the customer's needs is only half a year. The further handling tasks which have to be done in the future can only be estimated roughly.

Where a customer's production has manual working competitors, offers have to be given normally within one or two weeks after the enquiry. To produce a production plan, therefore needs a decision to be made within a week on whether or not the automatic system for workpiece handling can be used for the new type of workpiece.

Batch size. The time for running a batch is about one hour and cannot be prolonged for production reasons (this is also true for larger orders).

Manual cycle time. The manual cycle time (piece work) is about 2–3 seconds including the handling of the magazines. The cycle time depends mainly on the workpiece size.

Quality. The method of feeding the workpieces to the fixtures of the magazine greatly influences the quality of the coating process. A relationship exists between the manual cycle time and the quality of this joining process, leading to a conflict for the worker since higher output means poorer quality.

Cost of manual fixtures. This cost can be neglected.

Human aspects. Because of the short manual cycle time and the forced pace of the manual work, which cannot be avoided by ergonomic redesign of the workplace, automation of manual work is desirable.

System planning – conclusions
In planning an economic automated handling system, 'flexibility' and 'quality', as well as 'speed', are the important design parameters.

In particular, desired flexibility means the ability to cope with:

☐ Differing workpiece types and, as a result, handling quality. Also, new workpiece types have to be considered.
☐ Differing magazine types and their associated tolerances.
☐ Differing joining forces within a workpiece type.
☐ Hourly change-overs.

In addition, in order to win any one 'job' and thus stay competitive, the automated system has to prove that it is working with a new workpiece type within one week.

These difficult peripheral conditions were checked at the beginning of the project to detect whether it is possible to change the conditions to make the task easier. The following was attempted:

○ Prolonging the batch run-time.
○ Redesigning the magazine to increase quality and to decrease difficulties for automatic handling.
○ Standardisation of the workpiece spectrum.
○ Preoriented workpieces.

However, because of the high estimated costs only a small design modification of the magazine and a limitation of the workpiece spectrum (length from 10 to 50mm, diameter from 10 to 50mm) could be done.

System design

Workpiece orientation
For the automation of the feeding operation to the magazines it is necessary to orientate the workpieces. A comparison of the mechanical orientation of vibratory bowl feeders and the use of a workpiece recognition sensor was made. Tests clarified that both methods would work. A cost estimation (Table 1) of the alternatives resulted in:

Table 1 Cost estimation of the alternatives for the orientation of workpieces

Sorting with mechanical devices		Sorting with TV sensor	
Storing bin with unloading device	3,000 DM	TV sensor complete with interface	40,000 DM
Vibratory bowl feeder (including sound absorber) with exchangable orienting devices, unloading devices and positioning devices	8,000 DM	Storing and separating bin	8,000 DM
		Orienting gripper with three positioning axes	18,000 DM
Gripper with adjustable gripping width	5,000 DM		
Each additional workpiece specific orienting device	1,500 DM		
Each additional workpiece specific unloading and positioning device	1,000 DM		
Sum of the fixed costs	16,000 DM	Sum of the fixed costs	66,000 DM
Sum of the variable costs per workpiece type	2,500 DM		

☐ The system with the workpiece recognition sensor has its break-even point at 20 different workpiece types compared to the orientation system with the vibratory bowl feeder.
☐ The cost for the storing of unused tools (orientating devices, allocating devices) has to be considered.
☐ Obtaining trouble-free operation with a new workpiece takes longer with the vibratory bowl feeder.
☐ The use of the workpiece-recognition-sensor software makes it easy to install magazining systems.

It was decided therfore, that in this case a workpiece recognition sensor was the solution of choice.

System description
The system layout as shown in Fig. 3 was chosen because the time critical functions are well separated:

○ Feeding of workpieces to the recognition area of the workpiece sensor.
○ Recognition procedure.
○ Unloading workpieces from the supply conveyor belt by the assembly robot.
○ Placing workpieces on the removal conveyor belt by the assembly robot.

All these functions can be done in parallel, which results in an estimated cycle time of 1–1.5 seconds.
The planned system consists of the following essential components:

Separation/unloading bin – stores workpieces for about 1 hour batch running time and feeds workpieces to the supply conveyor belt.

238 PROGRAMMABLE ASSEMBLY

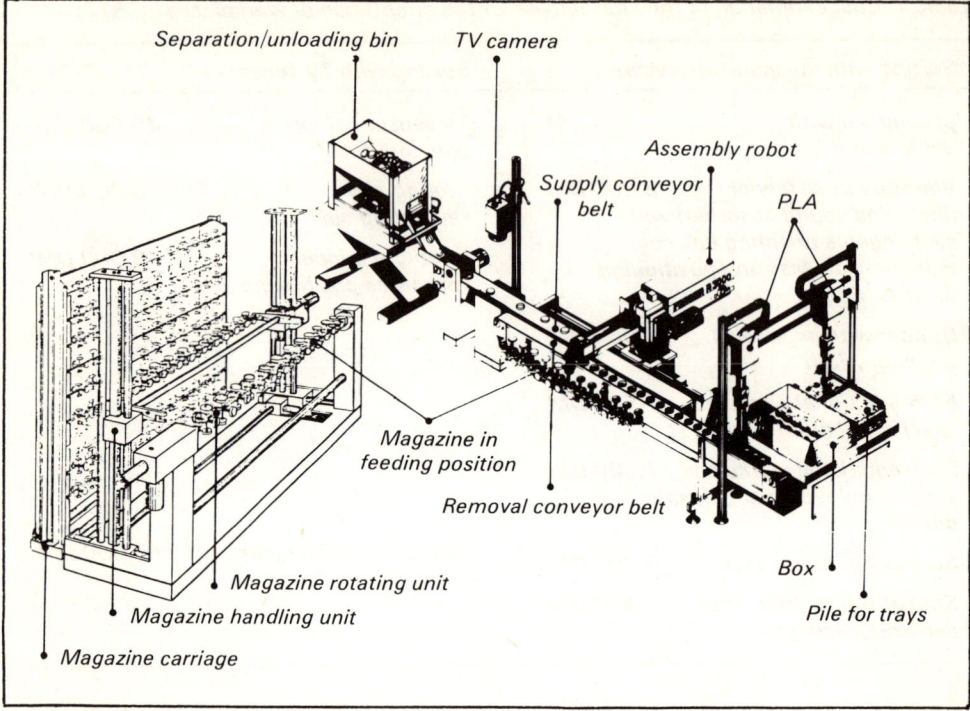

Fig. 3 Magazining system

Supply conveyor belt – transports the workpieces coming from the separation/unloading bin to the recognition area under the TV camera and then to the assembly robot. Workpieces that are not taken by the assembly robot are transported to a box at the end of the supply conveyor belt. An adhesive coating on the belt prevents workpieces from rolling. The conveyor belt is position controlled by the computer software system like an axis of the handling system.

TV camera with workpiece recognition system – is situated above a part of the supply conveyor belt where the transparent belt is illuminated. The silhouette of the workpieces lying on this part of the belt is sensed by the TV camera and the information subsequently processed by a microcomputer-controlled workpiece recognition system (type OMS from BBC Germany[1]).

Assembly robot – the assembly robot (DEA Pragma A3000) receives the coordinates of the recognised workpieces and is able to locate the workpieces lying on the supply conveyor belt. Using an orienting gripper the robot picks up the workpieces from the conveyor belt and positions them ready for feeding to the empty fixtures on the magazine. If a magazine full of coated workpieces is approached, first a fixture is emptied from the workpiece and placed on the removal conveyor belt, and an uncoated workpiece is fed into the now empty fixture.

The assembly robot has a double gripper: one working with a sucker cup which takes workpieces from the magazine and feeding them to the removal conveyor belt; the other gripper is able to orientate the workpieces and take them from the supply conveyor belt to the magazine.

In the prototype magazining system, the handling system has one arm with three linear main axes, two rotating auxiliary axes and one linear hand axis for the opening and closing of the gripper jaws. In future it will be necessary to use a second arm on the assembly robot with the same features in order to decrease the cycle time. This second arm will be mounted in the same way as the first arm[2].

Removal conveyor belt – takes coated workpieces and transports them to the so-called programmable loader (PLA). The removal conveyor belt is position-controlled in the same way as the supply conveyor belt.

Programmable loader – has two arms, each with two axes. One arm is able to take a row of workpieces laid in a pattern that fits the tray on the removal conveyor belt. The second arm takes trays from a pile and transports them to the box.

Magazine carriage – is transported manually to the magazining system and is able to store 20 magazines in exactly defined positions.

Magazine handling unit – loads and unloads the magazine carriage with magazines, and loads and unloads the magazine rotating unit. To decrease the cycle time it has a double gripper. The vertical axes of the magazine handling unit are position controlled, whereas the horizontal axes of the magazine handling unit are pneumatically driven and have three fixed positions.

Magazine rotating unit – has a double-sided fixture for one magazine. The unit rotates the magazines so that the assembly robot is able to feed workpieces to the magazine in the vertical direction.

Separation/unloading bin

The separation/unloading bin has the task of storing workpieces and of feeding the workpieces to the supply conveyor belt in such a way that they can be recognised by the workpiece recognition system. For the available sensor systems this means that the silhouettes of the workpieces do not touch (with some sensor systems workpiece silhouettes that do meet at one point can be recognised, but the time for the recognition is increased). The separation/unloading bin is shown in Fig. 4. It works in the following sequence:

☐ Filling the separation/unloading bin with workpieces.
☐ Start signal to the control of the separation/unloading bin.
☐ Gradual opening of the container flap.
☐ Sliding workpieces through the opening container flap.
☐ Signal from the light barrier by the first workpieces sliding out of the workpiece container.
☐ Closing the container flap after a workpiece specific delay.
☐ Falling out of the first workpiece of the group of workpieces that are sliding out.
☐ Clamping of the sliding workpieces by the container flap and the rubber stop.

By tests, it could be proved that by changing the delay and by arranging the light barriers in an appropriate way, the following results are possible:

○ Standby time from the start signal to separation of the workpiece is about 0.5–1 second.
○ The number of the fed workpieces at the same time is from 1 to 3.

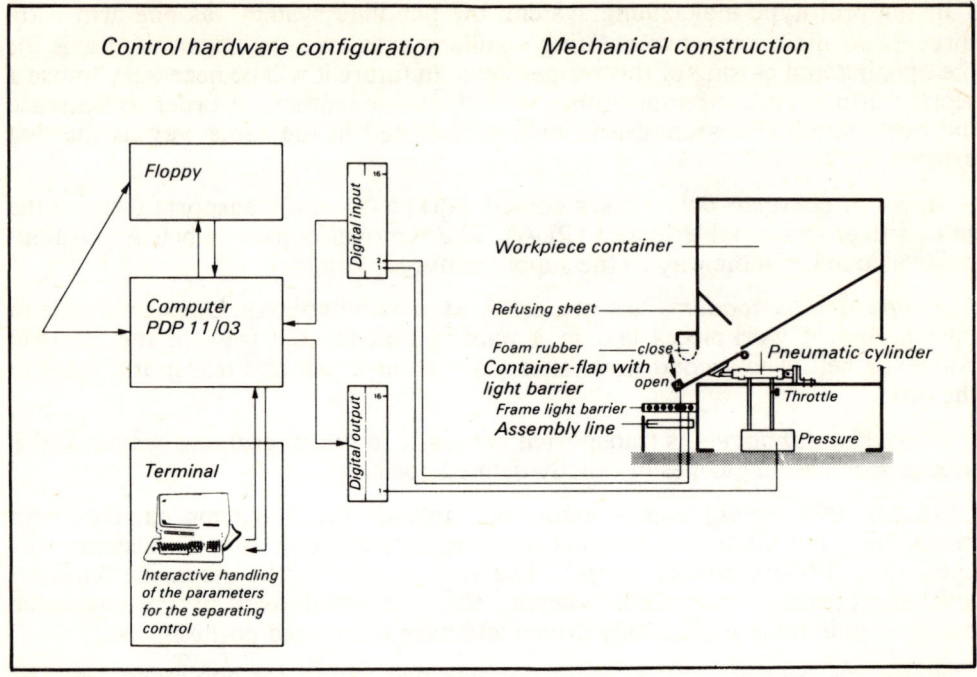

Fig. 4 System configuration of the separation/unloading bin

To make sure that the feed rate is constant the container flap is moved according to the light barrier signals and the time after the start signal. It is not necessary to use a buffer after the separation/unloading bin. The change-over for the separation/unloading bin consists of a change of parameters for the movement of the container flap, so there is no mechanical change-over necessary for the whole workpiece spectrum.

Sensor problems
Whether or not recognition of the workpieces is possible using available light was studied at the beginning of the project. We found out that, depending on the workpiece form, surface and colour, it was necessary to change the illumination of the scene. The effort for flexibly automated illumination and the expected low reliability led us to the conclusion that it is necessary to use a 'back' light for recognition. In this case a transparent conveyor belt is used which is lit up from below by a fluorescent lamp. By tests it could be clarified that even larger contaminations of the belt do not disturb the recognition. The TV sensor uses filter algorithms to avoid any wrong recognition results.

Using the 'back' light for illumination produces a new problem, however. It is not possible, because of the silhouette form, to detect whether a workpiece is lying on the conveyor belt with the bottom hole up or down. Therefore, it was checked which sensor-type could detect this. Depending on the required flexibility and the sensing distance which is needed because of collision reasons, we found out that the most simple solution would be the use of a reflective light barrier which senses the surface of the workpieces (Fig. 5).

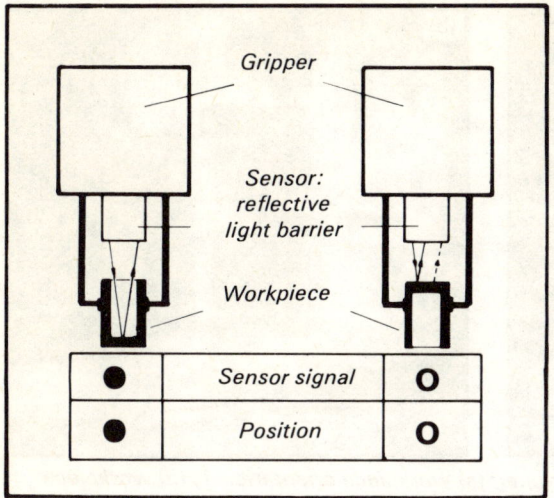

Fig. 5 Use of a reflective light barrier to detect the workpiece position

The sequence of operations is as follows:

☐ Feeding of the workpieces from the separation/unloading bin to the recognition area of the TV sensor by using the supply conveyor belt.
☐ Stop conveyor belt.
☐ Start signal to the workpiece recognition system for taking the image of the scene.
☐ Start the supply conveyor belt (the way between two positions of the supply conveyor belt depends on the size of the recognition area and the size of the workpieces).
☐ After the start of the workpiece recognition procedure the positions, orientations, and silhouette types of all workpieces lying in the sensor recognition field are transferred from the workpiece recognition system to the host. After this a new picture can be loaded in the memory of the workpiece recognition system. The whole recognition procedure lasts for about 0.3 seconds for each workpiece and is therefore fast enough compared with the cycle time of the handling system.
☐ The computer memorises the information of the workpiece recognition system, together with a variable, according to the part on the conveyor belt where the recognised workpieces are lying. The memory is able to store the information on more than 200 workpieces lying on the supply conveyor belt.
☐ The computer transfers the coordinates of the workpiece to be gripped to the assembly robot.
☐ Moving the assembly robot to the gripping position.
☐ Gripping the workpiece. During the gripping procedure the reflective light barrier in the centre of the gripper detects whether the bottom hole of the workpiece is up or down.
☐ After the workpiece has been taken off the supply conveyor belt, the gripper turns the workpiece in a way that the bottom hole is situated in the down position. The workpiece is fed in the fixture on the magazine by the handling system after finishing the orientation process.

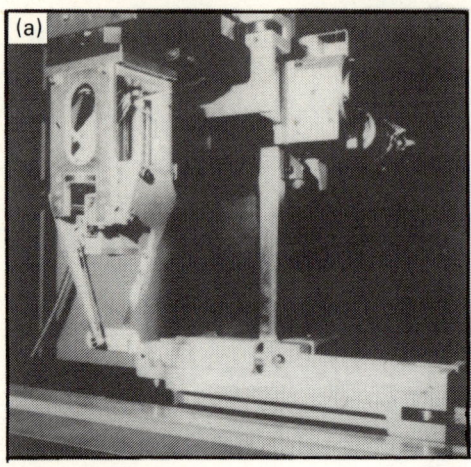 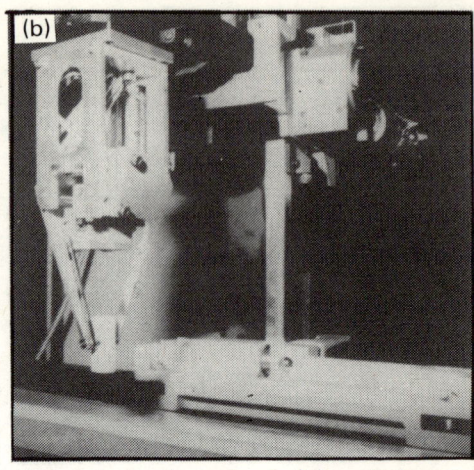

Fig. 6 Assembly robot with orienting gripper: (a) workpiece orientation 1, (b) workpiece orientation 2

Orientating gripper

The workpieces lie in their stable and indifferent orientation on the supply conveyor belt as they are falling out of the separation/unloading bin. After gripping it is therefore necessary to orientate the workpieces in a way that the bottom hole of the workpieces is in the upper position. The gripper developed to do this is shown in Fig. 6[3].

Since the workpieces are to be gripped with various forces, the measuring of the gripping forces is done by a sensor.

The following features are integrated in the gripper:

○ Tactile sensors to measure the joining and the gripping forces.
○ Positioning axes for closing the gripper jaws and turning the gripper jaws to orientate the workpieces.
○ Reflective light barrier to detect the orientation of the workpiece.

A summary of the sensors used for measuring purposes is given in Table 2.

Gripping sequence

The assembly robot is principally able to grip all the workpieces that are recognised on the supply conveyor belt, therefore it is necessary to decide in which sequence they are to be gripped. The sequence depends on:

□ the accessibility of the workpieces,
□ the size of the gripper jaws,
□ the opening of the gripper jaws, and
□ the distance between workpiece position on the supply conveyor belt and the desired position of this workpiece on the magazine.

Optimisation of the gripping sequence is done according to the following criteria:

○ avoiding collisions between gripper jaws and surrounding workpieces,
○ decreasing the cycle time by minimising the motion paths of the assembly robot, and
○ gripping 100% of the workpieces lying on the conveyor belt.

Table 2 Summary of the sensors used

Type of sensor	Sensor task	Measuring range	Resolution of the measuring system	Interface	Investment
Optical sensors					
TV sensors	recognition of the position and the orientation of workpieces	image size 80 x 100mm	workpiece position 0.2% of the image height	EIA RS 232 serial I/O 9600 BAUD	40,000 DM
Infrared reflective light-barrier	distance to workpiece surface	10–70mm distance	approx. 2mm	analogic 1 0–6 V	300 DM
Tactile sensors					
Magnetic sensor measuring the compression of a spring	measuring joining forces	0–10 N	0.2 N	analogic 1 0–2.5 V	500 DM
Strain gauge measuring the compression of a spring	measuring the gripping force	0–10 N	0.2 N	analogic 1 0–12 V	2,000 DM

Sensor parameters determination

Analogue sensors. The parameters for the analogue sensors (joining force, gripping force, distance to the workpiece surface) have to be done according to the type of workpiece that has to be handled. For each workpiece type a certain limit for the sensor input connected to a range for an axis position (e.g. the closing position of the gripper) has to be defined. This can be done by:

☐ calculation (e.g. the maximum gripping force can be determined from the workpiece specification),
☐ tests with a special test station,
☐ tests with the magazining system itself (teach-in), and by
☐ estimation on the basis of calculations or tests.

In our case, parameters are determined with a simple separate test station (e.g. determination of the maximum gripping force using a letter balance) and with the system itself. It is expected, however, that in the future the parameters for the sensors that are workpiece specific, can be estimated. The necessary correction of the parameters is then done during a test run of the system.

Workpiece recognition system. The workpiece specific parameters for the TV sensor are determined, nowadays, by showing the workpieces to the system. For complicated silhouette forms this procedure is practicable if the sensor determines the parameters itself. The used workpiece spectrum in this case, however, makes it possible to determine the parameters for workpiece recognition by describing the silhouette. This means that for programming the new workpiece it is only necessary to tell the workpiece recognition system the diameter and the length of the workpiece that has to be detected. This is done by the computer.

Therefore, special programming of the workpiece recognition for new types of workpieces is not necessary.

The workpiece recognition procedure is performed in the following way:

System start
○ Transferring the sensor programs and the workpiece parameters from the computer mass memory to the RAM of the workpiece recognition system (the workpiece recognition system has no mass memory).
○ Calibrating the TV camera to detect whether the camera is adjusted in the correct way, to correct non-linearities of the TV tube, to determine the conversion parameter for the calculation of the workpiece position, and to measure the distance between the coordinate systems of the sensor and the assembly robot. The calibration is done automatically using a plate with a hole pattern of fixed hole-centre distances.

Recognition of workpieces in the automatic cycle
□ Start of the image input of the workpiece recognition by the computer.
□ Start of the workpiece recognition procedure by the computer. At the end of the recognition procedure the workpiece recognition system creates a table in which, for every workpiece that is stored in the library, the x position, the y position, the rotating angle and the silhouette type are fixed.
□ Output of the table to the computer initiated by the computer.

Output of the workpiece recognition results is done after every recognition procedure. The computer memorises each table in its RAM.

Elements of the programming language for the use of sensors

In the assembly robot programming language there have to be elements that can be used for the calculations and the information exchange between assembly robot control and host, or the sensors. These language elements are included in the standard of the programming language HELP which has been developed by DEA of Turin or developed together with IPA, Stuttgart for this project.

System software

Fig. 7 gives a survey of the control structure used. The user programs are done in the higher real-time language HELP and in the OMS command language. The programs are developed off-line except a very small part which deals with the determination of parameters for the layout description.

Man-machine interface

To operate the system, non-skilled workers have to be used. They have to be able to make the change-over and to determine parameters for new workpiece types which are not fixed. Therefore a technical menu running on a VT 100 screen has been developed. This interactive menu makes sure that fog-free operation is possible.

Programming. To determine the parameters for control of the operating sequence with a new workpiece type, the operator chooses (after switching on the system) a certain operating mode at the screen. After this, the operator has to do simple things like measuring the diameter and the length of the workpiece or to remove parameters from the menu. If a parameter typed by the operator

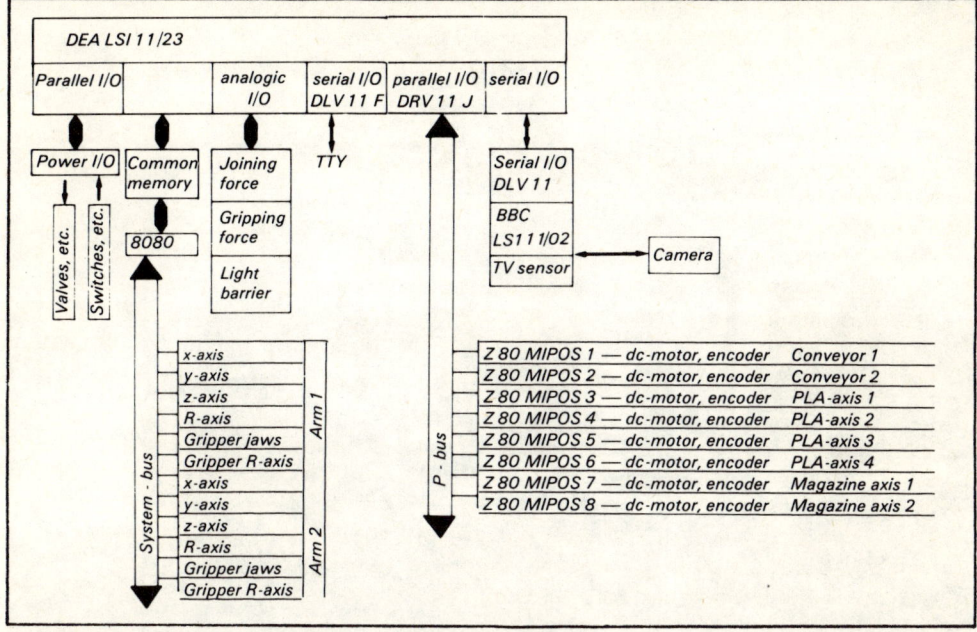

Fig. 7 Control structure of the magazining system

does not lie within the limits fixed by the program he will be asked for a new value. If the input for the parameters is finished, a single-step test run with slow speed is started to detect whether the input is correct. If the parameters have to be corrected this can be done by the operator during the run of the program at the screen. The sensors in the system are used to avoid collisions, especially with the joining force sensor. If an error occurs in the cycle run, an error message is printed on the screen. The time for programming a new workpiece is estimated at half an hour.

Change-over. The change-over procedure is as follows:

○ change the sucker cup for the assembly robot,
○ change the box,
○ change the pile for trays,
○ change the magazine carriage, and
○ load the separation/unloading bin with new workpieces.

If this is finished the operator starts the program by calling the number of the workpiece type.

Concluding remarks

A system has been presented that can be changed over to different workpiece types, joining forces and directions, magazines, and so on, very rapidly. The change-over is done mostly by changing parameters in the sequence program. The use of sensors for workpiece recognition promises to be more economic than the use of conventional mechanical automation equipment. The investment in electronic equipment is therefore reaching nearly 60% of the total investment.

To reach the desired flexibility and to use the combination of assembly robots together with sensors the following components had to be developed:

Fig. 8 Magazining system

☐ programmable separation/unloading bin,
☐ orientating gripper with integrated sensors,
☐ position-controlled conveyor belts,
☐ coupling of the control systems of an assembly robot, the workpiece recognition system and position control units with a computer,
☐ easy programming of the workpiece recognition system without teach-in by transmitting a geometrical description of the workpiece type.

The main tasks of IPA in the project were to develop these components partly in cooperation with the suppliers, and in the construction and the interlinking of the single components to a complex system (Fig. 8).

A noteworthy point with the design of the system was that the system consisting of assembly robots and workpiece recognition systems does not represent a flexible, programmable, complete feeding system. The above-mentioned developments have had to be made to complete the system.

References

[1] OMS optical measurement system. In, Proceedings of the 1st International Conference on Robot Vision and Sensory Controls, 1–3 April 1981, Stratford-upon-Avon. IFS (Publications) Ltd, 1981.

[2] Migliardi, G. and Bertolino, L. Assembly applications of the Pragma 3000. Proceedings of the 2nd International Conference on Assembly Automation. 18–21 March 1981, Brighton. IFS (Publications) Ltd., 1981.

[3] Graf, B. and Schmidt, I. Erkennungssysteme, Lageorientierungs-vorrichtungen und Teilezuführsysteme für die Automatische Montage Kongreß Verbindungstechnik VT '80, Köln 1980.

PROGRAMMABLE FEEDER FOR NON-ROTATIONAL PARTS

D. Pherson, G. Boothroyd and P. Dewhurst,
University of Massachusetts, USA

First presented at the 15th CIRP International Seminar on Manufacturing Systems,
20–22 June 1983, Amherst, USA

Abstract

A programmable feeder for non-rotational block-like parts is described. The feeder is being developed with applications to flexible robot assembly systems in mind. The development forms part of a broad study of flexible assembly at the University of Massachusetts.

Introduction

Since 1967 work at the University of Massachusetts Assembly Laboratory has been directed towards developing a practical systematic approach for efficient part and product design for ease of assembly and part feeder selection by applying group technology methods.

Group technology applied to assembly takes the form of classifying those design features that affect assembly tasks. The resulting classification or coding system then allows any part to be given a code based upon its geometry and several additional properties which affect its handling and insertion. Usually, therefore, when parts have the same code, they present the same assembly problems. Using this classification scheme, practical assembly data has been collected and grouped by part code and then reduced to the form of designer's handbooks and computer software[1, 2, 3]. To date, the information available from these handbooks is primarily for manual assembly and for automatic assembly operations using special-purpose equipment.

In the future, robot or flexible assembly systems are expected to satisfy some industrial needs better than either manual or dedicated systems[4]. Consequently, in the programme at the University of Massachusetts, group technology methods are being applied to accumulated data pertinent to flexible assembly systems. Conclusions drawn from this data will take the form of a design for assembly system applicable to robot or flexible assembly. An important aspect of this proposed system will be the design of parts so they are suitable for handling in programmable feeders. However, the feeders themselves must first be developed.

Programmable feeders

One criterion for the proper choice or design of a flexible assembly feeder is that it should be as flexible or adaptable as the robots themselves. These feeders should allow rapid adjustment or reprogramming for different parts in order to minimise set-up time.

Simplicity of programmable feeder design is being facilitated by limiting the application of each feeder to one class of part. Several different programmable feeders will be built; each feeder designed to orientate a different subset of all common part shapes. Taken collectively, then, these programmable feeders should be capable of feeding a wide range of parts.

Hopefully, the final design for assembly system will allow the designer to avoid those designs that cannot utilise the available feeders.

Other criteria intended for the programmable feeder are:

☐ low feed rate,
☐ low cost,
☐ no large part storage capacity, and
☐ no special-purpose delivery tracks.

These criteria arise because the cost of high speed feeding from large capacity hoppers cannot be justified in robot batch assembly.

Non-rotational parts feeder

The first programmable feeder developed at the University of Massachusetts is the programmable feeder for non-rotational block-like parts and is particularly suited for a well structured, robust, hierarchical flexible assembly system. Fig. 1 presents a structure chart illustrating this feeder's niche within a hypothetical flexible assembly control system.

The feeder, which is microprocessor controlled, is illustrated in Fig. 2; it is designed to handle parts which will lie stably on a moving belt and feed along an angled blade in one or more stable orientations. Parts must be no longer than 30mm and their overall height should be between 2 and 10mm. With these restrictions, the feeder can be quickly rearranged to feed a new batch of parts by

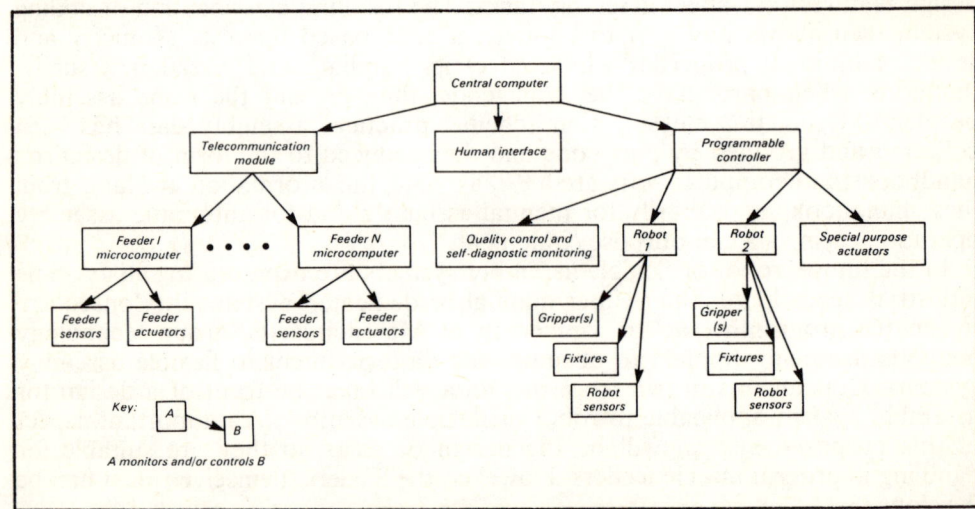

Fig. 1 Structure chart illustrating the control of a hypothetical flexible assembly system

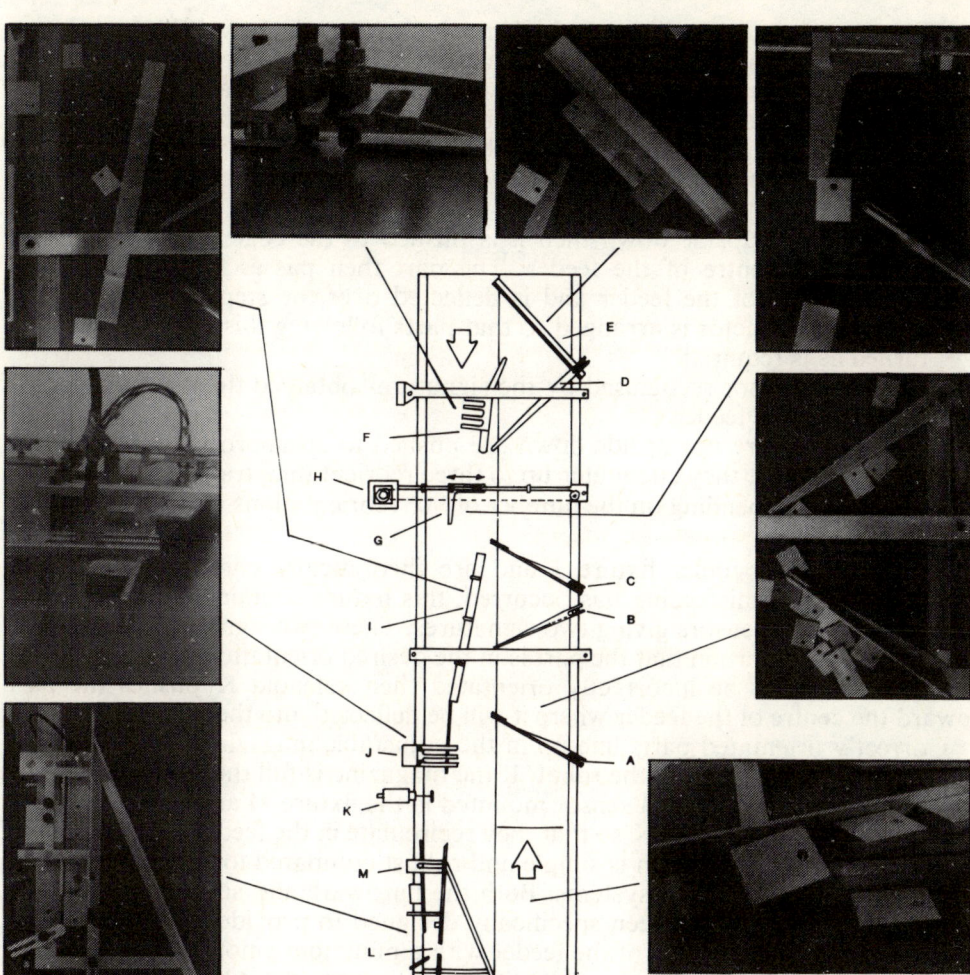

Fig. 2 Programmable feeder for non-rotational block-like parts

making minor mechanical adjustments and passing an example of the new part under its sensor during its teach mode.

The basic feeder consists of two parallel belts moving in opposite directions. The particular belt configuration and belt drive depicted here was developed in collaboration with AMP, Inc. One belt is lower at one end providing a pocket for the reservoir of parts. As the parts are carried along they encounter a blade A with a cut-out which allows one part at a time to pass. Excess parts or parts standing one on another are immediately deflected to the second belt to be returned to the reservoir. Parts passing through the cut-out are deflected towards the edge of the belt by blade B and are allowed to pass through to the blade C without restriction. Blades C and D perform the same functions as blades A and B respectively, and simply ensure that single-file feeding is obtained at the last blade E which deflects the parts on to the return belt.

The return belt is travelling at a higher velocity than the forward belt and, as a consequence, gaps occur between adjacent parts before they pass under the upper bank of four fibre optic sensors mounted on the frame F. At this point four signatures are obtained for the part and from these signatures one of the eight part orientations is recognised. The pusher G, which is activated by the stepper motor H, now positions the part on the return belt as follows:

○ If the part is upside down then it is pushed to the edge of the return belt nearest the centre of the feeder. The part then passes unimpeded to the reservoir end of the feeder and is deflected over the step between the two belts. The deflector is arranged so that parts following this path will be overturned as is required.
○ If the part is not recognised by the signatures obtained then it is pushed off the edge of the feeder.
○ Parts which are not upside down are pushed to an appropriate position on the belt so that they encounter up to three reorientating trips on the orientating blade I depending on the number of 90° reorientations required.

Parts now pass under fixture J and are theoretically correctly orientated. However, in case misfeeding has occurred, this fixture contains the lower bank of two fibre optic sensors giving two signatures. These two signatures are all that is necessary to ascertain that the part is in the desired orientation.

Should the part be incorrectly orientated then solenoid K pushes the part toward the centre of the feeder where it will be deflected into the reservoir.

Correctly orientated parts line up in the adjustable magazine device L where they wait to be grasped by the robot. If the magazine is full then the presence of the last part is detected by a sensor mounted in the fixture M and following parts are pushed by the solenoid K so that they recirculate in the feeder.

The feeder control system is simple and robust compared to those programmable feeders using camera systems. Both the hardware and software aspects of this control system have been specifically designed to provide clear and simple operation and maintenance of the feeder with a minimum amount of set-up time. Also, there is no intrinsic reason why the controller could not be applied to other mechanical orientating systems or be operated at significantly higher feed rates. The controller hardware can be broken down into six major sections as shown in Fig. 3, namely machine-human interface, power, actuators, photosensors, and computer.

The machine-human interface was a primary concern in the design of this system. Much consideration and effort has been devoted to making the controller as easy to use as the price range of this assembly system permitted. The machine-human interface consists of a front panel containing: calibrate, teach, and continue mode push buttons; calibrate, teach and run mode LED's; an orientation number display, two error code displays, a bar graph VU meter, and a power on/off switch and indicator.

The user may recalibrate or reteach the feeder merely by pushing one of the mode switches while the mode confirmation light indicates acknowledgement. The current analogue output levels of any of the six photosensors or three threshold levels may be observed at any time on the VU meter or on an attached oscilliscope screen by selecting the desired signal with the rotary selector. The orientation display cues the user for the next orientation in teach mode or it gives the last orientated part orientation number (1–9) in run mode. If an error occurs,

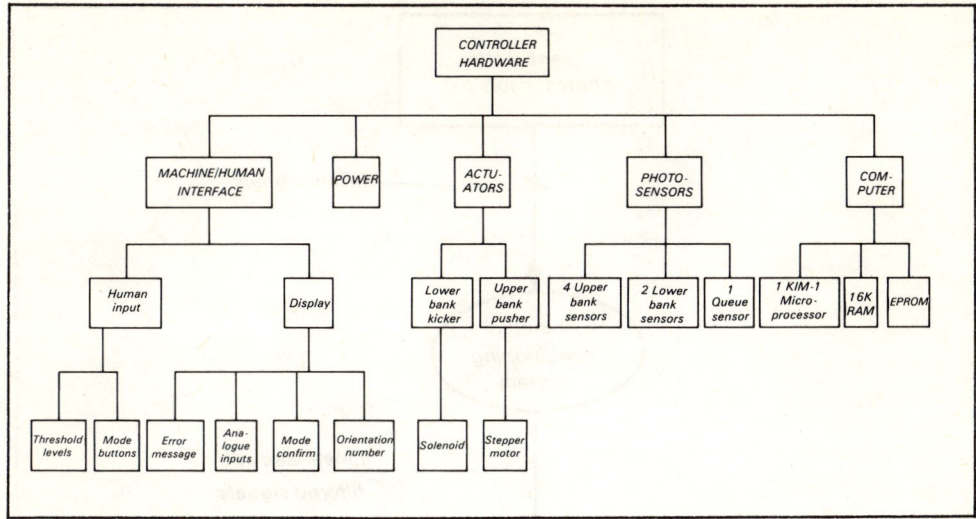

Fig. 3   Breakdown of controller hardware

say the sensor signal is too high or the part is too large to be handled with the current sampling rate, the error message code is displayed.

Seven Skan-a-matic S17103 sensors are used to obtain parts signatures and to monitor the number of parts waiting in the exit queue. The sensor tips are optical fibres with an annular outer sleeve transmitting incandescent light toward the part and an inner core returning the fraction of lamp light reflected from the part back to a phototransistor. A part being reorientated first encounters the four sensors comprising the upper sensor bank (F in Fig. 2). Four sensors are sufficient for reorientating many complex shaped parts and parts which may initially be upside down. The next sensor bank which the part would encounter is after the orientating blade and is referred to as the lower sensor bank (J in Fig. 2). This bank contains only two sensors since its only purpose is to confirm that the part is indeed in the proper final orientation. The last sensor is positioned in the exit queue (M in Fig. 2). The only purpose of this sensor is to allow the system to react when the number of parts in the queue has reached a maximum; it need only be connected to a conventional on/off control module.

The computer used in the controller is an Aim 65 8-bit microcomputer and operates with a 6502 processor. Memory expansion of 16K RAM, EPROM, and two 6522 versatile interface adaptors were subsequently added.

Two actuators are used: a lower bank solenoid (K in Fig. 2) and an upper bank stepper motor (H in Fig. 2). The lower bank solenoid has a one-inch stroke and the upper bank Slo-Syn stepper motor and controller actuates a pusher (G in Fig. 2). Power supplies for the stepper motor and solenoid are isolated from the remainder of the system.

Data flow in the system operation begins at the sensors (see Fig. 4). As a part passes underneath the upper sensor bank, samples of each signature for each of the four sensors are taken. This raw signal is then processed by an analogue signal conditioning and linearisation system. At the prescribed time, the data acquisition system samples a given signature and converts the sample into an 8-bit number between 0 and 255. This number is subsequently compared to three threshold values between which it falls. Those corresponding numbers are 0, 1, 2

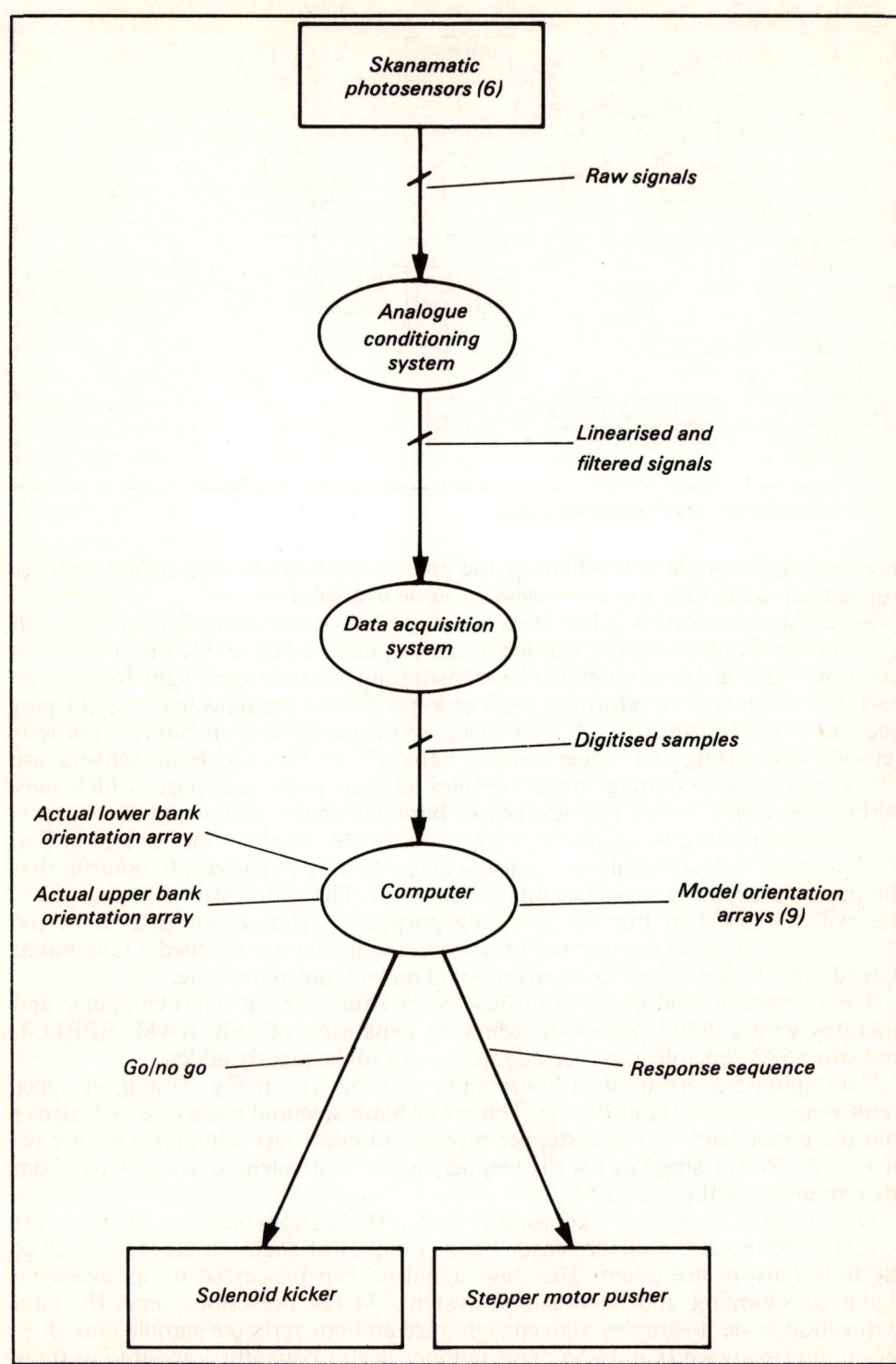

Fig. 4   Data flow of controller system

and 3, respectively; increasing as the threshold magnitude increases. The computer stores this value in one of three major data blocks: within the nine model orientation array, within a current lower bank orientation array, or within a current upper bank orientation array. At the proper time, the computer compares one of the current bank orientation arrays with one of the model orientation arrays. The best matching model orientation array dictates the manner in which the system will respond to the part currently passing under a sensor bank. All responses consist of either energising the solenoid or producing a prescribed set of motions of the upper bank pusher.

The treatment of photosensor signals described here is unconventional. Specifically, the raw signal from the sensors is converted from current to voltage and then logarithmically linearised with a Burr-Brown 4127 logarithmic amplifier. The next step is to filter the signal heavily with a first-order active filter. Commercial sensor vendors usually provide only nonlinearised on/off proximity control with their sensor products. The reason why more than one threshold level is used on a sensor output is that, for some simple parts, one threshold level provides insufficient orientation data for a signature. For instance if only one threshold level was used for a part having the cross-section shown in Fig. 5, the threshold level would be adjusted so as to pick up the highest signal level only. The signature would be completely symmetrical and consequently it would be impossible to determine whether the toe is on the left or on the right of the part as shown. So, with a high threshold level the orientation of this part cannot be detected. It the threshold level was set lower in order to detect the lower level of the analogue signature, again a symmetrical bump would result; only now it would extend for the whole length of the part. The only easy way to detect the orientation of a simple shaped part like this is to redesign the part or to upgrade the quality of the discretised signature; that is, to add another threshold level as shown in Fig. 5. Absence of material in the shape of the part is a string of logic O's, the lower level of the signature is a string of logic 1's, and the upper level of the signature is a string of logic 2's. Since, on a microprocessor it takes two bits to write a logic 2, one can add a third threshold level with no extra storage requirement since logic 3 is also two bits long in binary form.

Although each of the six sensors could have their own independently adjustable threshold levels, it was felt that 18 separate threshold dials would be rarely, if ever, needed and their inconvenience would be so great that it was decided to set only three threshold levels with potentiometer dials and let the same three levels apply for all six sensors. Note that the seventh sensor, mounted over the queue, has only to recognise the presence or absence of a part at the head of the queue and so only needs one threshold level.

In any signal processing operations where signals are stored, one objective should be to find ways to represent a signal as briefly as the operation's requirements allow. In the present design it was thought desirable to compress memory storage and consequently decrease the time penalty associated with processing large blocks of data, so that multi-task programming in real-time is easier.

As previously mentioned, every sample taken from every signature from every orientation is represented by the first threshold level below the digitised sample value. The possible threshold values consist of the set 0, 1, 2 and 3 consecutively from the lowest to the highest threshold level. In this way four signature samples may be taken, reduced to threshold numbers and stored as four two-bit samples per byte of memory. Each byte of storage for the upper bank of four sensors will

254 PROGRAMMABLE ASSEMBLY

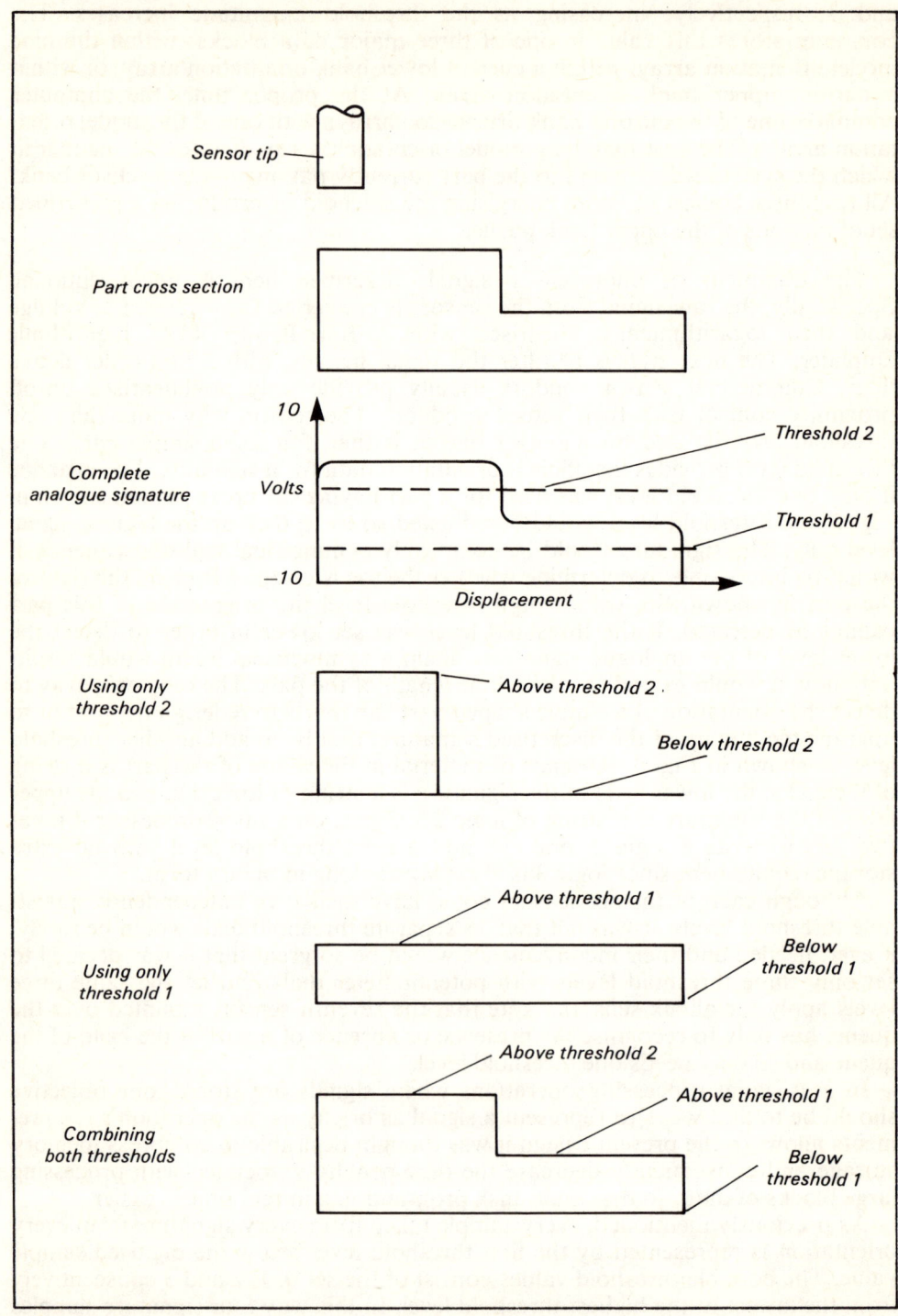

Fig. 5 Elimination of ambiguity for certain cases by using a signature representation based upon two threshold comparisons instead of one

contain the results of a sampling cycle of sensors (also numbered 0, 1, 2 and 3). For example, AABBCCDD would be one byte of RAM where AA is the threshold representation of the sensor zero sample, BB is the threshold representation of the sensor one sample, and so on. One page of RAM will store one orientation array consisting of 255 samples from each of the four sensors. The lower bank of two sensors will require only one nibble of RAM to store one cycle of sensor samples from that bank since it has only two sensors.

The calibration mode is entered upon system start-up or when a user pushes the calibrate mode button. In this mode, the versatile interface adapters are configured and the threshold levels are manually set. The computer loads the threshold levels and checks each sensor to see if its signal exceeds the upper limit (+10V) for input voltage into the data. The user adjusts the threshold dials in order to detect selected part textures (humps or dips in a signature) sufficient for the detection of orientation. Extraneous threshold levels may be disabled with a switch. The front panel VU meter or an oscilliscope may be used as visual aids to determine the best threshold level. When the calibration mode is left, the positions of the threshold dials may be disturbed since their input has already been loaded by the data acquisition system.

The SDM 854 data acquisition chip was obtained from Burr Brown and is configured to accept nine input signals: six sensor signals from the upper and lower sensor banks and three signals representing the three available threshold levels. At precisely the right moment, each sensor signal is digitised by this chip and sent to the Aim computer for processing.

Significant computer processing begins with the teach mode. Teach mode is entered by pushing the teach mode button. The microprocessor now searches for a non-zero input signal from the upper bank of four sensors. As the first orientation to be defined passes under the bank, points at equal intervals across all four signatures are sampled and digitised. Each signature sample from the data acquisition system is then read by the computer, reduced to a 2-bit threshold value and stored in one quarter byte of RAM. Loading into memory ceases when all sensors for the bank register zero signal levels. By the end of the teach process, eight signatures from the upper bank and one signature from the lower bank will be stored. These signatures become the models for all subsequent comparisons made during the run mode.

In the run mode, the sensors for both banks will begin to scan at the same rate as in teach mode. When a part passes under the upper sensor bank, a current orientation signature is stored. Now the computer tries to match the current orientation array with the model orientation arrays. First, comparisons determining the degree of mismatch between any given model orientation data array and the current orientation array will define error as the difference in magnitudes of the total number of bytes for the two arrays plus the number of mismatched bytes in a byte-by-byte comparison continued to the end of the shortest data array. This error value is stored in another array called the error array. The error array contains the final error from array-by-array comparisons of all the possible model arrays. Error comparisons for the lower sensor bank are similar but simpler since there is only one good orientation array. Finally, a response will be made according to which of the model arrays is the best match or if the user defined upper limit of mismatch error (rejection limit) is exceeded by all orientation comparisons.

The best match found by the comparison software will determine the proper response of the system to the part passing down the feeder belt. Possible responses for a part passing the upper bank were described earlier. Possible responses for a part passing the lower bank can be summarised as follows:

○ The part is properly orientated so no action is taken and the part is allowed to enter the exit queue.
○ The part is improperly orientated so it is rejected by the solenoid kicker and allowed to pass through the orientating system once again.
○ The exit queue is full so all parts are rejected and recirculated.

The upper bank response actually consists of a sequence of instructions to the stepper motor controller in the form of sequences of digital pulses on the forward or reverse channel of the stepper motor controller.

Concluding remarks

Although computing for this prototype is intended for one microprocessor, a multiprocessor system may be used instead. This configuration would reduce downtime due to failure of one computer since either bank can orientate parts, although the lower bank orientates passively. Also a double microprocessor system decreases real-time programming constraints and allows a more modular approach to later projects.

It is expected that the feeder described here will be easily integrated into an overall flexible assembly control system. Communication between the programmable feeder and other elements of the flexible assembly system would be minimal; being only part orientation/rejection reports, error messages, and basic on/off commands, etc. This should make interfacing relatively easy.

Acknowledgements

The authors wish to thank Professor J. Motherway for his helpful advice, T. Koff and D. Zenger who designed many of the mechanical devices, and D. Hammer who constructed some of the electrical hardware.

This work was supported in part by a grant from the National Science Foundation. Any opinions, findings and considerations or recommendations expressed in this publication are those of the authors and do not necessarily reflect the view of the National Science Foundation.

References

[1] Boothroyd, G. and Dewhurst, P. Design for Assembly Handbook. University of Massachusetts.
[2] Dewhurst, P. and Boothroyd, G. Computer-aided design for assembly. Assembly Engineering, February 1983.
[3] Boothroyd, G., Poli, C., and Murch, L. E. Feeding and Orienting Techniques for Small Parts. University of Massachusetts.
[4] Boothroyd, G. Economics of assembly systems. Journal of Manufacturing Systems, vol. 1, No. 1, pp. 111–127, 1982.

A COMPUTER CONTROLLED RECONFIGURABLE GRIPPER

A. Butcher and P. Fehrenbach,
GEC Research Laboratories/Marconi Research Centre, UK

First presented at the 15th CIRP International Seminar on Manufacturing Systems,
20–22 June 1983, Amherst, USA

Abstract

An important requirement of programmable assembly devices is fast operation. A gripper which can handle a variety of components helps in this by obviating the need to stop to change tools. If the gripper can also perform tests on the components to check that they are working prior to assembly, this also reduces the overall time required by ensuring greater reliability in the finished product.

The gripper presented achieves both of these aims for components with from two to sixteen dual-in-line pins in a research system for assembling circuit boards.

The gripper has eight pairs of opposing 'fingers' arranged in four groups for gripping the pins of the packages. Different sizes of package are accommodated by remotely selecting different combinations of the four groups. Each group has a 'thumb' which acts normal to the fingers and parallel to the pins of the package. The tip of each finger is electrically isolated from its support and wired to allow electrical tests to be performed on the packages.

The motors driving the gripper incorporate optical shaft encoders to provide positional feedback and the driving circuitry generates feedback signals proportional to the motor currents. These latter signals are used to determine the torque applied by each motor. Final alignment of the package over the holes in the circuit board is facilitated by optical fibres attached to each finger next to the pins being gripped. With a back-illuminated circuit board the fibres can sense whether or not the alignment is correct for insertion.

The gripper and sensing control are implemented using single board computers with various bus interface boards and some custom built hardware. The software for the gripper forms part of a multiprocessor multi-tasking control network which constitutes a programmable assembly station. The assembly station can take its instructions directly from a CAD system for completely automated production.

Introduction

In an assembly task, a variety of objects are brought together in a predefined manner. The difference between the objects' shape, size, etc., is usually sufficient to necessitate the use of several grippers to perform the assembly satisfactorily. If one gripper could be designed such that it could reconfigure itself as most, or all, of the grippers required, then this could offer savings in terms of time for assembly and cost of assembly hardware.

Another major requirement for grippers is that they should be able to supply as much information about their environment as possible, since this naturally enables a more controlled assembly. If information is available about the object being gripped, where the object/gripper is in relation to the assembly site, etc., then more of the assembly sub-tasks can be performed 'closed-loop' leading to better quality control of the product being assembled with all the associated advantages this provides. The gripper presented fulfils these requirements to an extent. It can reconfigure itself to cater for dual-in-line packages having from two to sixteen pins. Further, it provides control over the force applied to the object and feedback information relating to this, provides facilities for 'testing' of the package being gripped and is used as a platform for optical fibres which are used to enable precise positioning of the assembly site below the gripper.

System overview

The gripper itself and its controller form a small part of a research assembly workstation which is directly connected to a CAD system. The workstation comprises a network of multitasking processors, an anthropomorphic robot arm, the gripper, an X, Y, θ-table, a set of slide feeders and associated electronic hardware. The network has been designed such that interprocess communication is processor independent. Communication is achieved by a process (or task) sending a message to a process requesting one of its functions to be executed, using the accompanying parameters. The sending process then receives a message back at a later time which specifies success or failure and may also contain parameters relating to the achieved state.

To perform an assembly the CAD system sends an assembly description file to the assembly scheduler process in the workstation. Amongst the instructions in this file are the insertion instructions. These instructions are issued to the DIL insertion controller which coordinates all the processes necessary to insert a particular kind of DIL pack into the circuit board at the correct position. Three of the processes interacted with are: the gripper control process, the alignment process and the chip verification process. These correspond to control of the gripper, use of the optical fibres and use of the insulated finger tip contacts, respectively.

It is the workings of these three processes, their associated hardware and how they interact with the insertion controller which are to be described.

Gripper control and usage

Gripper and hardware description

The sixteen fingers of the gripper are arranged in two parallel rows of eight, with 0.1in. spacing along the rows, in four blocks having 1, 2, 4, 1 opposing pairs per block. There are also four thumbs which operate normally to the plane of action of the fingers, one per block (Fig. 1).

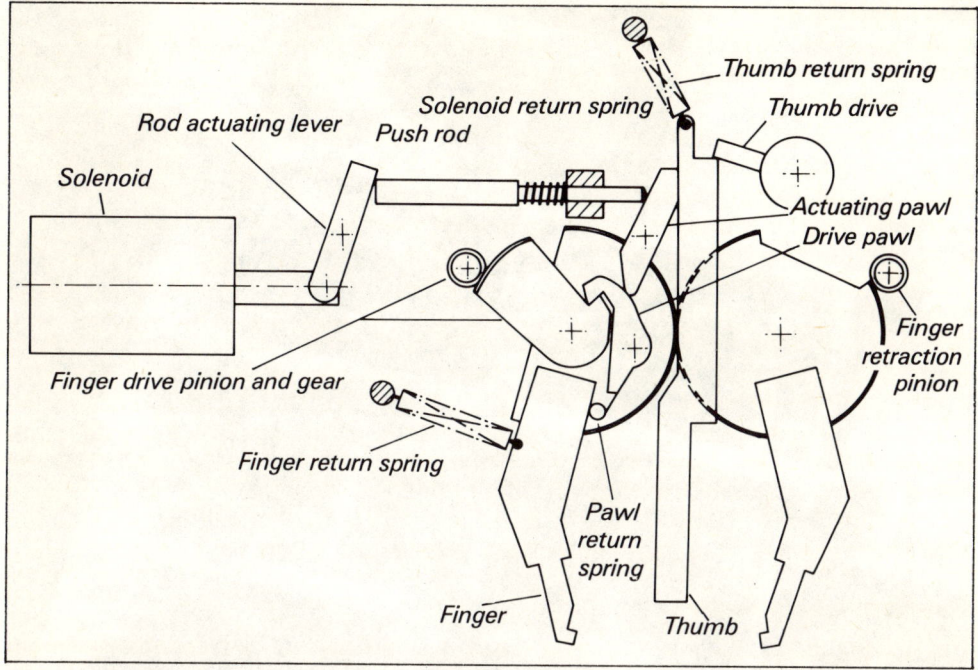

Fig. 1   Mechanics of the reconfigurable gripper

The finger blocks and their thumbs are selectable by energising the appropriate solenoids.

Movement of the finger blocks and thumbs is achieved by three dc motors. One motor drives the selected finger blocks, the second motor drives the unselected finger blocks and the third drives the selected thumbs. Each motor has an incremental optical shaft encoder with quadrature output for positional feedback (Fig. 2).

The tips of the fingers are electrically isolated to permit separate electrical connection to each pin by the chip verification process hardware. Also along each finger is a 0.01in. diameter plastic optical fibre used to detect light from the back-illuminated circuit board. The electronics for the sixteen pin diode detectors is also mounted on the hand and the resulting sixteen ±10V signals are accessible by the alignment controller via an A to D board (Fig. 3).

The free end of each motor drive shaft has a disc attached which has a radial slot in it. Absolute positioning of the motor is achieved by driving the motor until this slot is coincident with an optical proximity detector (the output is accessed by the motor controller) and using this as the datum.

Servo description

The motor servo loop is implemented as a three-term controller, the motor volts demand to the drive being given by:

$$V = K_1(p-d) - K_2(d-d_p) + I$$
$$I = I_p + K_3(p-d) \quad \text{when on}$$
$$= 0 \quad \text{when off}$$

where, V is the new volts demand, p is target displacement, d is present displacement, d_p is previous displacement, I is new integrator value, I_p is previous integrator value, and K_1, K_2 and K_3 are constants.

Fig. 2 Configuration for 8-pin DIP. One block of fingers is selected and three blocks deselected

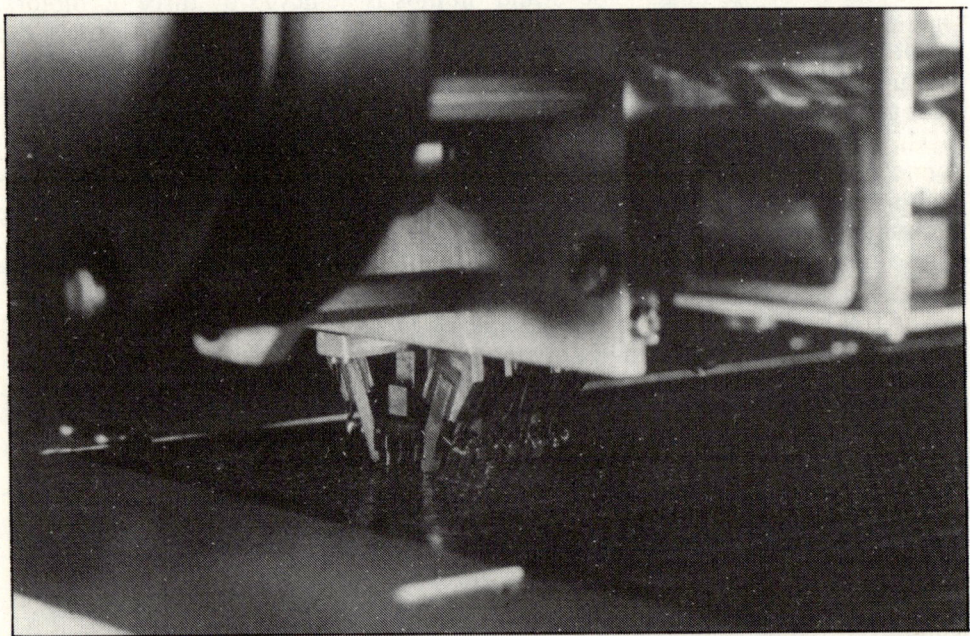

Fig. 3 The gripper during 16-pin DIP insertion. The thumbs are seen pushing the top of the package

As well as the volts demand, the driver circuit also receives a current limit demand from the controller. By monitoring the current taken by the motor it ensures that this current limit is not exceeded as a primary requirement and then if possible applies the required motor volts. Thus it has the characteristics of a programmable current limited voltage source.

The integrator is turned on towards the end of a move, if necessary, and is turned off at the start of a move. This ensures that when the move is completed, the motor displacement is that requested, or the motor current is at the requested limit-value.

The servo is software tuned to provide a minimum duration response to a step input, with no overshoot and a programmable undershoot. This software is at present separate from that of the controller in the workstation system but is implemented in the same network structure and is thus easily transferable. The servo is active at all times.

Moving the motors

The movement of the second motor relating to the unselected finger blocks is hidden from the user, so when requesting the fingers and thumbs to move, only parameters for motors one (selected fingers) and three (thumbs) are given. The request message to the move function of the controller contains destination positions and current limits for the motors. The controller uses the destination positions in calculating the positional error, but varies the applied current limits during the movement of the motors in the following manner. For the first few servo updates, the current limit is set to the maximum value. Thereafter, when the motor volts applied are such as to move the motor toward the destination, the specified current limit is applied, otherwise the maximum current limit is applied. This both facilitates starting and stopping, and limits the gripping (fingers) and pushing (thumbs) forces which the gripper can apply.

When the motors have finished moving, the achieved positions are compared with those requested and a decision is made as to whether the move was positionally successful or not. This success or failure is conveyed to the requestor, along with the actual positions and final update motor currents, in the response message.

Fig. 4 shows the actual selected thumb motor current for various limit values when the gripper grips a package at the feeder.

Reconfiguration

Reconfiguration is initiated when a request message for the reconfiguration function is received. The number of dual-in-line pins to be reconfigured for is included as a parameter of the message. The function controls the reconfiguration in the following manner:

☐ The motors are sequentially driven to their datum positions in accordance with the mechanical constraints imposed by the gripper design.
☐ The controller uses the 'number of pins' parameter as an index to a look-up table to establish the correct solenoid pattern and then energises the appropriate solenoids. With reference to Fig. 1 it can be seen that energising a solenoid has two effects. The rod actuating lever causes the push rod to push the thumb across (via the actuating pawl), such that the thumb drive will engage with the flat at the top of the thumb. The actuating pawl also causes the drive pawl and the finger drive gear to engage.

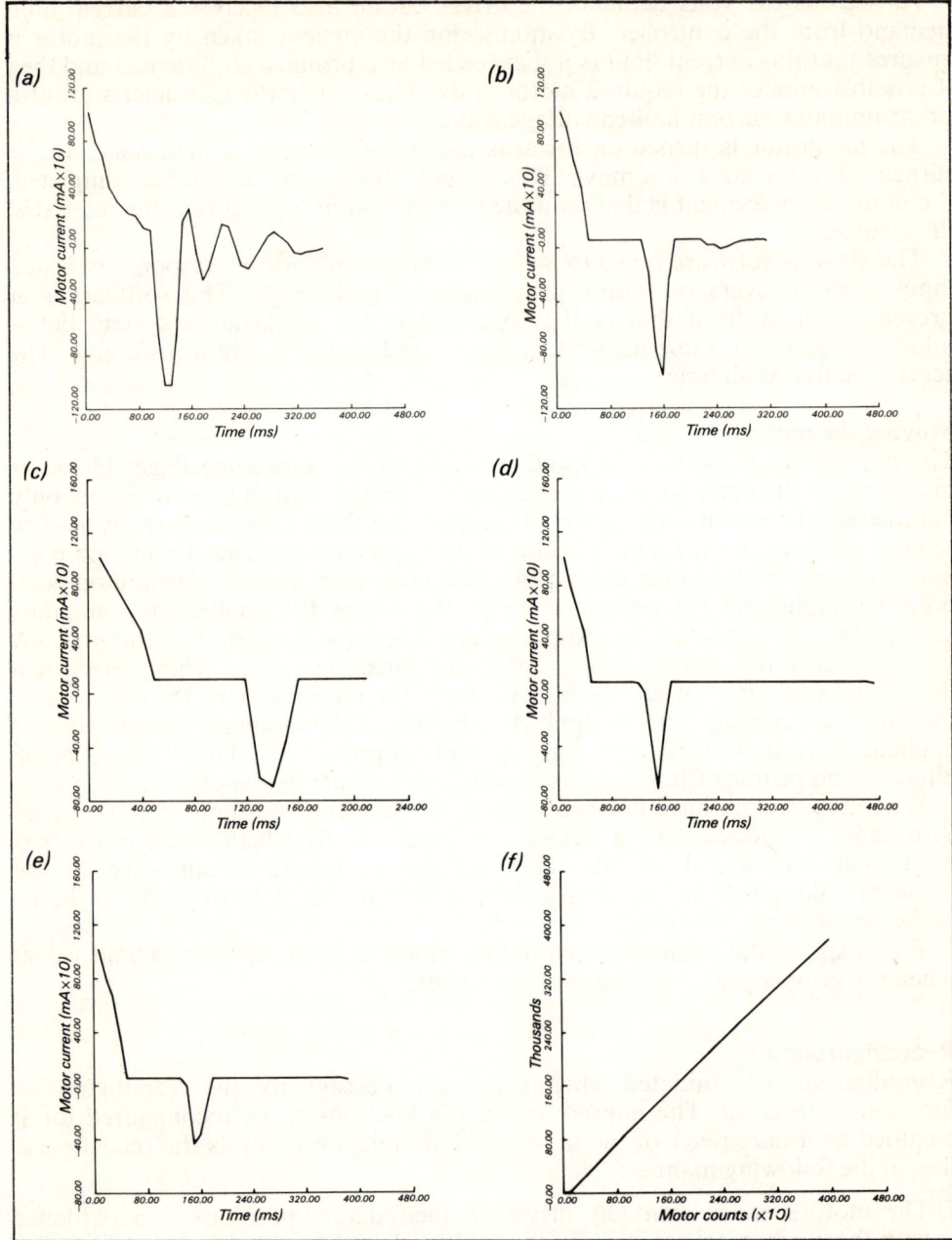

Fig. 4 The actual selected thumb motor current for various limit values when the gripper grips a package at the feeder. (a) and (b) show thumb motor currents when thumbs are driven to the top of DIP feeder with current limits of 900 and 50mA, respectively. (c), (d), and (e) also show thumb motor currents when servoing to feeder with current limits of 100, 75 and 50mA, respectively, but with DIP present. Hence, positionally the move fails, but current limit is applied. The higher the current limit, the further the thumbs extend given a constant system rigidity. (f) shows a motor count to extension calibration graph for the thumbs

- ☐ The selected fingers are now driven inwards by the finger drive pinion for a sufficient distance so that the finger retraction pinion no longer engages with the finger cog. The unselected fingers are driven outward such that they are clear of the work area.
- ☐ If all the above steps are successfully performed, then a success response message is sent to the requestor, otherwise a failure response is returned.

Reconfiguration of the gripper takes 2–3 seconds. However, the additional time seen by the system for reconfiguration is less than this, since reconfiguration is performed parallel with the manipulator moving to the slide feeder for the next component.

Gripping a package

Before the DIL insertion controller issues a request message to the motor move function of the gripper controller, it needs to establish several parameters of the package to be inserted. It does this by sending a request message, with the package identity as a parameter, to the system data base process and receives a response message with such parameters as package width and number of pins etc. It uses these parameters to calculate the finger grip position and motor current limit. The thumb motor parameters are target position at the feeder surface and current limit such that the motor will stall applying little force on the DIL package if it is present. The move response parameters are evaluated to check that the thumb motor has stalled and is applying its current limit, signifying that a package is present, and that the finger motor has reached its target position, thereby deforming the package pins to the correct separation for insertion into the circuit board.

Package insertion

The package insertion is performed by the gripper itself, the manipulator being used as a stable platform during this stage of the assembly. The insertion is performed in a sequence of three operations:

- ○ A request is issued to the move function of the gripper controller to extend the thumbs a small distance using a low current limit. This exploratory insertion is used to confirm that all the pins are located in holes without damaging the board or package.
- ○ Assuming a favourable response, a second move request is sent which opens the fingers and extends the thumbs to the insertion position with a current limit sufficiently high to achieve this.
- ○ If this is successful then a move request is issued to extend the thumbs to the surface of the circuit board with a low current limit to avoid damaging the package or board. If the insertion has been successfully performed then this last stage will confirm that the package is present on the board by the motor stalling and failing to extend the thumbs fully. It is possible that the package could have been dropped between the feeder and the circuit board in which case states 1 and 2 would have been successful so this final stage is a necessary check.

Package alignment

Before a package can be inserted it must be correctly aligned over the holes in the circuit board. In a perfect system alignment would always be correct initially, but in practice it is necessary to remove errors introduced by both the robot and the manufacture of the circuit board. This is achieved by moving the X, Y, θ-table which supports the circuit board. The table movements are controlled by the inputs from the optical fibres on the gripper in order to maximise the light entering each fibre. The fibre inputs provide 4096 levels of light which give sufficient discrimination to locate the centre of a hole in the circuit board. The fibres are not used merely as binary sensors. The two dimensional response of a fibre to a hole depends on the hole diameter and the height of the end of the fibre above the board, which is a constant 1mm; thickness of the board is a less important variable. A response to a typical hole is shown in Fig. 5.

The table moves incrementally under servo control and the direction and magnitude of each step is determined by the change in light input at the fibres during the previous step. Large numbers of steps, typically 0.1mm in length in the x or y directions, are combined to form a search pattern or hill climb or both to maximise the light. Incremental rotations about a vertical axis through a point on the board determined by the light inputs ensure that the alignment angle is correct. A typical path followed by the table when searching for a hole and then

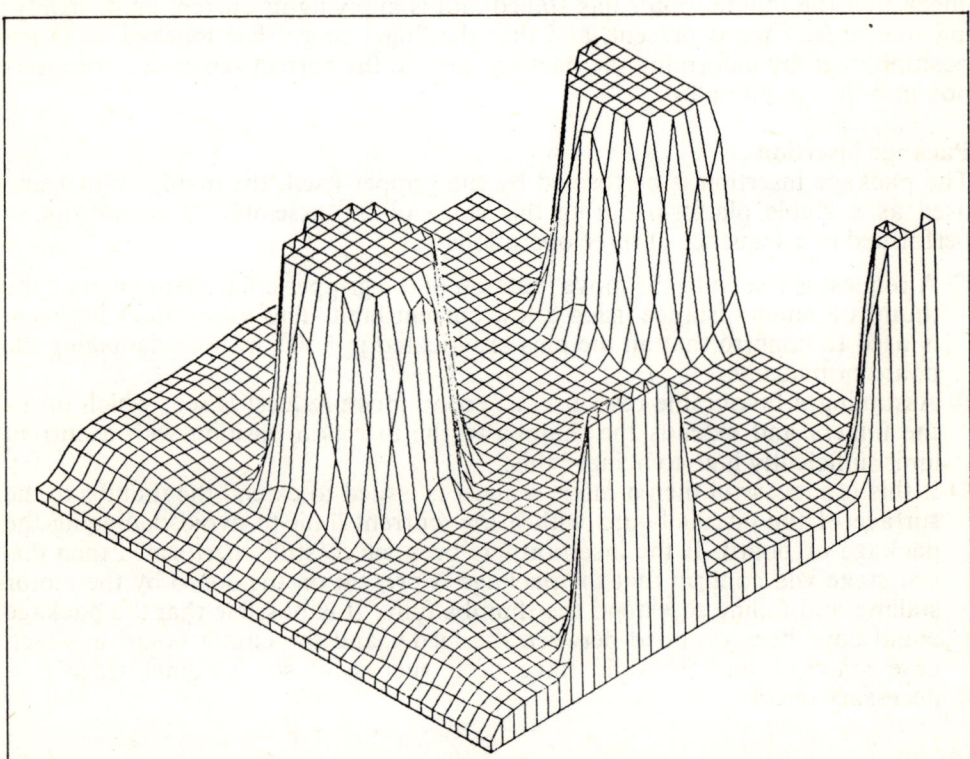

Fig. 5 Typical optical sensor response. This scan at a grid interval of 0.1mm shows a 1.0mm diameter hole in a circuit board and parts of three similar holes on 0.1 in. centres. The lowest horizontal level on this isometric view represents no light input to the optical fibre

hill climbing to its centre is shown in Fig. 6. The response of the process when called is either success or failure. If a failure is responded then insertion of the package does not proceed.

Package verification

Verification of a package can begin immediately it has been removed from the component feeder and proceeds as it is moved towards the circuit board. By applying various voltages to appropriate pins dependent on the type of circuit in the package, and measuring the voltages at other pins and the currents flowing, it is possible to check (a) in which of the two possible orientations the package is held in the gripper and (b) that the circuits in the package are functioning satisfactorily. The result of test (b) determines whether insertion of the package

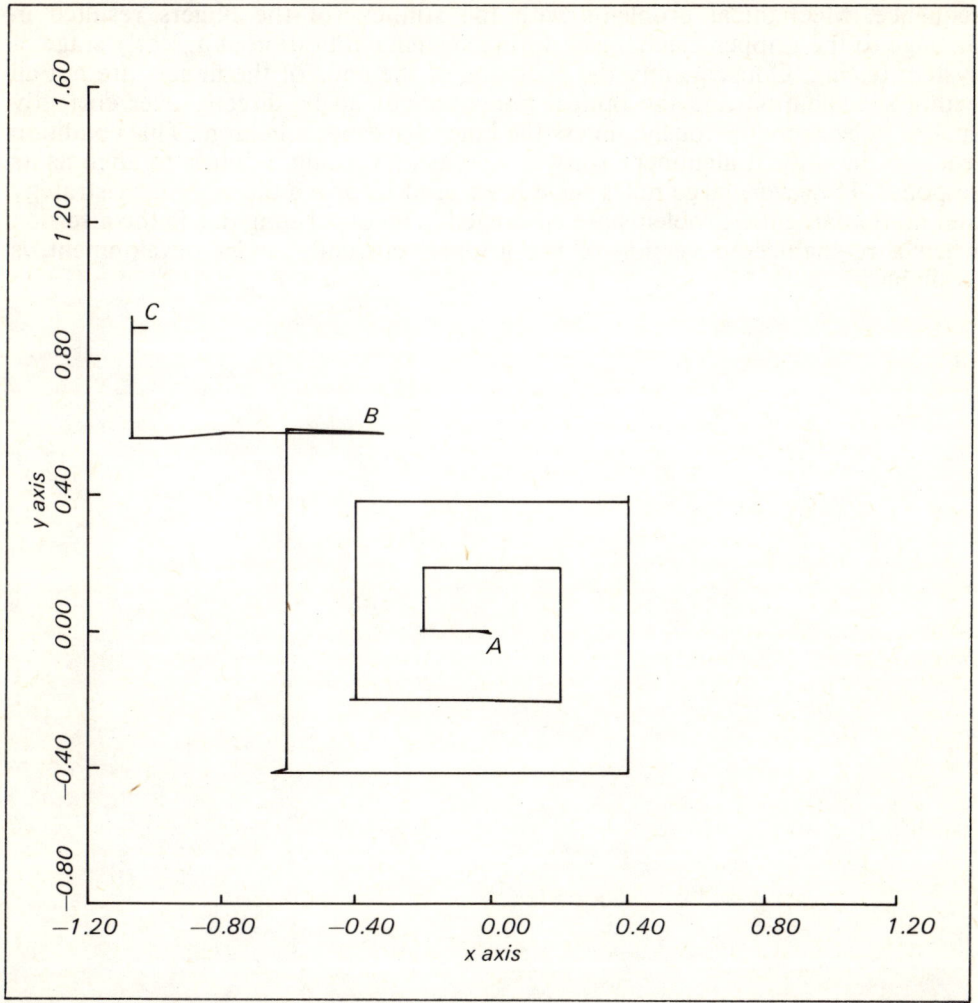

Fig. 6 Typical table path during alignment. Starting at point A the X,Y,Θ-table searches to produce a change in the input to one of the fibres. This is detected at point B. The table then moves to maximise the light entering the same fibre, which it achieves at point C. A rotation of the table may follow but cannot be easily represented on a plot of this form

goes ahead or the gripper is diverted to drop the package into a reject bin. The result of test (a) determines the position to which the gripper is moved for the insertion stage if the package is working properly. The two tests are performed as quickly as practicable to obtain reliable results and return their responses before the decisions based upon them need to be made. They thus add nothing to the assembly time.

Concluding remarks

Use of the gripper in the assembly station so far has shown that it is capable of meeting its design aims. Insertion of dual-in-line packages of several sizes has been successfully achieved under controlled operation using a sample circuit board with specially toleranced holes.

The optical alignment has yet to be fully incorporated in the insertion sequence. Mechanical problems with the stiffness of the fingers resulted in damage to the gripper mechanism during system calibration at an early stage of system testing. Consequently the positions of the ends of the fingers are not all within specification and the optical fibres cannot all be directly over correctly spaced holes simultaneously, unless the holes are especially large. This condition leads to the optical alignment software always returning a failure to align as its response. However, large holes have been used to prove the alignment strategy and no fundamental problems are envisaged in incorporating this in the insertion when a re-engineered version of the gripper, currently under development, is available.

THE FLEXIBLE PARTS FEEDER WHICH HELPS A ROBOT ASSEMBLE AUTOMATICALLY

T. Suzuki and M. Kohno, Hitachi Ltd., Japan

First published in *Assembly Automation* (February 1981)

Abstract

A flexible assembly system comprising a flexible part-feeding machine and a robot has been developed. It has been proved that the part-feeding machine developed is adaptable for the changing shape of the parts. A simplified matching method, which identifies the posture of the parts in less than 1 ms by a microcomputer, has been found practical. Introductory experiments have been carried out, and further experiments are being performed to improve the system towards more practical applications.

Introduction

The automation of assembly has been applied mostly to mass-produced goods such as home appliances, watches and car components. It has almost reached its culmination by means of so-called hard automation, which consists of many single-task working stations.

However, the recent trends of increasingly diversified tastes of customers have been forcing manufacturers of such products to adopt more flexible assembly facilities, which are adaptable to frequent changes of products with short changeover times. The flexible assembly machine is also needed to improve the productivity of small batch production, which has hitherto been less automated by conventional manners.

Under such circumstances, industry has found industrial robots as the most effective tool for handling parts and assemblies, and some new concepts of flexible assembly systems and elementary technologies have so far been reported. These include flexible part feeding devices[1], universal grippers[2,3], and new concepts of assembly configurations[4,5]. However, they are still at laboratory stages and have not yet been successfully applied to practical production at factories.

A development project for a flexible assembly system for small mechanical products such as portable tape recorders has been carried out at the Production

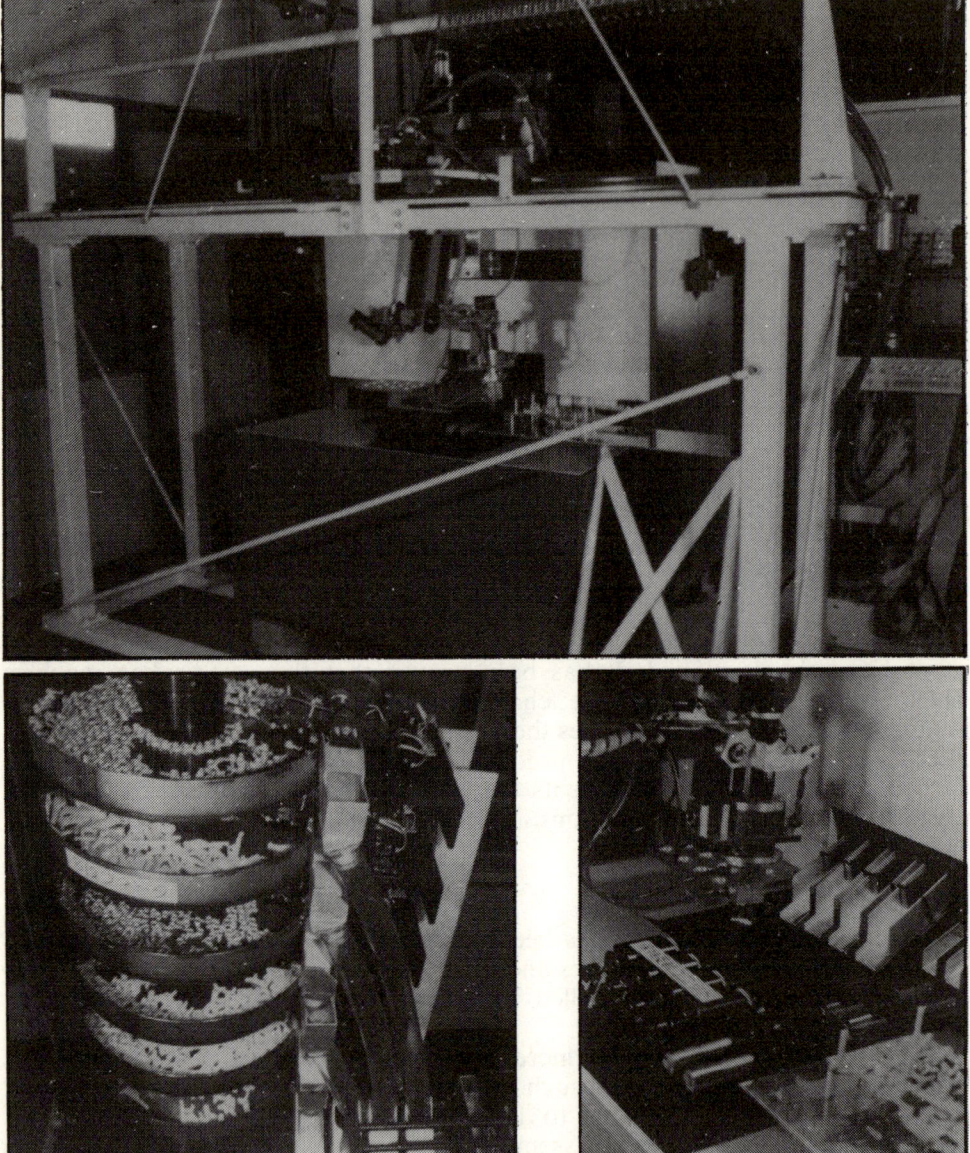

Fig. 1 Views of the flexible assembly machine showing the multilevel bowl feeder, visual system and robot arm

Engineering Research Laboratory of Hitachi. A pilot system has been developed, and early experiments performed.

The system comprises a jointed-arm robot and a programmable part feeding machine equipped with vision.

Hardware

The system concept is that several parts are supplied and assembled at one station where the robot picks them up in a programmed sequence. This is par-

FLEXIBLE PARTS FEEDER 269

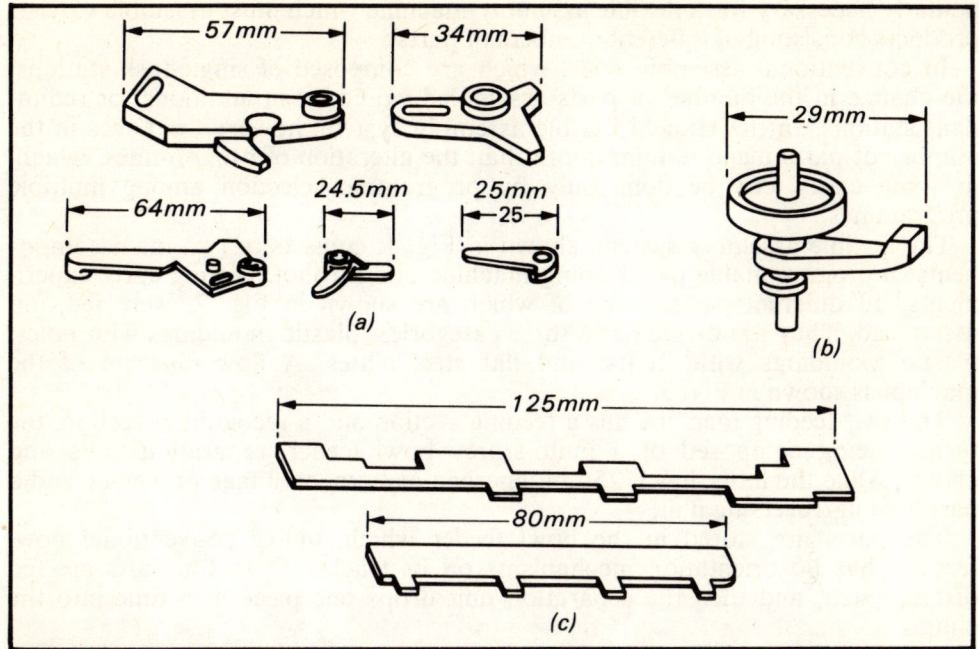

Fig. 2 Parts handled by the programmable assembly complex fall into three basic groupings: (a) mouldings with holes, (b) mouldings with shafts and (c) pressed flat steel plates

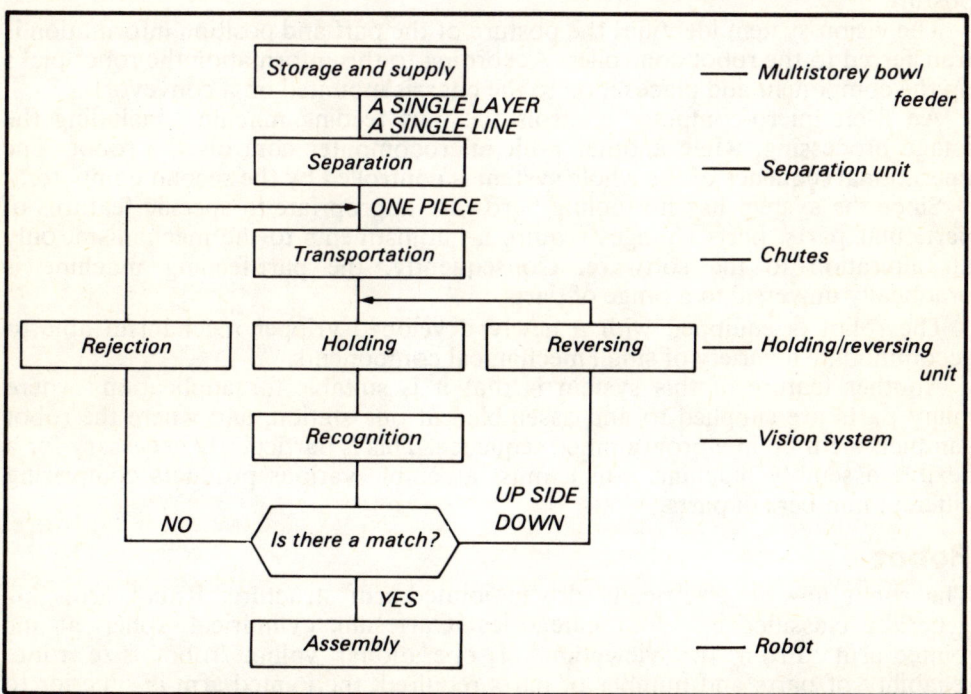

Fig. 3 Flow diagram of part-feeding and assembly operation shows the six stages of the process

ticularly necessary for a flexible assembly-machine which must assemble various products consisting of different numbers of parts.

In conventional assembly lines, which are composed of single-task stations, the change in the number of parts assembled would mean additional or redundant stations. In the Hitachi flexible assembly system, however, changes in the number of parts mean nothing more than the alteration of programmes, which, in some cases, can be done only by programme selection among multiple programmes.

The flexible assembly system, shown in Fig. 1, consists of two main components: a programmable part feeding machine and a robot. During early experiments, 13 different parts, some of which are shown in Fig. 2, were fed and assembled. They are divided into three categories: plastic mouldings with holes, plastic mouldings with shafts, and flat steel plates. A flow diagram of the machine is shown in Fig. 3.

The part feeding machine has a feeding section and a recognition section, the former being composed of a multi-storey bowl feeder, separation units and chutes, while the latter has a 256-bit linescan camera, an image processor and a part holding/reversing unit.

The parts are stored in the bowl feeder which, unlike conventional bowl feeders, has no orientation mechanisms on its tracks. Thus, the parts are fed disorientated, and then the separation unit drops one piece at a time into the chutes.

At the bottom of the chute, the separated part reaches the part holding/reversing unit. In this unit, two dimensional pushers press the component against two-dimensional datum planes, and the part settles into a predictable stable posture.

The vision system identifies the posture of the part and position information is transferred to the robot controller. According to this information the robot picks up the component and places it on to the chassis mounted on a conveyor.

An 8-bit micro-computer controls the part feeding machine, including the image processing, while another 8-bit microcomputer controls the robot. The operational sequence of the whole system is controlled by the second computer.

Since the system has no tooling hardware appropriate to specific features of particular parts, parts changes require no adjustments to the mechanism, only an alteration to the software. Consequently, the part-feeding machine is practically universal to a range of parts.

The robot is equipped with a newly developed gripper mechanism able to accommodate a variety of small mechanical components.

Another feature of this system is that it is suitable for applications where many parts are supplied to and assembled at one station, and where the robot can pick them up in a programmed sequence. This is particularly necessary for a flexible assembly machine, which must assemble various products comprising different numbers of parts.

Robot

The robot has an electrically driven jointed-arm structure. Robot arms are generally classified into four categories: Cartesian, cylindrical, spherical and jointed-arm. From the viewpoint of operational volume/robot size ratio, reliability of parts and number of parts required, the jointed-arm is superior to the others. In terms of control on the other hand, jointed-arms require the most complicated coordinate calculations.

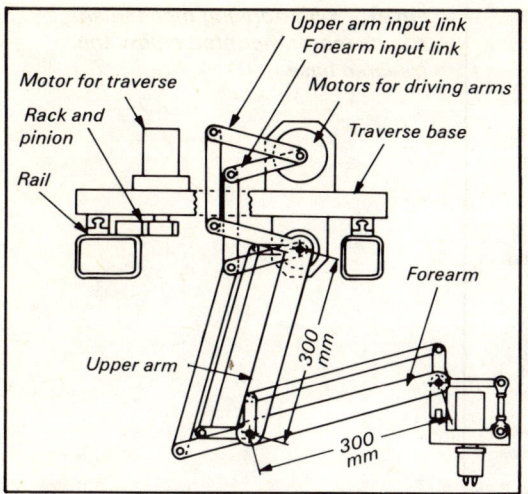

Fig. 4 Arm-driving mechanism has the motor above the traverse base

However, the recent development of microcomputers and the reduction in their cost has enabled the jointed-arm robots with fairly high functions to be made at moderate cost.

Robot drives fall into three categories: pneumatic, hydraulic and electric. The pneumatic still lags behind the others in the capability of its servo control and is used only for simple, fixed-operation robots employed as a low cost actuator. The electric drive has an advantage that power supply is widely available. The advantage of the hydraulic drive is that it can generate high power from a small actuator, but it requires intensive maintenance. For this flexible assembly system, the robot handles only small components and conventional dc servo motors are powerful enough to drive the robot.

To broaden the robot's operational volume, the arm is suspended downwards from a traverse base which can move on a pair of rails. The arm has three degrees of freedom: shoulder rotation (rotation of the upper arm), elbow rotation (rotation of the forearm), and traverse.

A wrist mechanism with three degrees of motion has been attached to this arm to give six degrees of freedom as a whole, excluding finger movement.

The mechanical configuration of the robot arm is based on parallelogram linkage mechanisms, as shown in Fig. 4. Two dc motors, driving the upper arm and the forearm, are attached to the upper side of the traverse base to reduce the weight of the arms. Driving forces are transferred to the arms through parallelogram linkage mechanisms.

The base, on which the robot arm is mounted, travels on a pair of rails and is driven by a rack and pinion mechanism from a dc servo motor.

Wrist

The mechanism for the wrist motions is also based on parallelogram linkage (Fig. 5). The motors for these motions are attached to the underside of the traverse base, and two series of parallelograms link the movement of the motor to the end of forearm.

The two-degrees-of-freedom spatial linkage mechanism (Fig. 6), is built at the end of the arm. This mechanism generates bend and/or yaw motions of the wrist according to the movement of the two linkages.

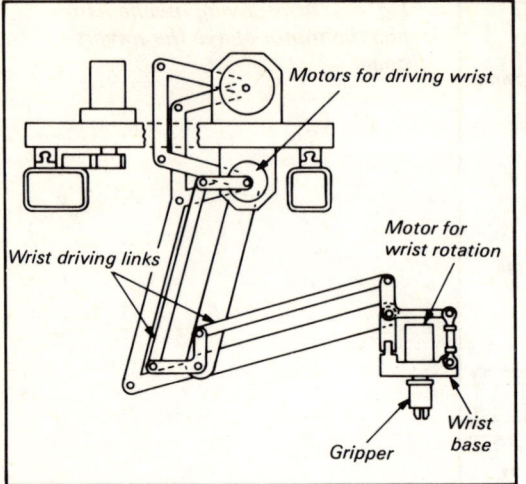

Fig. 5   Wrist-driving mechanism has its motor mounted below the traverse base

When the input links move in the same direction, the wrist bends, as indicated by black arrows in the figure. When the input links move in opposite directions to each other, or when the two input links move with a phase difference, the wrist yaws, as shown by the hatched arrows. The third motion of the wrist, rotation, is generated by a dc servo motor attached to the wrist base.

Since the motors for driving the wrist are fixed to the traverse base and the motions are transferred through the parallelograms, the wrist maintains its posture against the floor regardless of the arm position, unless the wrist motors are activated. Consequently, the posture and motion of the wrist can be controlled independently of the arm motions.

Fig. 6   Wrist mechanism uses a simple linkage to achieve motions of yaw and bend

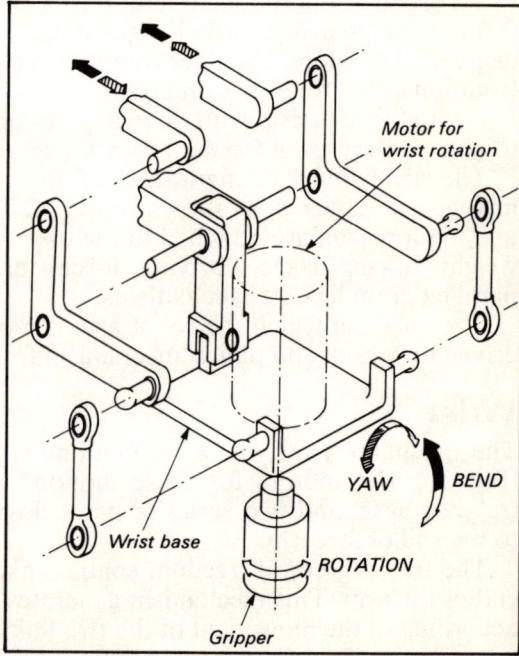

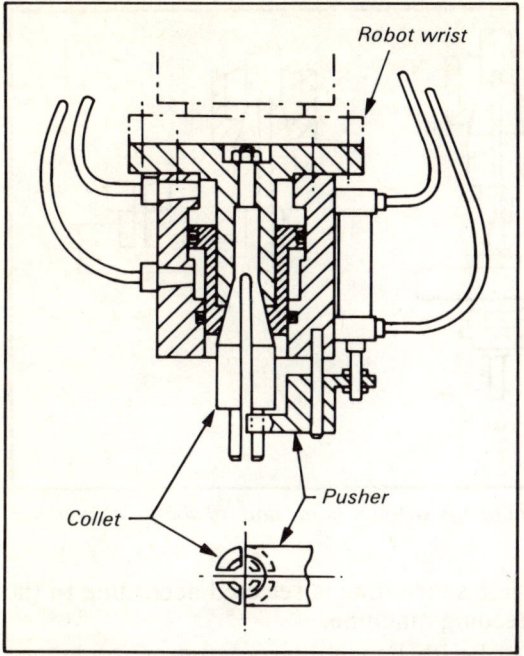

Fig. 7 Cross section through the gripper shows the collet and pusher used for collecting components

Gripper

The key factor of a flexible assembly robot is design of suitable grippers which accommodate a variety of mechanical parts. One route to coping with part differences is to prepare several grippers corresponding to the parts and to change them in the course of assembly operations. This technique has been applied to some experimental assembly systems[8,9], but it requires time to change the grippers.

Since there are no mechanical hands which are as adaptable as human hands in handling varied parts and assemblies, it is better to limit the variety of parts and design a gripper mechanism adaptable enough to handle the limited range of parts.

A gripper capable of handling all these parts without re-tooling has been designed and built; the cross section is shown in Fig. 7. It comprises a collet chuck and a double-acting pneumatic cylinder.

The parts are gripped as follows (Fig. 8): mouldings with holes are gripped in the bore by the external diameter of the collet (Fig. 8a); mouldings with shafts are gripped at the shaft by the internal circle of the collet (Fig. 8b); flat plates are gripped at slits between the collet segments (Fig. 8c).

Control

An 8-bit microcomputer controls the robot. The programmable part feeding machine has its own microcomputer to process visual information while sequential operations of the whole system are controlled by the robot's microcomputer. Both computers are linked to each other through their digital input/output interfaces, and all possible locations and postures of the parts at the part holding/reversing unit are stored in the robot microcomputer as subroutines of

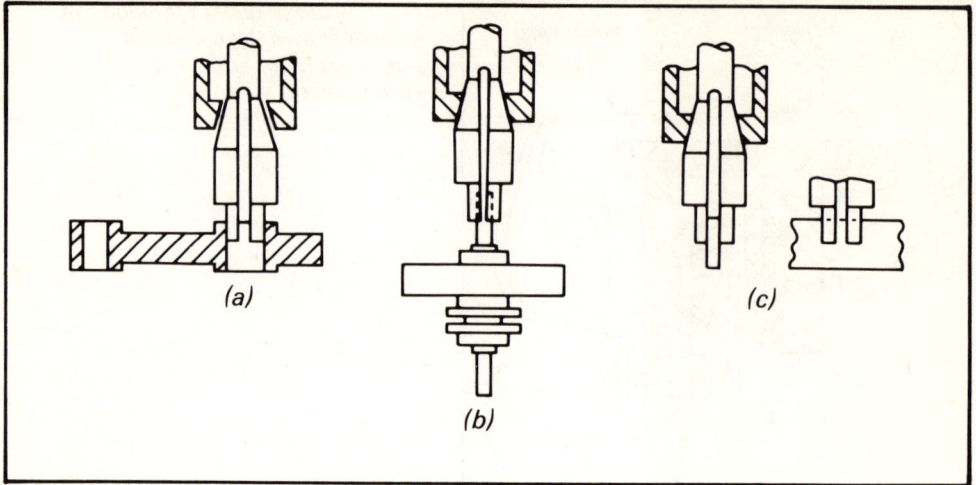

Fig. 8   The gripper can be used to hold mouldings with (a) holes and (b) shafts, as well as (c) flat plates

the motion programmes. The appropriate subroutine is selected according to the coded information sent from the part feeding machine.

The robot uses point-to-point control with linear interpolation. Circular arc paths are approximated by small segments of straight lines, and the circular interpolation can be added to the current software if necessary.

Precise operations, such as the assembly of small mechanisms, generally require very accurate motions. Accurate linear motions are needed particularly for peg-hole insertion, the basic operation in assembly, and they save labour when long straight motions of the arm need to be taught. If the linearity is accurate only a point at each end of a straight path is necessary.

To improve accuracy of linear interpolation, software servoing, in which the computer is located within the feedback loop of the servo system, has been applied to the robot control. Fig. 9 shows the block diagram of the system. At end sampling, the position and velocity signals are fed back to the microcomputer, which calculates the next outputs (i.e. the velocity signals to the servo amplifiers) from the current values, thus eliminating the deviations which are caused by the change in the dynamic characteristics of the robot mechanism and, therefore, are inevitable in conventional servoing. The sample cycle is 20 milliseconds.

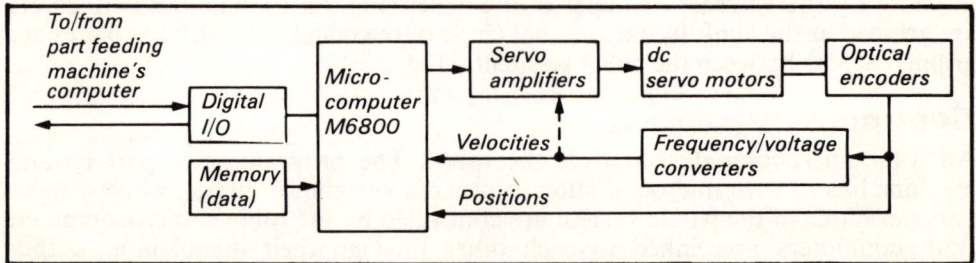

Fig. 9   Block diagram shows how software servoing is linked into the part feeding machine's computer

Fig. 10   The six-axis robot with its gripper is suspended from overhead rails

A special feature of this control system is that the velocity feedback signals are taken from the incremental optical encoders, also used to take the position feedback signals, through the frequency/voltage converters. Tacho-generators are not therefore required.

Some specifications of the robot are listed in Table 1, and Fig. 10 shows a view of it.

Part feeding

The part-feeding machine uses a multistorey bowl feeder for compactness. The feeder has five bowls and each bowl contains a different part. The function of the multistorey bowl feeder is confined to feeding parts one at a time and in single file to the separation unit. Thus each bowl is equipped with a wiper and a dish-out. The wiper and the dish-out can accommodate some change in the shape of the parts; they can be adjusted for larger changes in shape, if necessary.

Table 1   Specifications of the robot

Configuration	Jointed arm
Degrees of freedom	6
Maximum tip velocity	1000 mm/s
Arm length	
upper arm	300 mm
forearm	300 mm
Operation range	
base traverse	1800 mm
upper arm	+40°,−25°
forearm	+30°,−40°
wrist bend	±45°
wrist yaw	±30°
wrist rotation	±180°
Load capacity	2 kg
Positioning repeatability	±0.3 mm
Drive	dc servo motors
Controller	microcomputer HMCS 6800

276 PROGRAMMABLE ASSEMBLY

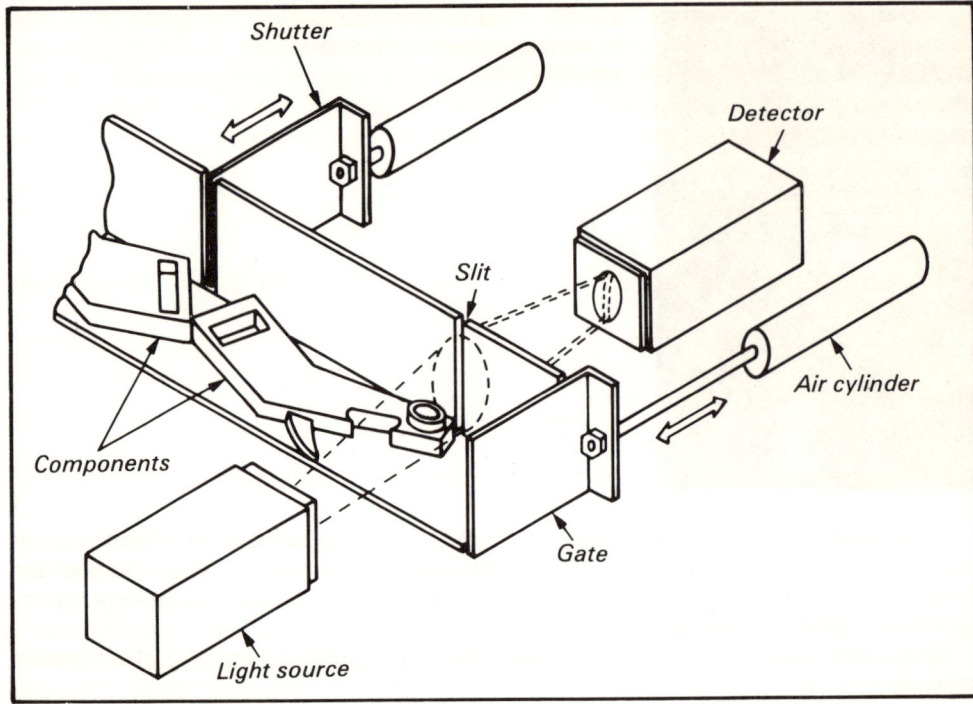

Fig. 11 Separation unit isolates one part to be assembled from another

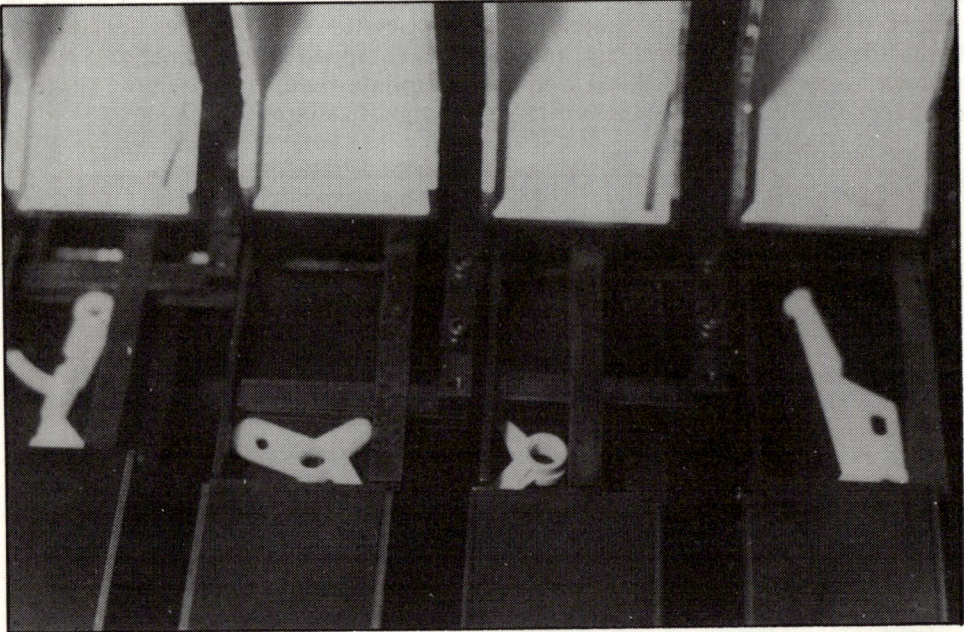

Fig. 12 The holding/reversing unit enables components to be identified or turned over if they are the wrong way up. Components can also be rejected if they are of the wrong type

The separation unit is at the exit of each bowl (Fig. 11). The parts enter into the separation unit, where the leading edge of the first part is detected by the photoelectronic sensor. The leading part of the line is then separated by the shutter from the others. Next, the gate is opened by a signal from the microcomputer and the separated part is dropped into the chute.

In this unit, the following conditions are required to ensure separation:

☐ to limit the area of detection by passing the beam through a slit,
☐ to detect by adjusting the strength of the beam.

The separation unit has almost the same facility as the bowl for accepting parts of different shape.

The part then drops down chutes from the separation unit into the holding/reversing unit. The chutes consist of vinyl tubing and aluminium square tubing. They are cheap and adaptable for size within a limited range.

The holding/reversing unit (Fig. 12) has three functions:

○ From the chute the part is pressed against the two-dimensional datum planes, firstly in the x direction, and then in the y direction, by a pair of comb-shaped pushers. These put the part into a known position to be easily identified by the image processor. Being held by the pushers, the part is checked by the robot without fail.
○ When a part is positioned the wrong side up, it is pushed into a turning case, which vertically rotates the part 180° and pushes it back to the datum planes. Then the first routine is repeated and the part is identified again. By this means all the correct parts entering the unit can be used without rejection regardless of their postures.
○ If the wrong kind of part is found in the unit the bottom plate opens and the part is rejected.

The unit can handle any part providing it is smaller than the rectangular area formed by the pushers and datum planes. It is possible to deal with five different types of part sequentially with one unit.

Vision

The vision system comprises a Reticon LC600 solid-state linescan camera and an image processor controlled by an 8-bit microcomputer (Fig. 13).

The camera travels over the holding/reversing units by means of a feed screw mechanism. An optical encoder is attached to one end of the screw and provides the image processor with the camera's location signal.

The traverse movement allows the linescan camera to compose two-dimensional images while the image processor only inputs images of the area where parts are expected. The camera's optical array consists of 256 elements spread over 70mm, providing a resolution of 0.3mm in the y direction. The screw lead is 93.3mm and the optical-encoder generates 2,000 pulses/rev. The resolving power for x direction is therefore about 0.05mm.

These resolutions are sufficient to characterise the different parts to be assembled. The analogue video output of camera is compared to a threshold level – which is programmable in 256 grades – to produce a train of binary pulses. These pulses are temporarily stored in a buffer memory and then only useful data are transferred to the microcomputer.

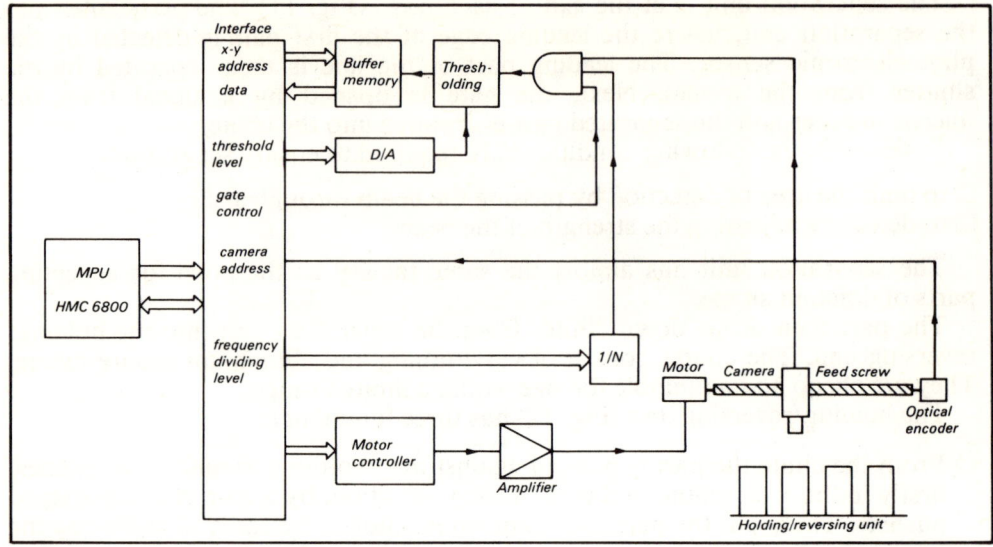

Fig. 13 Block diagram of the vision system

The method used to identify component posture is simple and practical. It is shown in Fig. 14. Since the sequence of assembling components has already been programmed, the part in question is already known. The component is in any case restricted to several postures in the holding/reversing unit.

In this example there are four possible postures of the component (Fig. 14b), which when superimposed compose Fig. 14c. If the three points (1, 2, 3) are distinguished into black and white, the posture is recognised as shown by Table 2.

In conclusion (using n points for reference), 2^n postures at maximum are recognised, including the case of no component being placed. The reference points should be set away from the edges of the components in order to avoid errors. Also to avoid errors caused by dust or noise, adjacent pixels around a

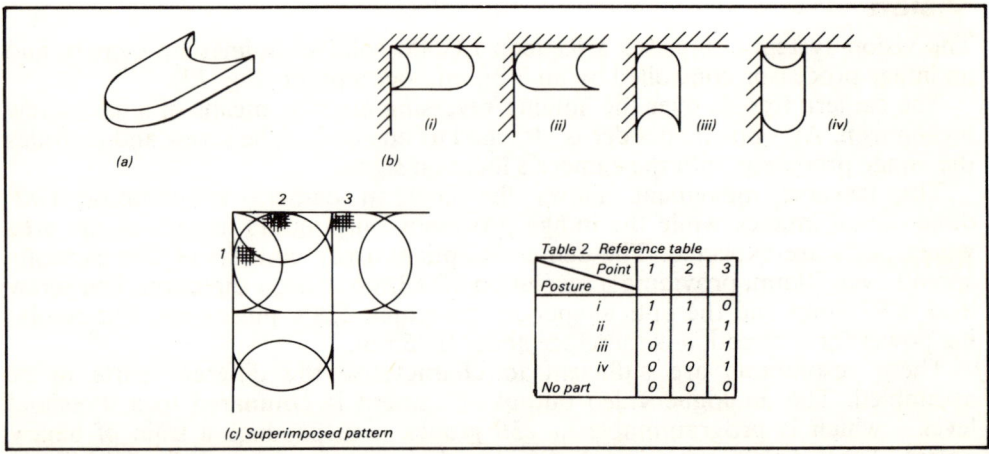

Fig. 14 Simplified parts matching system: (a) example of part, (b) possible postures of the part, and (c) superimposed pattern. The posture is recognised as shown in Table 2.

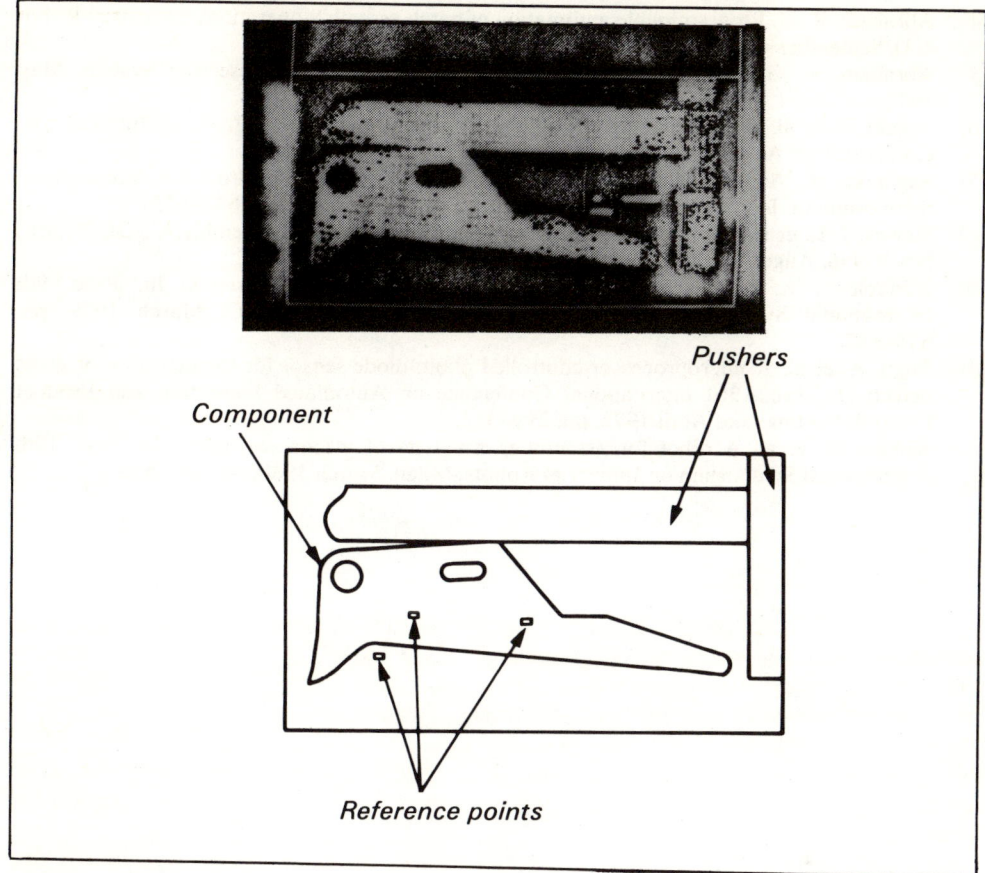

Fig. 15 Binary image of a typical part positioned in the holder/reverser

reference point are examined. If seven or more pixels appear in the same colour, black or white, the reference point is regarded as this colour. The time required for this process of the 8-bit microcomputer is less than 1ms, which does not influence the cycle time of the assembly operation at all.

The whole system has been tested through experiments of an assembly operation of a home appliance mechanism. In this experimental operation, five kinds of parts are supplied and identified completely. The feasibility of this sort of flexible part-feeding system has therefore been proved. The binary image of a part, which is held by a holding/reversing unit, is shown in Fig. 15. The three intensified dots in the figure are the reference points.

References

[1] Warnecke, H. J. et al. Pilot work site and industrial robots. In, Proc. 9th International Symposium on Industrial Robots. Washington DC, March 1979, pp. 71–86.
[2] Okada, T. and Tsuchiya, S. On a versatile finger system. In, Proc. 7th International Symposium on Industrial Robots, Tokyo, October 1977, pp. 345–352.
[3] Hirose, S. and Umentani, Y. The development of soft gripper for the versatile robot hand. In, Proc. 7th International Symposium on Industrial Robots, Tokyo, October 1977, pp. 353–360.

[4] Abraham, R. G. Programmable automation of batch assembly operations. Industrial Robot 4(3), September 1977, pp. 119–131.
[5] Abraham, R. G. State-of-the Art in Adaptable-Programmable Assembly System, May 1977.
[6] Suzuki, T. et al. An approach to flexible parts feeding system. In, Proc. 1st International Conference on Assembly Automation, March 1980, pp. 275–286.
[7] Sugimoto, K. An approach to structural synthesis of robots. In, Proc. 9th International Symposium on Industrial Robots, Washington DC, March 1979, pp. 641–655.
[8] Nevins, J. L. et al. Exploratory Research in Industrial Modular Assembly. CSDL Report, No. R-996, August 1976.
[9] Kondoleon, A. S. Cycle time analysis of robot assembly systems. In, Proc. 9th International Symposium on Industrial Robots, Washington DC, March 1979, pp. 575–587.
[10] Pugh, A. et al. A microprocessor-controlled photo-diode sensor for the detection of gross defects. In, Proc. 3rd International Conference on Automated Inspection and Product Control, Nottingham, April 1978, pp. 299–312.
[11] Kohno, M. et al. A robot for assembling a variety of mechanical parts. In, Proc. 10th International Symposium on Industrial Robots, Milan, March 1980, pp. 501–510.

DEVELOPMENT OF A FLEXIBLE PARTS FEEDING SYSTEM

K. Azuma, S. Hara and K. Hironaka,
Mitsubishi Electric Corporation, Japan

First presented at the 15th CIRP International Seminar on Manufacturing Systems,
20–22 June 1983, Amherst, USA

Abstract

The needs for flexible assembly systems have been steadily increasing in recent years. Many attempts have been made to develop a flexible assembly robot with sensors, but a practical one is not yet available.

In this paper, a method to increase flexibility of an assembly system is discussed. The most dedicated process in assembly operation is parts feeding, because an ordinary parts feeder must perform feeding and orientating functions at the same time.

Here, the function of orientation was separated from a feeder, and it was performed by a combination of a vision sensor and a robot. By reviewing product design, a practical parts feeding system was developed successfully.

Introduction

If assembly operation is divided into work elements, the operation time for each work element is usually several seconds. It is comparatively easy to develop an automatic workhead dedicated to a single work element, so an automatic assembly line which consists of such workheads is often employed, when the production volume is enough to justify the line economically. However, the recent market needs have been diversified, and an automatic assembly line has been required to cope with smaller production volume and frequent change of product design. In order to satisfy this requirement, it is desired that a workhead should cover more than one work element. A well-known approach is trying to develop a robot with sensors which can handle and assemble many kinds of workpieces[1,2,3]. Various vision algorithms have been investigated to detect one workpiece from a pile of workpieces in a bin. In particular, a number of studies have been made on estimating the position and orientation of workpieces which have a finite number of states on a plane. Considering gripper design and vision algorithm together, efficient acquisition was accomplished by a special vacuum cup hand and a simple vision algorithm which finds central portions of smooth surfaces[4]. However, these approaches are still in their early stages and have not yet been applied to practical production. Our approach has been to increase flexibility of a workhead by removing defects that make the workhead dedicated, and the following procedure has been taken:

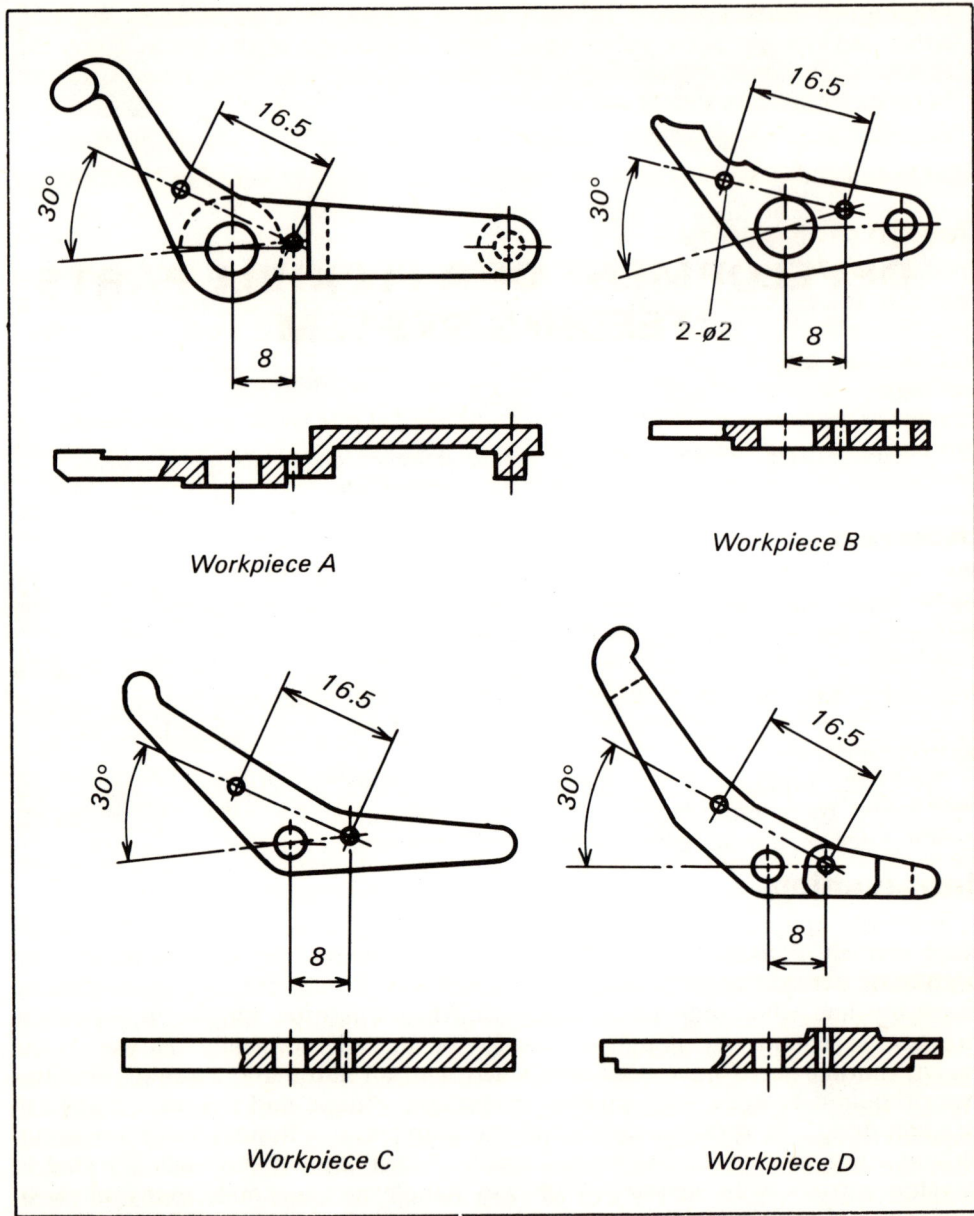

Fig. 1   Examples of the workpieces

- □ *Design review.* Design technique for automatic assembly is not firmly established, and there is yet room for improvement in product design to ease automatic assembly. Especially since universal robots are not practical for the present, it is necessary to standardise workpieces or to add common features to several kinds of workpieces for simplification of vision algorithm and gripper mechanism.
- ○ *Parts feeding.* Vibratory feeders are most commonly used to feed and orientate mechanical workpieces, but they are far from flexible. Feeder track and

orientating devices should be matched to a specific workpiece. In addition some workpieces cause jamming at the fine toolings of the track. Here, the function of orientation was separated from a feeder, and performed by a combination of a vision sensor and a robot.

Our intention is to develop a practical system; that is, a technically feasible, reliable and economical system, which can be applied to practical production.

Design philosophy

Production volume of domestic appliances is fairly large, and automatic assembly is desired. However, considering frequent change of product design, a fixed assembly line is not adequate. In the field of domestic appliances production, the tact time which is often observed is from 10 to 20 seconds. As stated previously, operation time for one work element is several seconds, so the number of workpieces which could be assembled at one workhead was estimated approximately as three. Video cassette recorders (vcr) consist of many mechanical workpieces, and as they are considered to be a typical example, plastic workpieces of a vcr were selected as an object of the flexible parts handling system. Some examples of the workpieces are shown in Fig. 1. As they have complicated features, conventional parts feeders inevitably become dedicated. In order to increase flexibility, the functions of feeding and orientating were separated. That is, a vibratory feeder only feeds workpieces one by one regardless of orientation, a vision sensor detects the position of a workpiece and a robot grips the workpiece according to the information.

Thus the construction of feeders became simple. Each component of the system should be as simple as possible to apply to actual production. The design of workpieces was reviewed at first, and two common holes were added as shown in Fig. 1. This is very useful for simplifying the vision algorithm and gripper mechanism. In order to reduce the tact time, the camera was fixed and the scanning range was masked to see only one workpiece at a time. High speed material handling was also required and a five-bar-linkage robot was employed.

System structure

The flexible parts feeding system consists of four vibratory feeders, a stage with reverser and rejecter, a vision sensor, and a 5-bar-linkage robot (see Fig. 2). A photographic view of the system which was applied to the actual line is shown in Fig. 3. Each vibratory feeder (A, B, C, D) drops workpieces A, B, C and D, respectively, on to the stage almost one by one. It starts running by a command from the main controller, and stops when a workpiece passes through photoswitches which are set at the exit of each bowl. The stage is divided into three blocks (A, B, C). Workpieces A and B are dropped on to the blocks A and B, respectively, and workpieces C and D are dropped on to block C.

The vision sensor consists of an 8-bit microcomputer, about 50 standard ICs, a solid-state linescan camera, and appropriate lighting equipment. The vision sensor identifies a workpiece on the stage, and transfers the positional information to the robot controller if it is placed face up. If the workpiece is placed face down, the stage reverses it, and if there are more than one or unsuitable workpieces on the stage, the stage rejects them.

A cross section of the robot gripper is shown in Fig. 4. It has two fingers: finger 1 is a 1.9mm diameter pin, and installed at the rotational centre of the

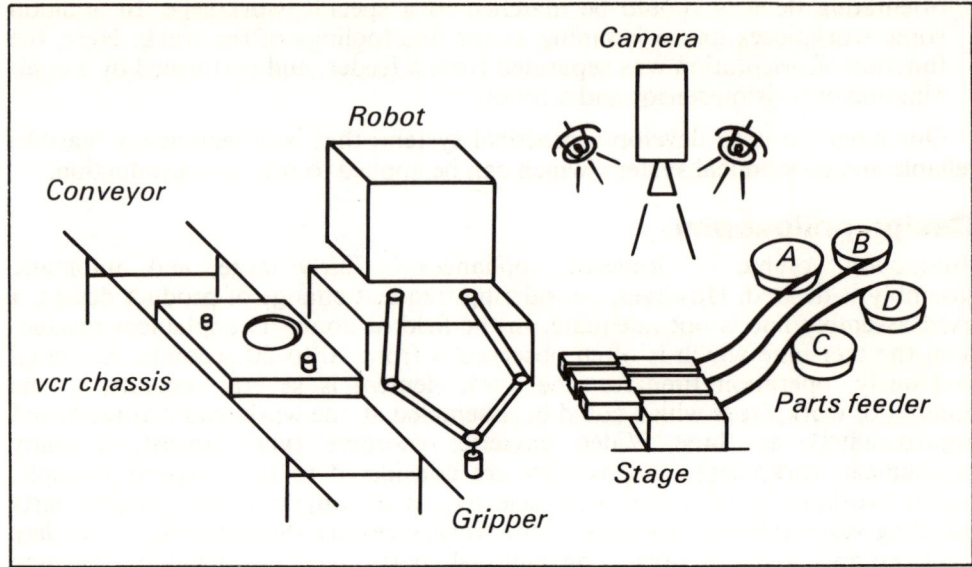

Fig. 2   Outline of the system

gripper. It assures relative positioning accuracy between a workpiece and the gripper. Finger 2 is a 1.0mm diameter pin. As it is pressed against a stopper by a piston, it keeps a constant pitch from finger 1. When they are inserted into the two holes of the workpiece, the pneumatic system automatically changes the piston into the other position, then finger 2 is pulled by a spring to hold the workpiece. The control system has a kind of hierarchical structure, and the main controller supervises a vision sensor, a robot, feeders and the stage.

Fig. 3   Photographic view of the system

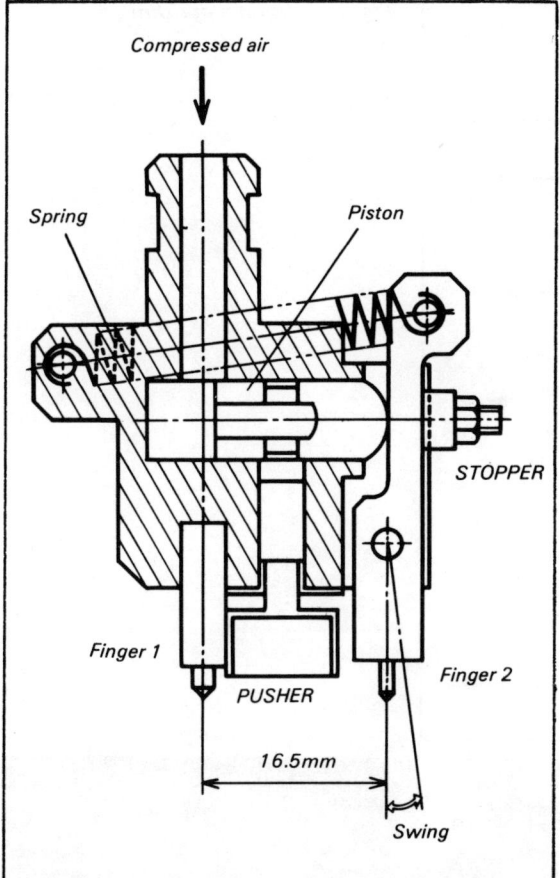

Fig. 4 Cross section of the gripper

Vision algorithm

The vision algorithm is simplified by extraction of holes. As shown in Fig. 5, distances between the left edge of the frame and the edge points of the object are measured. These edge points are gathered and recognised as lines. In these lines, a pair of lines, the length and the width of which are much the same, is regarded as a hole. The coordinates, diameters, distances, and relative angle of the holes are calculated. No. 1 hole and No. 2 hole are identified on the basis of the teaching data. Whether the workpiece is faced up or down is identified by the sign of angle α. When there are none, more than one, or unsuitable workpieces, they are eliminated through area measurement. The positional data of the two holes are transferred to the robot controller after correcting the distortion of the vision frame. The resolution is 0.5mm/pixel (242 x 256 pixels) and the time for identification is within 0.4 seconds. The maximum number of templates stored in the vision sensor is five, and the teaching-by-showing method is employed.

5-bar-linkage robot

A 5-bar-linkage robot has a fixed linkage bar and four moving linkage arms (Fig. 6). This robot has four degrees of freedom, two in the x-y plane, and has rotational and vertical motion of a gripper. In this configuration the arms are

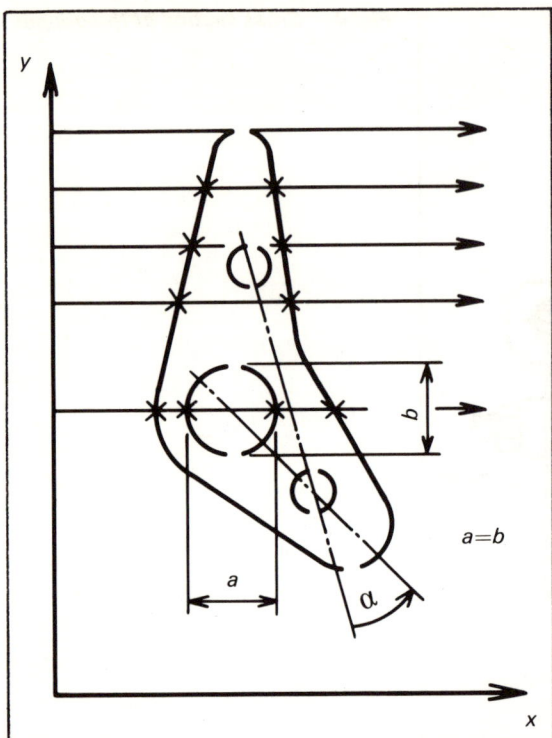

Fig. 5 Vision algorithm

Fig. 6 5-bar-linkage robot

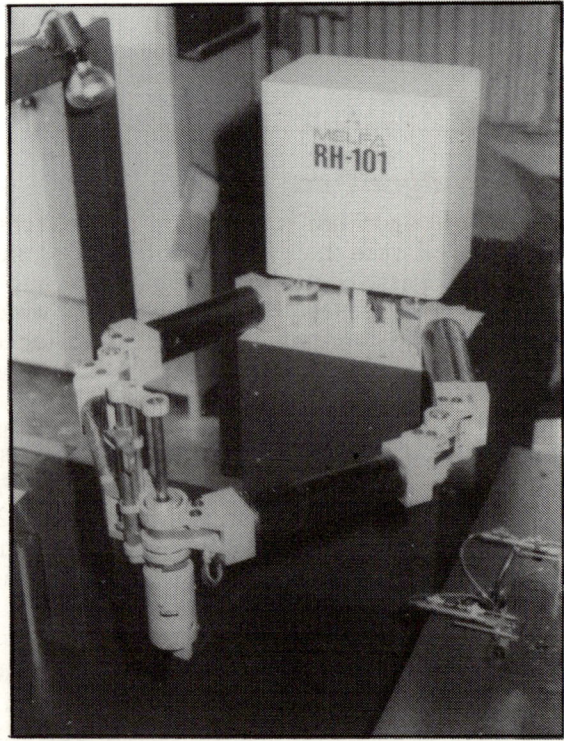

very rigid laterally, and are light-weighted because all the driving servo motors are stationary. Therefore the robot is adequate for high speed and high accuracy operation. The maximum handling load is 3kg, the maximum velocity is 2.5m/s, the maximum working area is 1370 x 780mm, and the repeatability is ±0.03mm.

Error estimation for acquisition

The robot controller calculates the target position and orientation of the gripper, on the basis of the positional data of the two holes transferred from the vision sensor. If both tips of the two fingers of the gripper are positioned within the two holes of workpieces, the gripper can insert the fingers into the holes with the aid of the chamfers and the lateral passive compliance. The maximum permissible deviation of the fingers from the holes is within 1.0mm. An estimation of the deviation is shown in Table 1.

Table 1 Error estimation

Item	Value (mm)
Resolution of the vision sensor	$\pm 0.25 \times \sqrt{2}$
Calibration error of the coordinate system	0.1
Eccentricity of the gripper	0.1
Calibration error of transformation	0.2
Pitch error of holes	±0.16
Maximum deviation	0.91

The resolution of the vision sensor is $\pm 0.25 \times \sqrt{2}$mm, because the diagonal length of one pixel (0.5 x 0.5mm) is regarded as the maximum deviation. Both the robot frame and the vision frame have some distortion errors due to coordinate transformation, or installation error of the camera, and they are corrected to be within 0.1mm. The gripper has eccentricity of 0.1mm because of its lateral passive compliance. The calibration error due to the relative coordinate transformation can be $0.25 \times \sqrt{2}$mm, but this error is improved by a statistical parameter identification to be 0.2mm. The target position of finger 1, which is at the rotational centre of the gripper, is the No. 1 hole of a workpiece, and the maximum deviation is 0.75mm. In addition, plastic mould workpieces usually have dimensional error due to non-uniform contraction, and the pitch error of holes is ±0.16mm. This pitch error is added to the deviation of finger 2 from the No. 2 hole. So, the value of the maximum deviation of the fingers from the holes was estimated to be 0.91mm, and this value is less than the maximum permissible deviation value of 1.0mm.

Assembly

The workpiece acquired by the gripper is transferred onto the vcr chassis on the conveyor and assembled one by one. The positioning accuracy required for assembly is very high. Since some kinds of workpieces have holes and projections, as shown in Fig. 1, it is required to insert the pegs of the chassis into the holes of the workpiece and the projections of the workpiece into the holes of the chassis. However, this positioning accuracy depends on the repeatability of the robot and the positioning accuracy of the chassis. All workpieces were assembled successfully with the aid of the chamfers at the tips of the above-mentioned pegs and projections, and the lateral passive compliance of the gripper.

References

[1] Hagihara, S. et al. Automatic assembly of components with flexible wire. Proc. of the International Conference on Production Engineering, Tokyo 1974.
[2] Suzuki, T. et al. An approach to a flexible part-feeding system. Proc. 1st International Conference on Assembly Automation, March 1980.
[3] Kohno, M. et al. A robot for assembling a variety of mechanical parts.

Chapter 6
ECONOMICS

Assembly is a complex process and consequently it is not easy to analyse costs and justify its implementation. The three papers in this chapter present some of the techniques now available.

EQUIPMENT JUSTIFICATION FOR AN ASSEMBLY SYSTEM

C-S. Ho,
Unimation, Inc., USA

First presented at the 15th CIRP International Seminar on Manufacturing Systems,
20–22 June 1983, Amherst, USA

Abstract

The upper bound of the cost of various equipment used in an assembly system is analysed based on the effect of the equipment on the assembly time, downtime and state transition time. In some cases, the labour cost for the same work is utilised to estimate the acceptable cost of equipment. Designing the programmable equipment with the state transition time in mind can make the programmable assembly system more competitive, when the average batch size is of the order of magnitude of the critical batch size. The basic functional requirements of each equipment are discussed.

Introduction

An assembly system is a system that positions, aligns, holds, secures and manipulates input parts and subassemblies in a specific sequence to satisfy the relative position, kinematic relationships and physical properties at the junction of parts defined on the assembly drawing[1]. For such a system, the input parts are presented to the workhead or operator at each station by using the parts feeders, magazines or bins. Each workhead or operator has a gripper or tool to perform the assembly operation. The assembly stations are linked by the conveyor or transfer mechanism. The assembly cycle at each station can be either synchronised or asynchronised. If a workhead is used at a station, then it can be a special purpose machine, a pick-and-place arm, a programmable workhead or a robot.

A programmable assembly system is an assembly system that can assemble different products through the program control. The capital investment in this type of system is relatively high but insensitive to the product change as compared with the dedicated assembly system. The production cost in this system is more stable than that of the manual assembly system. In general, the programmable assembly system is suitable for small and medium batch size production.

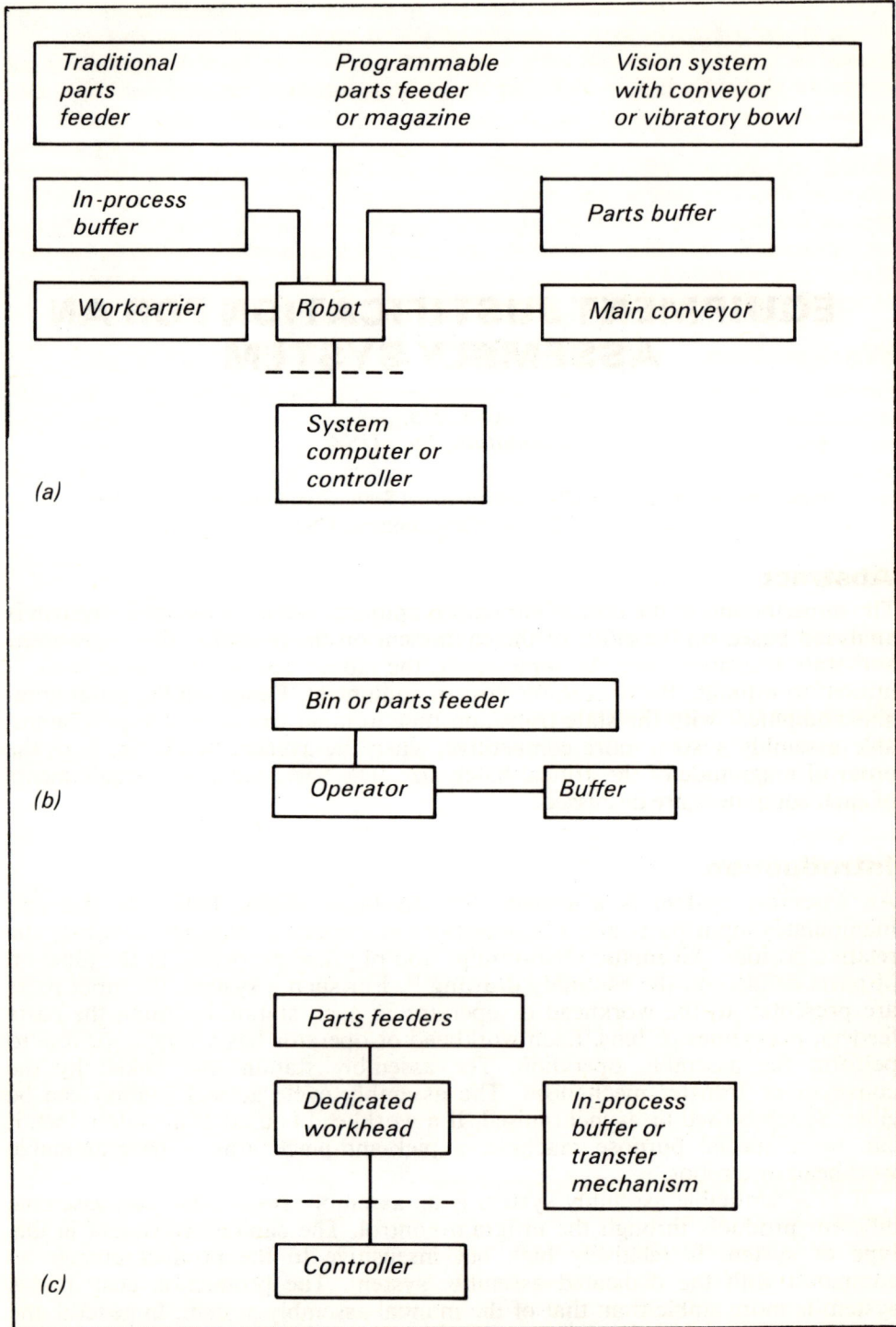

Fig. 1 Basic assembly stations: (a) robot assembly system, (b) manual assembly system, and (c) special purpose assembly station

The selection of the assembly system is mainly an economic decision. The design of an assembly system should be based on the same consideration. The economic analysis does not only identify the economic feasible domain of an assembly system but also identifies the critical technical development that can expand its feasible domain[1,2].

The justification of which equipment should be used in an assembly system should be based on the contribution of this equipment to the system productivity. The use of one particular equipment may reduce the production time or the defective rate of an assembly system. A system designer can always use information like this to estimate the maximum acceptable spending on particular equipment. An equipment designer can use the same information to design the right equipment for a specific system.

Workstation

Fig. 1a is a block diagram of a programmable workstation. The robot is capable of tracking objects on a moving conveyor indexed by an encoder. If the objects are randomly placed on the conveyor with the right side up and no touching between parts, then a vision system can be used to estimate the centroid and orientation of the objects. The use of sensors for defects checking and tools such as remote centre compliance for parts mating can further improve the performance of a programmable workstation. The use of buffers for parts can ease the difficulty due to unsteady parts supply and minimise the change-over time.

Figs. 1b,c are block diagrams of the manual workstation and dedicated workstation, respectively. These have been thoroughly discussed in previous work[1,2].

Workstation performance

The performance of a workstation depends on the speed and precision of the workhead, the quality of the parts, the assembly process and the peripheral equipment such as parts feeders, transfer mechanisms, vision systems and inspection apparatus. In general, the productivity of a system cannot be improved by simply increasing the speed and accuracy of the workhead. A designer has to have an overall view of the whole system before he makes the decision.

Another important performance characteristic of a programmable workstation is the state transition time, i.e. the time needed to start, restart, stop and recover the system. The change of the product of a system is also a state transition activity. The state transition time for product change is also called the change-over time. The effect of the state transition time is analysed later.

When a group of workstations are linked to make a production line, the performance of each workstation is no longer independent of each other[3]. The performance of a production system is analysed in the following section.

Assembly system performance

The production time, T_p, of a K-station indexing machine isolated by in-process buffer storages from other segments of the system, is given by:

$$T_p + (N N_s T_c + N K N_s P_q T_d + T_{st})/N$$

or,

$$T_p = N_s T_c + K N_s P_q T_d + T_{st}(1 + N_a P_q)/N_b \qquad (1)$$

where N is the number of acceptable assemblies in a batch, N_a is the number of parts in an assembly, N_s is the number of operations performed at a station, T_q is the time for one assembly operation or machine cycle time, P_q is the ratio of the number of assembly operations causing downtime to a number of other operations, T_d is the average downtime, T_{st} is the average state transition time per batch, and N_b is the batch size. If the whole production is synchronised, then $K N_s = N_a$.

Eqn. (1) is also true for the free transfer line, if each station is isolated by using T_d/T_c or more buffer storages[2,3] and $K = 1$. Accordingly, the equation is true for a hybrid production line consisting of I segments of K indexing stations with sufficient buffer storages between segments.

By equating the second and third terms in Eqn. (1) and solving for N_b:

$$N_b = T_{st}(1+N_a P_q)/(K N_s P_q T_d) \qquad (2)$$

Eqn. (2) shows the critical batch size for which the effect of the state transition time on the production time is as important as the effect of the faulty parts or other breakdown modes. For convenience, the N_b given by Eqn. (2) is called the first critical batch size, N_1.

By equating the first and third terms in Eqn. (1) and solving for N_b:

$$N_b = T_{st}(1 + N_a P_q)/(N_s T_c) \qquad (3)$$

This is called the second critical batch size, N_2. Since $N_s T_c$ is always greater than $K N_s P_q T_d$, N_2 is always less than N_1.

Fig. 2 shows how T_{st} and N_b affect the production time, T_p, for a typical assembly system.

For a manual assembly system, the change-over time alone normally varies from 900 to 2,700 seconds. The production time can increase dramatically for small batch production if planning and training are poor. For a manual assembly system, T_p is given by:

$$T_p = N_s T_c (1+P_q) + T_{st}(1+N_a P_q)/N_b \qquad (4)$$

The first and second critical N_b are:

$$N_1 = T_{st}(1+N_a P_q)/(N_s T_c P_q) \qquad (5)$$
$$N_2 = T_{st}(1 + N_a P_q)/(N_s T_c) \qquad (6)$$

For the dedicated assembly system, T_{st} is extremely large. This makes the dedicated assembly system unfeasible for small and medium batch size production.

For the programmable assembly system the state transition time, T_{st}, is one of the major factors in the evaluation of programmability.

The above analysis for T_p is relevant as long as $T_p \geq T_q$, where T_q, the required production time to meet the annual production volume, is given by:

$$T_q = 0.72 P_e/V_s \qquad (7)$$

where P_e is the plant efficiency and V_s is the annual production volume per shift in millions. If $T_p < T_q$, then T_q should be used as the production time. In the latter analysis:

$$T = \max(T_p, T_q) = T_p \qquad (8)$$

is assumed for the assembly cost estimation.

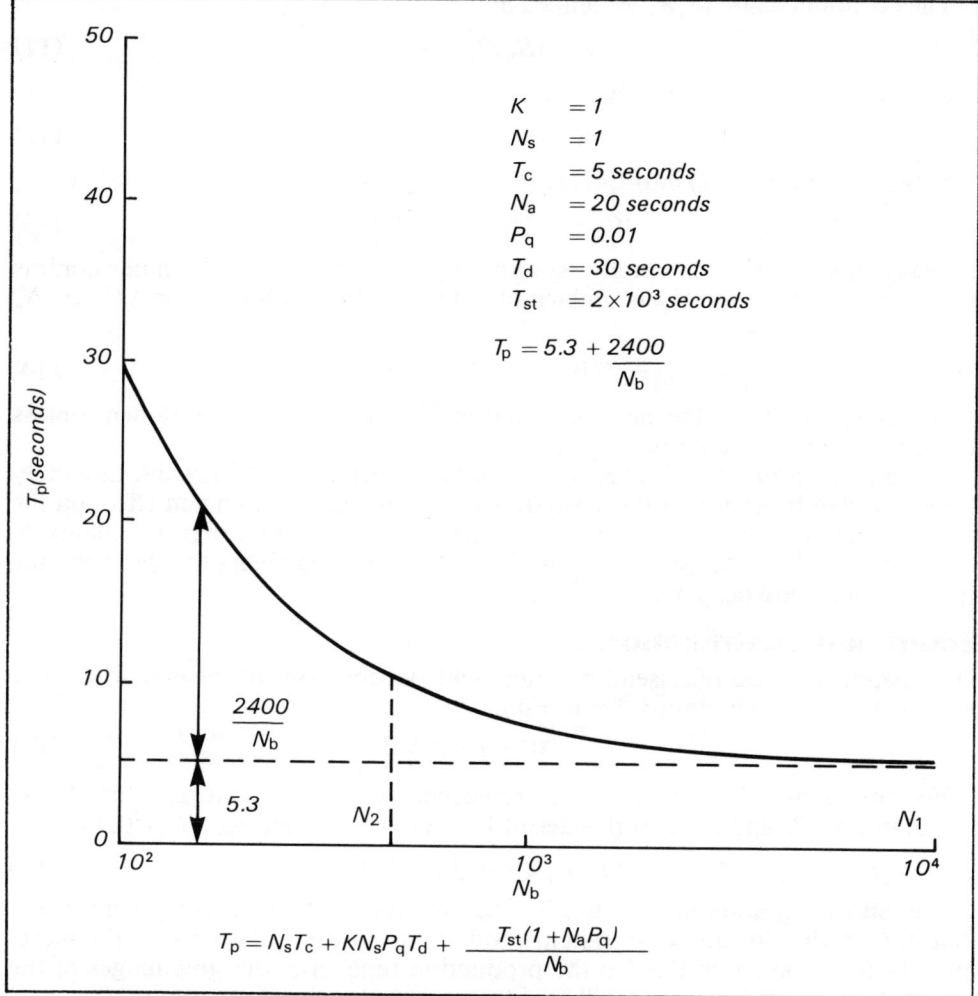

Fig. 2 Effect of batch size on production time

Economics of assembly system

Formulation of the assembly cost and equipment cost is basically the same as the work presented elsewhere[1,2]. For the continuity of the analysis, the basic equations are rederived. The detailed formulae for each system and the nomenclature are shown in Appendices A and B.

The assembly cost, C_a, can be expressed as:

$$C_a = TW \tag{9}$$

where T is defined by Eqn. (8) and W is the rate for production including the labour and equipment costs, that is:

$$W = W_t + C_e W_a/(S_h Q_e) \tag{10}$$

where W_t is the total rate of labour, C_e is the equipment cost, W_a is the annual wage of one operator, S_h is the number of shifts and Q_e is the capital investment that is equivalent to one operator on one shift[1,2].

The economic climate, R_i, is defined as:

$$R_i = (S_h Q_e)/W_a \tag{11}$$

Substituting Eqn. (11) into Eqn. (10):

$$W = W_t + C_e/R_i \tag{12}$$

Substituting Eqn. (12) into Eqn. (9):

$$C_a = T(W_t + C_e/R_i) \tag{13}$$

Using the manual assembly system as the reference, the dimensionless assembly cost per part, C_d, is obtained by dividing both sides of Eqn. (13) by $N_a W_a T_a$:

$$C_d = T(W_t + C_e/R_i)/(T_a N_a W_a) \tag{14}$$

Appendix B shows the necessary parameters to estimate the dimensionless assembly cost of each assembly system[1].

For a given product, C_d is used to identify the economically feasible assembly system. It can be seen that the state transition time has a significant effect on the economic picture of a programmable assembly system. The programmability of a programmable workhead or a robot is not fully utilised unless the tools and the system control are designed properly.

Equipment justification

The justifiable value of assembly equipment depends on the system where the equipment is to be employed. From Eqn. (9):

$$C_a = W \Delta T + T \Delta W \tag{15}$$

For the justification of the use or replacement of equipment, ΔC_a has to be less than zero. Multiplying both sides of Eqn. (15) by R_i and set $\Delta C_a < 0$, then:

$$\Delta W R_i < -W R_i \Delta T/T \tag{16}$$

The above equation means that the incremental value of an equipment is less than the product of the total system production cost and the ratio of the increment of the production time to the production time. Any design changes of the system that satisfy Eqn. (16) will lead to a cost reduction. Eqn. (16) can be used to estimate the upper bound of the cost of an equipment. Dividing both sides of Eqn. (16) by $W R_i$:

$$\Delta W/W < -\Delta T/T \tag{17}$$

Inspection equipment

Due to the accuracy, precision and reliability of the inspection equipment, 100% inspection does not guarantee that 100% of the 'acceptable' parts are really acceptable. Other factors contribute to the uncertainty, such as the fuzzy definition of the measurement or the inadequacy of the equipment to measure the desired feature.

Assuming $T = 3$ seconds, $K = 1$, $N_s = 1$, $P_q = 0.01$, $T_d = 30$ seconds and $T_{st}(1+N_a P_q)/N_b = 0$, then the amount of money that can be spent on the inspection equipment without increasing the cost can be obtained from Eqn. (17). The total inspection expenditure is $W R_i/10$ or $O(W R_i/10)$. (O means 'the order of magnitude of'.) Obviously, the indexing line will allow for more investment. In that case, the use of the in-process buffer storage becomes a potential alternative.

If $P_q=0.001$, then 1% of the total investment can be spent on the inspection equipment. The average penalty of having bad assemblies should be added to the above estimation, if the information is available.

In general, the assembly system design is product dependent. The system cannot be designed properly unless knowledge of the assembly to be assembled is available.

Programmable workhead

The speed of a programmable workhead depends on the rigidity and inertia of the workhead. The continuous path planning also affects the speed. For instance, the PUMA 550 robot can save approximately 0.75 seconds for a pick-and-place cycle, if it is in the continuous path mode instead of the point-to-point mode. The continuous path mode eliminates the unwanted stops on the path.

The typical assembly time for a programmable workhead varies from 3 to 5 seconds, if the time to transfer a workpiece in and out of the workstation is not included. For the indexing or free-transfer line, the time to transfer a workpiece in and out of a station may take 0.5 – 2 seconds. The time is determined by the slew speed, V, and the maximum acceleration, A. If $S \leqslant V/A$:

$$T_{tr} = (S/A)**0.5 \qquad (18)$$

where S is the distance and T_{tr} is the transfer time. If $S > V/A$:

$$T_{tr} = S/V + V/A \qquad (19)$$

Many people believe that the production cost can be reduced by increasing the speed of the workhead. However, this is not always true. If the speed of the workhead is too high, then the queue of machines waiting for service starts to build up, and accordingly, the production cost will increase. For a free-transfer line, the increase in ratio of $T_d:T_c$ may cause the buffer isolation to break down and result in an increase on the downtime. Obviously, there is an optimal production time for a given assembly system. The optimal production time is a function of the quality of parts, the cost of buffer (in-process storage), the average time to fix the malfunction and the labour rate. The parts with higher quality always cost more[3].

The optimal speed for the workhead can be estimated by simulation or analysis, if the basic parameters mentioned are known. A stochastic model for a group of indexing lines has been built. The profit rate is optimised by using dynamic programming.

There are other considerations in the selection of a programmable workhead, such as the force and moment required to manipulate the workpiece, the reach, the reliability and the programming language. The programming language is very important to the user. The controller that allows the user to delete, create and transfer the program during the assembly cycle can reduce the software change-over time to zero. It is important for every component in the assembly system to be able to prepare for the change-over before current production ends.

Parts feeder

The parts feeder is always a serious design problem for the assembly system. A major proportion of the downtime is due to parts jamming in the parts feeding system. The reasonable cost of a parts feeder can be estimated by equating the costs of manual handling and automatic feeding. For a dedicated parts feeder, the upper bound of the cost of a feeder, C_f, is given by:

$$C_f \leqslant W_a T_h R_i F_f / [N_p(S_v + R_d)] \tag{20}$$

where T_h is the manual handling time[1], F_f is the required feed rate, S_v is the style variation index (Appendix A) and R_d is the design variation index (Appendix A). The dedicated parts feeder is not suitable for programmable assembly because of the lack of flexibility and the long change-over time. For a programmable parts feeder, the upper bound of the cost of the feeder, C_{pf}, can be expressed as:

$$C_{pf} \leqslant W_a T_h R_i F_f \tag{21}$$

The programmable parts feeder has to be designed with the minimum change-over time in mind.

Parts presentation and inspection with vision system

If a vision system and other hardware are used for parts feeding and inspection, the cost of such a system, C_v, is bounded by:

$$C_v \leqslant W_a T_h F_f R_i + O(W R_i)/(10 N_a) \tag{22}$$

The second term in Eqn. (22) can be estimated more accurately if all parameters in Eqn. (1) are known.

The vision system identifies and locates an object by using a template or statistical data of various geometric features. The process of establishing data for recognition is called training. The training process can be done off-line to reduce the change-over time and it can also be simplified or eliminated by theoretical analysis of the digitised image[6]. Industrial vision systems cannot identify objects that are touching each other. There are other restrictions to each individual vision system, such as the contrast between the background and the object. The vision system has to satisfy all functional requirements as well as economic justification before it can be justified.

Gripper

The quick-change gripper is needed for the change-over period and on-line tool replacement in case of failure. If several parts have to be picked up by the same workhead, then the gripper has to be designed to pick-up them all. Changing grippers between assembly cycles is not practical. The constraint of the cost of a general purpose gripper is the product of the number of different parts to be handled and the cost of a dedicated gripper.

The mechanical gripper requires up to 0.2 seconds delay for the fingers to respond to the grasp or release command. The vacuum tool does not require the delay for the response. However, the vacuum tool is not suitable for all parts.

The gripper with parallel fingers can secure the part better than the one with toggle fingers. The tangent of the half angle between the two toggle fingers has to be less than the coefficient of friction. Design of the finger tip can affect the reliability of gripping. The use of the notch can eliminate the relative positional and angular error between the part and the workhead (Fig. 3). This is the only design where a designer can expect an improvement on accuracy with negligible expenditure.

Work carrier

An assembly which does not require a work carrier is most suitable for programmable assembly. If the time to reprogram a programmable carrier is no more than the production time, then the upper bound of the cost of a programmable carrier is given by:

$$C_{pc} \leqslant N_p S_v C_c \qquad (23)$$

where C_{pc} is the cost of a programmable carrier, N_p is the number of products and C_c is the cost of a dedicated carrier.

System computer

The use of a system computer can reduce the state transition time to 10 seconds by coordinating various activities of an assembly system. For an average batch size of 500, $T_{st} = 1000$ seconds and $T = 10$ seconds, the upper spending bound on the system computer is 20% of the present system production cost [Eqn. (17)]. For a system with $N_b = 200$, $T_{st} = 1000$ seconds and $T = 10$ seconds, this increases to 50%.

The system computer does not only coordinate various production activities, but also plays a role as the interface between the operators and a complex production system. The computer can make the system easy to operate and prevent human errors.

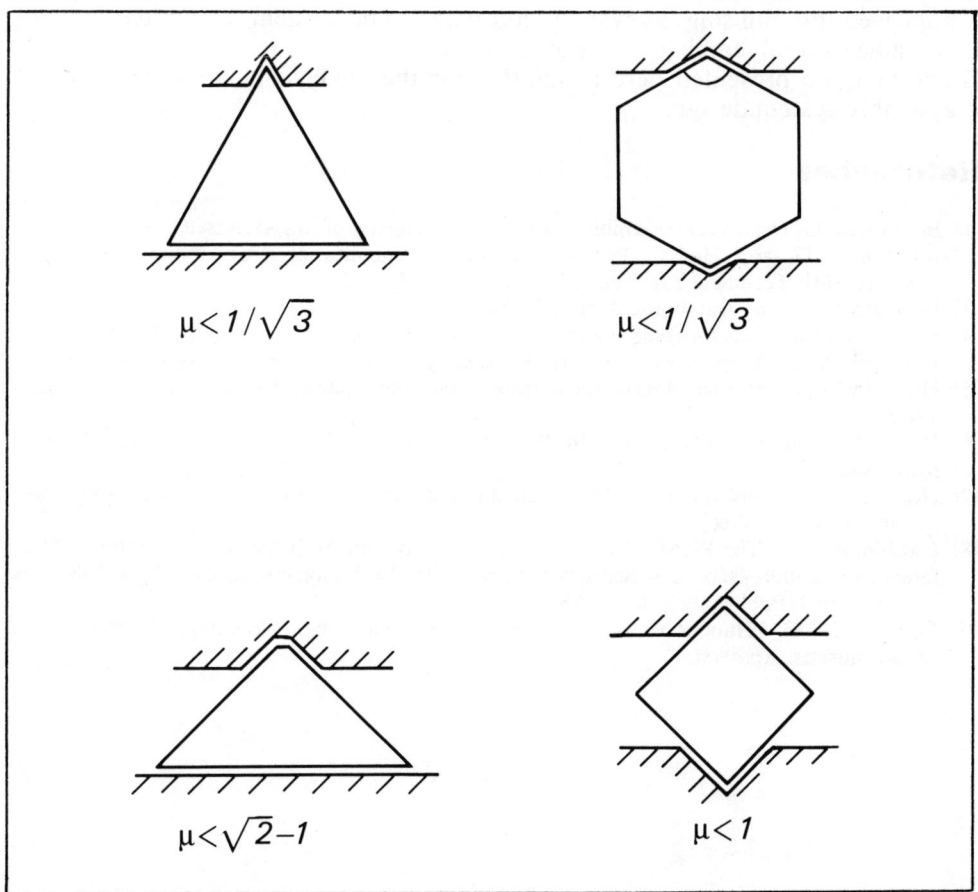

Fig. 3   Notches on finger tips: μ is the maximum coefficient of friction allowed for self-align action to occur

Conclusions

The important points concluded from this study can be summarised as follows:

- The justifiable value of an equipment depends on:
    the effect of equipment on production time,
    the original production cost,
    other alternatives such as using human operators,
    utilisation of equipment (e.g. quick-change hands), and
    reliability (e.g. notches on finger-tips).
- The effect of the state transition time should be considered for small batch assembly operations [$N_b = O(N_1)$].
- The design of an assembly system is product dependent. The best assembly system that can possibly be designed is also a measure of the quality of the assembly and its design.
- The production cost does not always decrease as the speed of the workhead increases.
- The small positional and angular errors can be filtered out if the gripper is designed properly.
- The state transition response and the man–system interaction can be improved by utilising a system computer. The system computer is very valuable to small batch production systems.
- The analysis presented here is suitable for the analysis in the early stage of assembly system design.

References

[1] Boothroyd, G. Design for Assembly Handbook. University of Massachusetts, Amherst.
[2] Boothroyd, G. and Ho, C. Performance and Economics of Programmable Assembly Systems. SME Technical Paper AD77–720.
[3] Boothroyd, G. et al. Automatic Assembly. Marcel Dekker, 1982.
[4] Ho, C. Error analysis for robot assembly systems. In, Proc. 4th International Conference on Assembly Automation, Tokyo, October 1983, pp. 243–254. IFS (Publications) Ltd., 1983.
[5] Hill, J. W. Dimensional Measurement from Quantized Images. SRI Project 4391, August 1980.
[6] Ho, C. Precision of vision system. In, Proc. IEEE Machine Vision Workshop, pp. 153–159, May 1982.
[7[Ho, C. Effect of linkage errors on manipulator accuracy in non-joint oriented coordinate system (to be published).
[8] Carlisle, B. et al. The PUMA/VS-100 robot vision system. In, Proc. 1st International Conference on Robot Vision and Sensory Controls, Stratford-upon-Avon, UK, April 1981, pp. 149–160. IFS (Publications) Ltd., 1981.
[9] Boothroyd, G. Handbook for Automatic Feeding and Orienting. University of Massachusetts, Amherst.

Appendix A – Nomenclature

Equipment costs:
- C_a — cost of assembly
- C_b — cost of buffer spaces or transfer device per station for a free-transfer machine
- C_c — cost of work carrier
- C_e — cost of equipment
- C_f — cost of feeding device and delivery track
- C_g — cost of gripper per part in assembly
- C_m — cost of magazine
- C_{pf} — cost of programmable feeder
- C_{pc} — cost of programmable workcarrier
- C_{sc} — cost of system computer
- C_t — cost of transfer device per station for an indexing machine
- C_u — cost of universal gripper
- C_w — cost of special purpose workhead
- Q_e — capital expenditure to replace one assembly operator on one shift

Production
- F_f — required feed rate
- P_e — plant efficiency
- N — $N_b/(1 + P_q)$
- N_1 — first critical batch size
- N_2 — second critical batch size
- N_b — number of products in one batch
- N_i — number of operators additional to supervisor (in-line machine)
- N_r — number of operators additional to supervisor (rotary machine)
- N_s — number of operations performed at a station
- V_s — annual production volume per shift
- S_h — number of shifts
- T_{tr} — transfer time between two stations
- W — total rate for production, $W = W_t + C_e W_a/(S_h Q_e)$
- W_a — annual wage of one operator
- W_s — annual wage of one supervisor
- W_t — total wage

Product
- P_q — ratio of faulty parts to acceptable parts
- N_a — number of parts in the assembly
- N_d — number of design changes in first three years
- N_p — number of different products to be assembled in first three years
- N_t — total number of parts available for building different product styles
- R_d — N_d/N_a, design stability
- R_i — $S_h Q_e/W_a$, investment potential
- S_v — N_t/N_a, style variation index

302 PROGRAMMABLE ASSEMBLY

Time

T max $\{T_p, T_q\}$
T_a manual assembly time per part
T_c time of one assembly operation at a given station
T_{co} change-over time
T_d machine downtime due to faulty parts
T_h manual handling time
T_p production time
T_q required production time to meet the annual production volume
T_m manual assembly time per part with mechanical assistance
T_r assembly time per part with robot or programmable workhead
T_w assembly time per part with special purpose workhead
T_{st} average total state transition time per batch

Appendix B – Modified version of UMASS model

Special purpose indexing line

$T_p = T_w + N_a P_q T_d + T_{sti}(1+P_q)/N_b$
$W_t = N_r W_a + W_s; \qquad N_a \leqslant 6$
$W_t = N_i W_a + W_s; \qquad N_a > 6$
$C_e = N_a\{S_v C_t + N_p[S_v C_c + (S_v + R_d)(C_f + C_w)]\}$

Special purpose free-transfer line

$T_p = T_w + P_d T_d + T_{stf}(1 + P_q)/N_b$
$W_t = N_i W_a + W_s$
$C_e = N_a[S_v(1 + T_d/T_w)(C_b + N_p C_c/2) + N_p(S_v + R_d)(C_f + C_w)]$

Free-transfer line with programmable workhead

$T_p = N_s(T_r + P_q T_d) + T_{stp}(1 + N_a P_q)/N_b$
$N_s = T_q/(T_r + P_q T_d)$
$W_t = N_i W_a + W_s$
$C_e = [C_r + (1 + T_d/T_r)C_b]/N_s + N_p[(S_v + R_d)C_m + C_g + (1 + T_d/T_r)C_c/(2N_s)]$

Two-arm robot assembly centre

$T_p = N_a(T_r/2 + P_q T_d) + T_{str}(1 + N_a P_q)/N_b$
$W_t = W_s$
$C_e = 2C_r + N_p[C_c + N_a C_g + N_a C_m(S_v + R_d)]$

Manual assembly line

$T_p = N_s T_a(1 + P_q) + T_{sta}(1 + P_q)/N_b$
$N_s = T_q/[T_a(1 + P_q)]$
$C_e = N_a(2C_b + N_p C_c)/N_s$

If the feeders are used:

$T_p = N_s T_m(1+P_q) - T_{stm}(1+P_q)/N_b$
$N_s = T_q/[T_m(1+P_q)]$
$C_e = N_a/N_s[2C_b + N_p C_c + N_s N_p(S_v + R_d)C_f]$

Universal assembly centre
$$T_p = N_s(T_r/2 + P_q T_d) + [T_{stu}(1+P_q)/N_b]$$
$$N_s = T_q/(T_r/2 + P_q T_d)$$
$$W_t = W_s$$
$$C_e = N_a/N_s(2C_r + 2C_u + C_c) + N_a C_p S_v$$

AUTOMATIC OR MANUAL ASSEMBLY? BOUNDARIES OF ECONOMY AT MIDDLE OR LOW BATCH PRODUCTION

D. Elbracht and H. Schacher, University of Duisburg, West Germany

First presented at the 3rd International Conference on Assembly Automation and 14th IPA Conference, 25–27 May 1982, Boeblingen, West Germany

Abstract

The labour costs of selected industrial companies are prognosticated to 1990. Investments, for the future economic acceptance of replacing human power by automatic production result from this. In competition with countries of low wages, investments for automation have to be much lower.

The possibilities and borders of economic assembly under conditions of investments are discussed. The problems of small batch assembly are commented on mainly in this field. Economical borders within the production of automotive engines are discussed, comparing them with structures of manual, semi-automatic and automatic assembly. This comparison leads to the result that in the laboratory, distinctly lower assembly times may be reached – especially by methodical training of sensomotoric handling – than industry is now calculating. From this we can deduce quite different economic borders for automatic assembly.

In this context the possibilities of using flexible assembly structures with the help of inductively controlled industrial trucks within the production of automotive engines are discussed.

Finally, some possibilities of economic small batch assembly with the assistance of industrial robots are detailed. Mobile and inductively controlled industrial robots (MOBIROBS) and their districts of economic work are discussed.

Introduction

All of us who are involved in assembly know that the automation of assembly procedures is very difficult and is mostly overcome only with tremendous expense in time and money. On the other hand our impression is that the permanent increase of labour costs gives no alternative but to automate production and, above all, assembly.

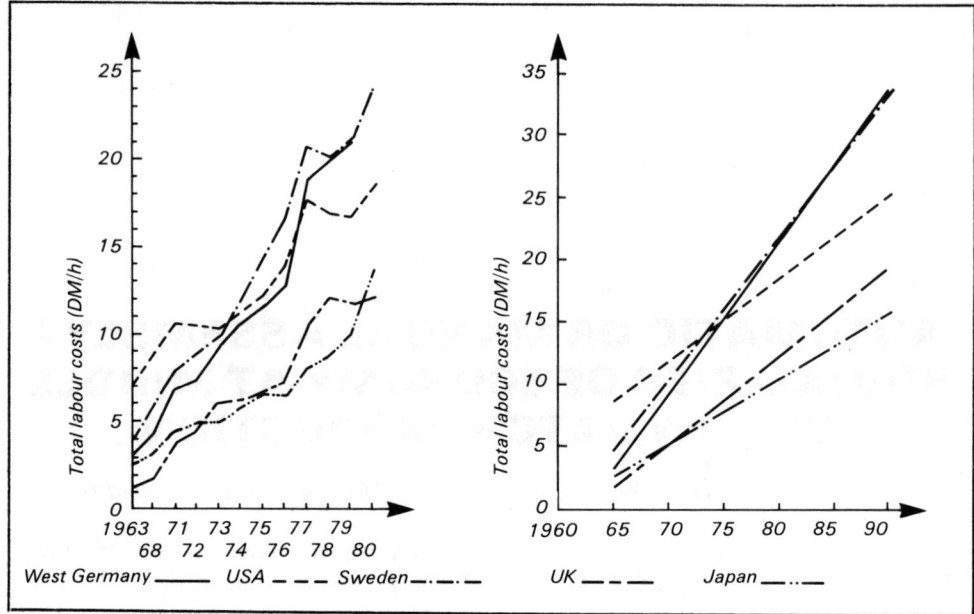

Fig. 1 The international development of labour costs since 1963 and a linear extrapolation to 1990

If we look closely at the efficiency of manual, mechanised or automated assembly, there are very clear boundaries concerning the different steps of an economic automation with respect mainly to middle and small batch production. This problem is analysed in detail.

Cost dependence

The cost of assembly essentially depends on labour costs, not only because of the usually very high proportion of manual work within the assembly process, but also for its automation – the economy of which depends on wages.

Fig. 1 shows the international development of labour costs since 1963, from which a prognosis up to 1990 is taken. In West Germany, it is expected that wages will not increase as much as in the past ten years. Nevertheless that cannot be said for all the highly industrialised countries, especially Japan and the USA.

The limit of investment for substituting human hands with automated production can thus be deduced. The investments needed to replace one worker on one- or two-shift work can then be calculated. Total costs from these investments should be much lower than the costs for manual work – including the expense for the equipment needed in the manual workplace.

Economy and rentability with respect to future wages

Fig. 2 shows that today in West Germany the greatest amount acceptable for automation of a workplace being used in one shift is about DM 78,000 when calculated with two years return on investment (ROI). Using the same workplace in two shifts this amount increases to about DM 140,000. If scales of rentability are used as, for instance, the automotive industry is using, the investments acceptable for overcoming one workplace are much lower. Targets for

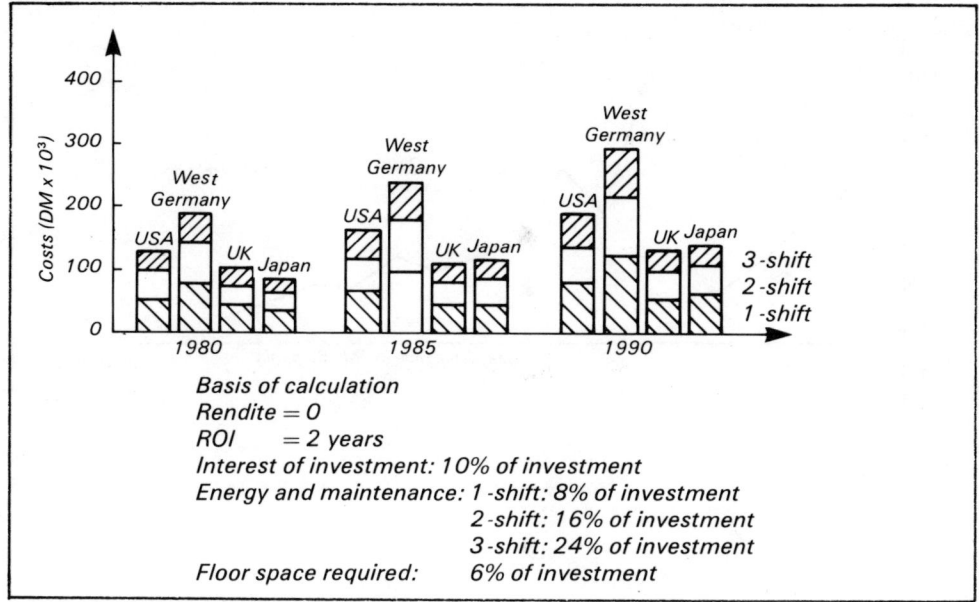

Fig. 2   Limit of investment for replacement of one human hand

rentability in the automotive industry are normally about 25% including two years ROI.

A second conclusion can be derived from Fig. 2: different industrial countries are forced into different investments. For instance the Japanese are able to invest only about DM 36,000 for replacement of one workplace in one-shift production and some DM 65,000 when working in two shifts (calculated with two years ROI). Supposing that a German producer wants to compete with a Japanese producer in automation, he must be able to do this with the investment for automation that corresponds to the Japanese conditions. Otherwise there is no chance to produce with comparable costs.

On the other hand prices of machines and equipment for automation are also increasing each year, but not in proportion with wages. Consequently, we will economically master a somewhat higher degree of automation in the coming ten years compared with our situation today. Evidently this will be applicable for all countries with rising rates of productivity, especially Japan.

The special position of assembly in production

Assembly has had up to now, and will also have in future, a prominent position in all intentions for rationalisation. The three reasons for this are:

☐ The share of human power in assembly is much higher than in most of the different manufacturing processes. This is because of the high advantage of man in sensomotoric work in relation to mechanised production.
☐ The replacement of human power is therefore possible in those jobs only by relatively high investments.
☐ The present tendency of individual and small batch production is strengthening this problem. If we need a flexible automatic assembly machine for different tasks it must have a much higher degree of complexity. Hence the cost for automation is increasing once more.

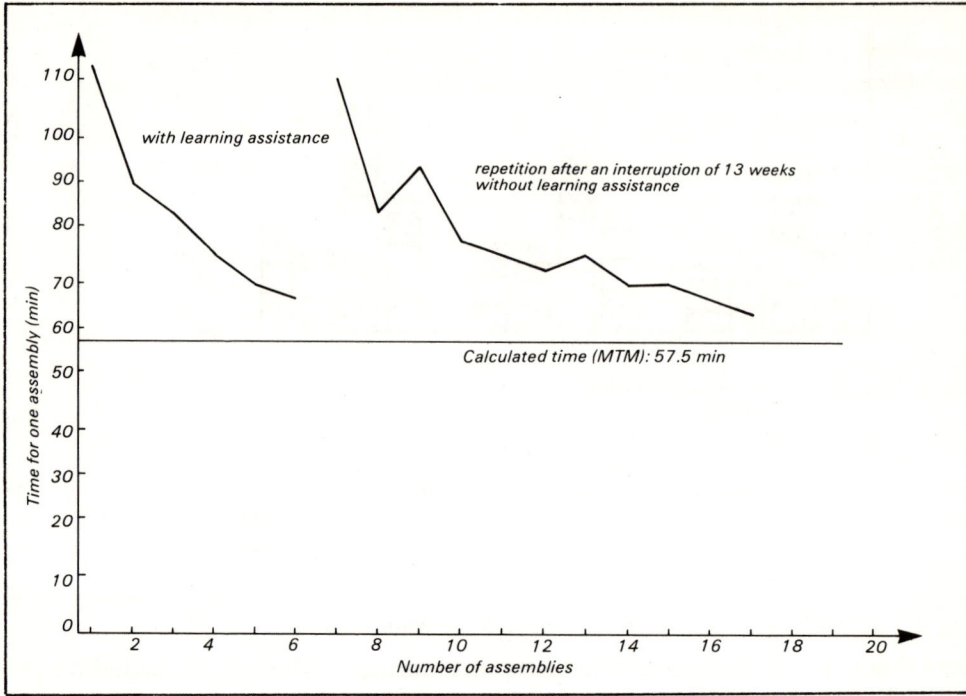

Fig. 3   Results of training in the manual assembly of a complete engine

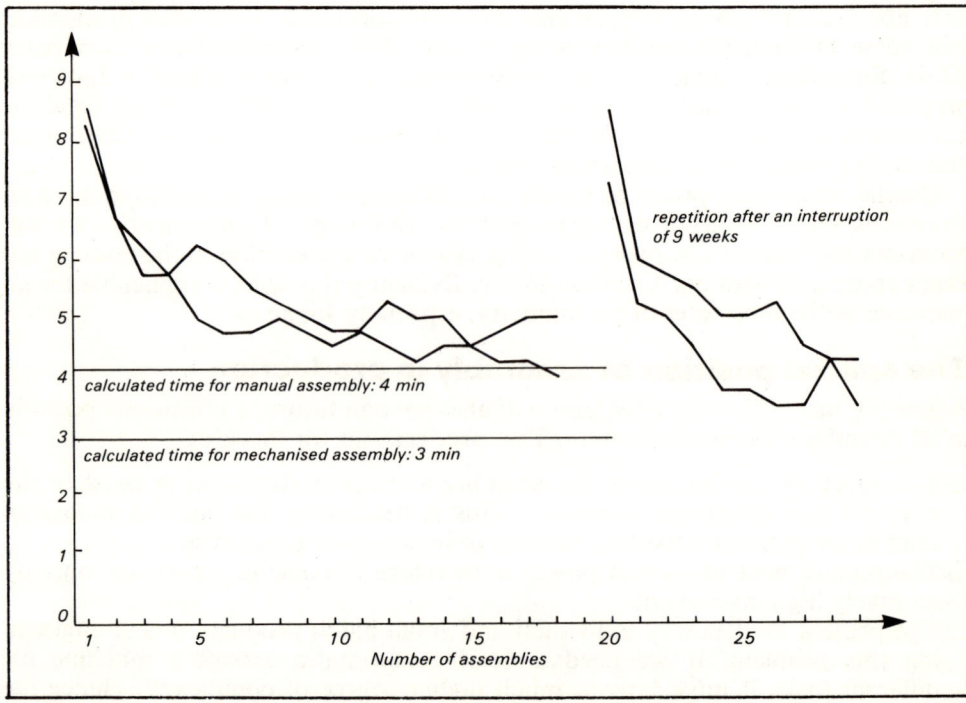

Fig. 4   Progress in learning the assembly of an electrical contactor

Possibilities and boundaries of economic assembly

Economic methods of assembly

Installations for assembly may be manual applications (assisted by tools), semi-automatic stations (partly mechanised), or automatic stations. Manual assembly is proved best at small batch production. High batch production is realised as a combination of the three types of assembly installations.

The following criteria are important for assembly automation, generally, and especially for differentiation of small batch production from high batch production:

○ Content of work (assembly time for one unit).
○ (Possible) degree of automation.
○ The usability of automatic assembly stations.
○ The usability of industrial robots.

Content of work in manual assembly

There has been a trend towards a greater content of work (set by legislation) since the beginning of the 1970s, however there is neither a human nor a commercial interest to enlarge the content of work.

Let us look, for instance, at the assembly of automotive engines. In a well known humanisation project, the workers preferred to assemble a whole engine on their own instead of sharing the work. Instead of the usual 30 seconds work content, in this project workers decided on a content of about 80 minutes. We surely all agree that those large work contents do not correspond with the aims of humanisation. Let us have a closer look at the facts.

Every year a German automobile company produces 1.5 million engines in 230 different types. The annual production numbers of the variants differ from 10 to 400,000, but these 400,000 are far from identical. A typical motor family may itself be subdivided into some 30 variants. Depending on the market, special engine types can change from 10 to nearly 60% of total production volume within only two years. When different variants are to be assembled a lot of training is required, especially in the case of variants seldom produced. Fig. 3 shows the problems of training when assembling those engines in our laboratory. After an interruption of three months, the same training efforts as before are required.

Fig. 4 shows that the training time is shorter when using mechanised assembly installations. The reason for this can be that the number of moving parts is much lower, with the result that less conscious effort is required and it would be reproduced with smaller mental effort after longer periods of interruption.

Furthermore, when frequently changing assembly jobs, intensive training in the laboratory significantly reduces the time necessary for training (Fig. 5). When an experienced 'teacher' gives hints on the forthcoming work (e.g. on the tool to use), calculated time for assembly of an engine may be reached after 10–15 assemblies, including a content of work between 6 and 12 minutes.

Comparable with these results is the experience of industry, where a training time of 8–24 hours is needed for a content of work of about one minute.

From our method laboratory results we think that, because of the training time, contents of work greater than ten minutes must be out of economical interest. They also depend on the frequency of change of series, the complexity of work and the expected interruption until repetition of this work.

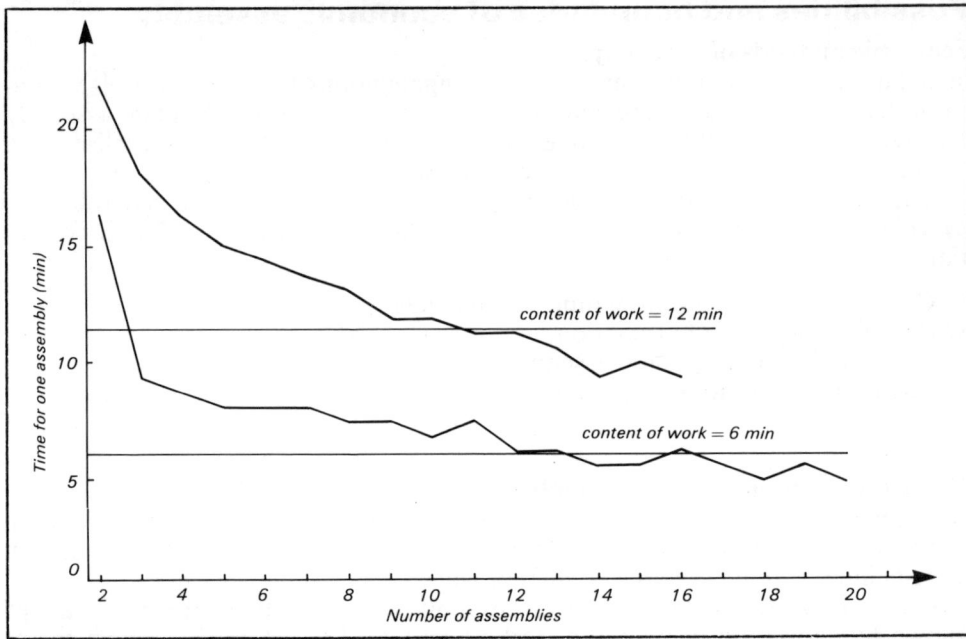

Fig. 5   Progress in learning by methodical teaching assembly of automotive engines with different content of work

Limits of the degree of automation

The economical conditions for automation compared with manual assembly are now discussed using three examples.

The influence of using assembly devices in comparison to manual assembly by simple tools is shown in Fig. 4. In this test every person has repeated the same assembly nearly 30 times. A comparison is made of the performance by using assembly devices with mechanical aids (e.g. pneumatic screwdrivers) and manual assembly on a desk. This example shows that the content of work in assembly depends on different facts. We do not recommend to choose a content of work as large as possible because of supposed profits in humanisation. The content of work has to be greater than the minimum set by legislation but lower than the boundaries given by economic calculations. These economic calculations must include the investment for workplace, the performance expected and the costs of training.

As previously mentioned, the boundaries for automatisation of assembly are given by the investments, the complexity of assembly, the lot sizes and so on. This is explained using the next examples.

Consider a batch test made on the assembly of electrical contactors. For this test an assembly system, recommended by Bosch, working with palletised carriers on a double-belt conveyor system is used. The initial costs for the whole system were:

Basic equipment:	DM 36,300
Special equipment for three stations:	DM 34,800
Workplace arrangement for three stations:	DM 15,960
60 palletised carriers:	DM 36,000
	DM 123,060

The conveyor system improves the performance by almost 10% (Fig. 4). Some calculations for economical comparisons can now be made. Regarding the 10% improved performance, the maximum investment should be DM 42,000 (including two years ROI, three workstations and two shifts; calculation method as above).

A second example uses the assembly of automobile engines. The automotive industry is now projecting the assembly of engines in sections. In the first section, automatic assembly stations assemble a so-called 'basic motor'. Then several manual and partially mechanised sections complete the motor. In the same way the company tries to assemble components automatically like cylinder heads. The problem with these procedures is the inflexibility of the installations to cope with:

☐ Changes in construction of the engine.
☐ Changes in production quantities (quantities have to be guaranteed).
☐ High initial costs (in this case between DM 200,000 and DM 250,000 for one station).
☐ Technical problems on time sequences as low as 11 seconds (the problem is the accuracy of position in the automatic station).

Flexible assembly structures can supersede the inflexibility of automated assembly sections. Different kinds of assembly stations may be connected, in different sequences, using automated wire-guided vehicles (Fig. 6). These assembly stations may be manual, partly mechanised or automatic. The economical and technical problems of such flexible assembly structures using inductively guided vehicles are:

Fig. 6   Flexible assembly structure using automated wire-guided vehicles

PROGRAMMABLE ASSEMBLY

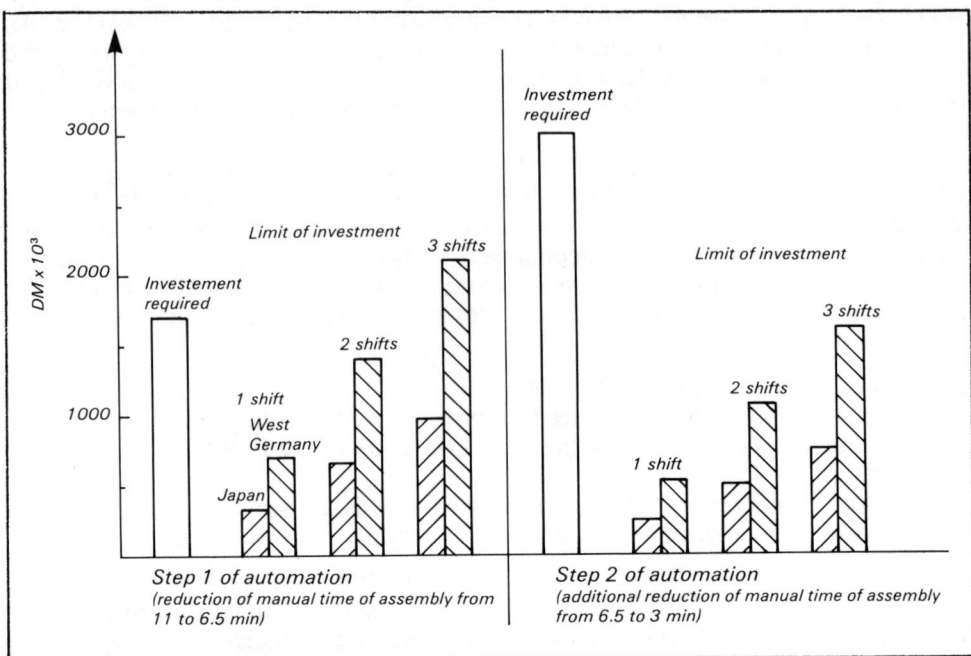

Fig. 7 Limit of investment and investment required for automation of assembly of automotive engines in Japan and West Germany

○ The accuracy of positioning the vehicle in the automatic stations and the accuracy of positioning the workpiece-carrier by handing it over to the stations.
○ The expensive equipment of several parallel workplaces with the desired high degree of automation.

Economic comparison of selected assembly methods of automobile engines
Some economic comparisons between different assembly steps in the assembly of automobile engines (work content about twelve minutes) are now made:

☐ Manual assembly of engines (experimental results).
☐ Calculated times for these manual assemblies (MTM method).
☐ Calculated time of automotive industry for these assemblies.
☐ Assembly time needed for automatic assembly.

Fig. 7 shows that in our example step 1 of automation becomes economic only when two shifts are in use. For the assembly path mentioned only 6.5 minutes out of this assembly are today made manual in industry. Investments for this first step of automation should be some DM 1.7 million. Also shown in the figure is the limit of investment for West Germany and Japan for one-, two- and three-shift operation.

The second step of automation needs an investment of about DM 3 million. The part played by manual assembly would be reduced to three minutes. However, there is no rentability. In addition, Fig. 7 shows that investments for rationalisation must be evaluated differently in different countries. Step 1 of automation may be economical in West Germany when working in two shifts, but not in Japan. On the other hand it is recognised that significant changes in

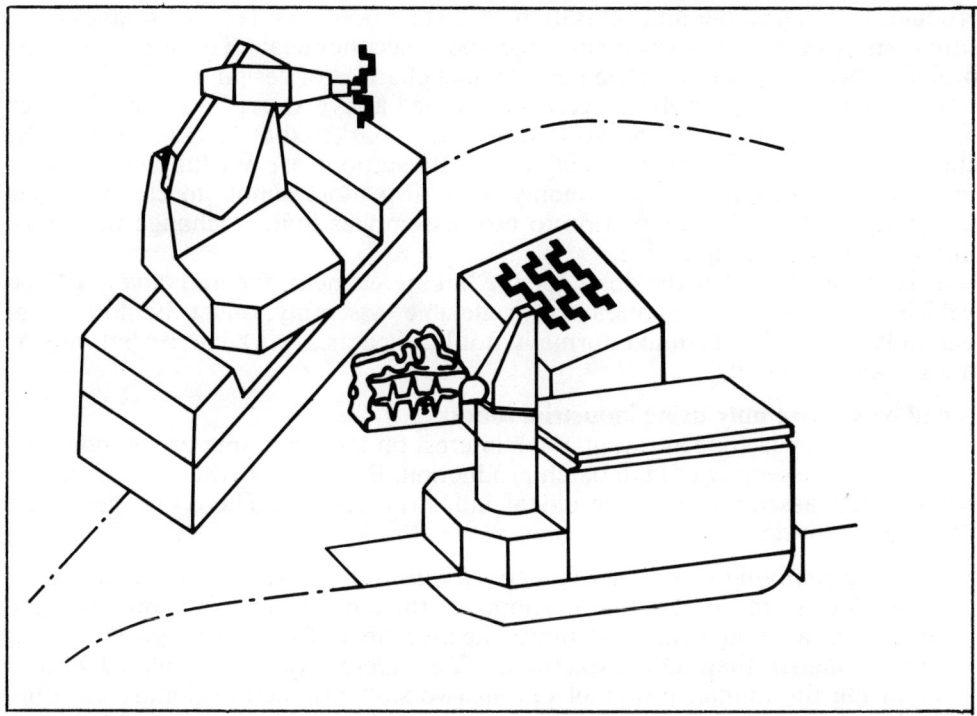

Fig. 8 Mobile industrial robot (MOBIROB) works at the assembly of an automotive engine

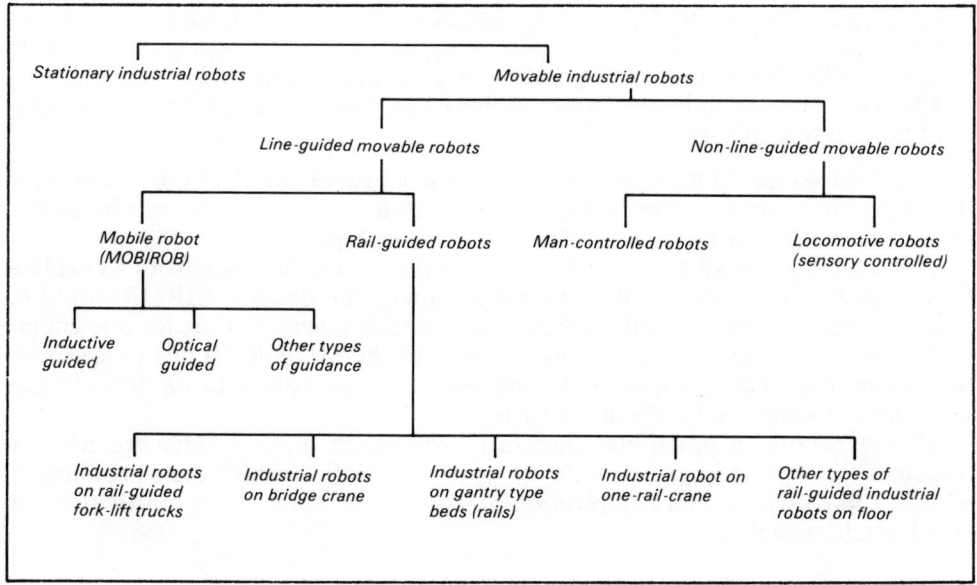

Fig. 9 Industrial robots for assembly

production programme and utilisation of installations can lead very quickly to situations where the investments become uneconomical. To overcome this problem there is only the solution of avoiding changes of design.

For the flexible assembly structures discussed above, the risks are much lower if there are variations in the structure of the market that lead to production changes, with the effect that special automatic stations are not fully used resulting only in a partial loss of economy. It is also more simple to change single automatic stations for adaptation to new assemblies than to change the whole inflexibly linked assembly line.

It is self-evident that the costs for the linking element, the inductively guided vehicle, by making structures for a flexible assembly, must be looked at carefully. Therefore it should normally not be used as a work carrier but only as a transportation unit.

Small batch assembly using industrial robots?

Research today is concentrating much interest on the increasing use of industrial robots in the assembly of high batch production. But what are the possibilities for small batch assembly with the aid of industrial robots? There are, generally, three alternatives:

○ *Robot assembling at one workplace without any workers.* The advantage of this comes from the fact that economy of the industrial robot is only guaranteed when working nearly full time. The investment for this configuration may not be higher than DM 140,000 in West Germany (as calculated above) replacing the human power of one in two-shift production (industrial robot and equipment).

○ *Robot assembling with one or more workers.* For instance only half a human hand is substituting the investment (as calculated above) of the industrial robot and equipment should be lower than DM 70,000 at two-shift production. For instance, this may be a robot installation with only three axes and low intelligence.

○ *Robot assembling at different workplaces.* In this configuration the robot should be installed only for complex work at different workplaces and which is very difficult for men and requires a large amount of time. There is a second alternative for the industrial robot in this case, when replacing remaining parts of human workplaces.

Fig. 8 shows an example of a mobile industrial robot (MOBIROB) working in the assembly of an automotive engine. Some different types of mobile industrial robots, especially for use in assembly, are given in Fig. 9.

The working area of the MOBIROB is limited by the distance from workplace to workplace. It is easy to calculate, for instance, that one MOBIROB would be able to work at more than 50 workplaces, if the full assembly time for one engine is 80 minutes and at each of the 50 workplaces the MOBIROB has to assemble for one minute. The distance from one assembly workplace to another, in this example, is thought to be about ten metres.

Mobile robots in assembly, therefore, will reach a very extensive area of possibilities, especially in the field of semi-automatic small batch production. Mobile robots will have a key position on the way to automation, particularly in small batch assembly.

AUTOMATED ASSEMBLY CAN EQUATE WITH SHORT PAYBACK PERIODS

A. E. Owen, Tony Owen MBA Ltd., UK

First presented at the 3rd International Conference on Assembly Automation and 14th IPA Conference, 25–27 May 1982, Boeblingen, West Germany

Abstract

Automated assembly projects require management approval before implementation. This approval equates with financial viability whereby a payback period of less than two years is usually required. Frequently, the only benefit to be quantified is that resulting from the transfer of human operatives, which unfortunately does not always satisfy the specified payback period. This paper examines *total* benefits that can accrue from automated assembly and illustrates how stochastic quantification can be applied to these benefits, such that an acceptable payback period can be achieved.

Introduction

The true potential of assembly automation can only be assessed if all of the advantages and disadvantages of the proposed system are quantified. Any predictive type of proposal must take into account the degree of uncertainty of each of the attributes considered and therefore it *must* be analysed both objectively and stochastically, such that the ultimate decision can be made in the light of all known and assumed data[1].

The theme of this paper is an examination of this philosophy and the application of the same to a theoretical situation so as to show that when the total benefits of automated assembly are given quantitative values, a more complete financial argument can be laid for (or against) the proposal.

It is generally forgotten or ignored that the manpower saving anticipated from a project is only one of a number of cost savings that can be expected from the correct application of automated assembly. When non-labour benefits are quantified the resulting savings may be far in excess of those resulting from the reduction or removal of the labour production element and permit payback periods well within the near mandatory two years[2].

Manufacturing industry and assembly

With the exception of the monopolistic industries, the survival of a company in the manufacturing industry is dependent upon that company satisfying the market demand for a given product, at a given price and quality. The demand for the product is neither constant nor guaranteed, but varies according to the market forces. The fickleness of the market place and the now necessarily short life cycle of any product, make it necessary that the goods are both available for purchase at a moments notice and are manufactured in an efficient manner such that their market price is competitive[3].

It is documented that 53% of the manufacturing time and 22% of the total labour incurred in a typical product is expended in some form of assembly tasks. Hence a company's viability is enhanced through using efficient assembly methods that result in low reject/scrap rates, a quality level that satisfies the product's requirements without being too lax or tight, and an optimal expenditure of consumables, whereby waste is balanced against speed and ease of assembly[4,5].

Assembly tasks can be done by human operators, flexible automation or hand automation. The limitations of the first mode are those of performance variation, inconsistency, fatigue, etc. The second and third modes vary in their cost, speed of operation, operational cost per unit produced, and production level for viability.

Flexible automation is slower than hard automation because it has not been specifically built for that particular assembly task, but rather for a more generalised group of tasks. However its purchase and running costs are often much less than that of hard automation, and similarly it requires a lower production level for viability. Hard automation is specifically designed and built for the particular assembly task(s). As such it is an efficient production unit and will produce a large number of assemblies at high speed and minimal marginal cost. It cannot be simply changed to process assemblies other than those for which it was designed, therefore when those products are no longer required the assembly system can only be scrapped.

The ultimate choice between the three modes of assembly (human operators, flexible automation or hard automation) is best made using some form of econometric graph, whereby the cost of production per unit can be compared over a range of production quantities. The quality levels attainable using the different assembly methods should also be compared and any cost savings resulting from these should be taken into consideration[6].

Basic considerations

The manufacture of any product incurs three distinct cost groups:

☐ *Fixed costs* – These are the overheads that are proportioned by management to the machine; the depreciation of the machine and all other costs that do not vary with the production quantities.
☐ *Variable costs* – These are the directly incurred costs that are experienced in the assembly of each and every product that is manufactured by the machine. Typically these costs are: the material; labour if applicable; energy and other expendable costs that vary directly with the quantity of products manufactured by the machine.
☐ *Hidden costs* – These are costs that pertain to:
scrapped materials and energy, wasted labour incurred through scrapped products, lost sales revenue from scrapped products, costs incurred in refurbishing/replacing products that have failed within the guarantee period, costs incurred in reworking or salvaging below quality products, costs of

inspection facilities because of low competence level of human operatives, lost sales revenue because of inefficient production methods resulting in low throughput, and lost opportunity costs because of monies tied up in inventory as a cover for fluctuating production levels.

In general, fixed and variable costs reverse their importance as indicators of the factory costs when the method of assembly changes from manual to automated. With manual methods the incurred cost for the workplace tends to be low compared with that required for automated assembly. The variable costs of manual assembly are those of labour and materials, whereas for automation there is only the material cost and no labour component. The expended energy used by the assembly unit will be lower for the manual system than the automated system and allocation of the energy cost to the fixed or variable sector is very dependent upon the company's costing system. The hidden costs tend only to apply to the manual assembly systems, albeit that it should be remembered that an automatic system that has been incorrectly set or that develops a fault can generate far more rejects per unit time than can any human operative.

The above shows that for high volumes of output, the factory cost of products assembled by automated methods is lower than those assembled using human operators. This cost advantage leads to increased contributions to profit and the ability to reduce the sale price in response to market pressures.

The cost elements for manual, flexible automation and hard automation are ranked according to magnitude:

Method	Fixed costs	Variable costs	Labour	Energy	Hidden costs
Manual	3	1	1	3	1
Flexible automation	2	2	–	2	2
Hard automation	1	2	–	1	2

Whilst fixed and variable costs are well known and quoted as parameters against which new projects are assessed, hidden costs tend only to be assessed (if at all) in qualitative rather than quantitative terms. Let us now look into these hidden costs more fully and try to indicate the degree of magnitude of benefit that could occur.

Scrapped materials

A prime source of reject products is the production variance of the human operator when employed as a production element. It is well known and documented that the human assembler exhibits emotion, fatigue and fallibility which results in variable performance, output and quality, inconsistent and non-predictable in absolute terms. By careful analysis of the reasons for which the products are deemed to be scrap, it should be possible to identify the root cause, isolate those that derive solely and completely from human error, and thus determine the advantages in terms of reduction of rejects if the human operator was removed from the process[7].

Another cause for the production of reject products is the product design itself, since unless a product has been designed for automated assembly, there are often a number of options in terms of the assembly possibilities – obviously only one will be correct – but, if it is possible to assemble a product incorrectly or damage some component within that product through incorrect assembly, then according to Murphy's Law it will happen[8].

There are a number of 'normal' production engineering reasons that can result in scrapped materials. Examples are: incorrect setting of the machine; too tight a tolerance band compared with the process tolerance; and too high a quality requirement compared with what is really sufficient for the product. Rejects that are attributable to these reasons can often be reduced by re-examining the parameters of rejection, since if, for example, the acceptance band can be increased to include $\pm 2\sigma$ rather than $\pm 1\sigma$ of the products assembled, then the resultant increase in acceptable products will be approximately 41%.

Irrespective of the cause of scrapped materials, the net result is that there is an often irredeemable loss in terms of material, labour and energy costs that have been expended in processing a product that is now deemed to be worthless. It should also be remembered that there is an additional cost in terms of the loss of revenue because the new reject products cannot be sold.

When a product fails within its warranty period, it is necessary for the manufacturer to replace or refurbish it, with the total cost being borne by the manufacturer.

Again statistical analysis is of benefit, since if components of a specific quality level are incorporated or if reduncancy is built into the design of the product, then the probability of a failure within the guarantee period can be predicted. Therefore the cost saving from a process that increases the implicit reliability can be laid against the costs that would be necessary to obtain that reliability. The decision whether to replace or refurbish depends upon the salvage value of the failed product, the cost of refurbishing, and the contributions to profit that would be lost if the decision is to replace from a product in stock.

Cost of inspection

If a production process results in products of varying quality, there is a need for an inspection station, such that those products that are deemed to be unsatisfactory are recognised and removed from the system.

When the process has a consistent and acceptable quality level, the viability of inspection as a production task must be challenged. Here we are talking about inspection to determine the quality and functional level of the products as opposed to the need for inspection stations in an automated system to check that a desired condition has happened before a subsequent operation.

It follows that any inspection costs must relate to the number of inspectors involved, the probability of reject units being accepted and the resultant cost and embarrassment to the company for that erroneous acceptance. If the *total* cost of these rejects exceeds the cost of inspection, then inspection as an external task to the production process *must* be considered.

Increased production levels

The human operator is inefficient as a production element, since he/she requires frequent breaks for re-energising (food and drink), time for cleansing of the system (washroom visits), and also the interaction of other humans in order to function rationally (social function). The human operator has a varying time based operational level (production varies throughout the day), and requires payment even when not working (coffee breaks, washroom visits, illnesses, vacations). Their cost as a production element is also constant (piece work), hence there is usually no direct benefit to be gained by a production run of large numbers when employing humans[9].

The machine as a production element is more efficient. Its energy supply is (usually) continuous, the properly designed machine is self-cleansing in a continuous mode, and it does not require social interaction (although it sometimes functions within a cell requiring interaction with a central controller). Its level of production is consistent throughout time and it does not incur costs when it is not processing products. Its illnesses tend to be short-lived (dependent upon the design of the unit) and also the cost of the machine is usually paid off in a time that is far less than its potential operational life. This last statement is more true of flexible automation, where its tasks can be changed, than of hard automation where the life cycle of the product being processed is often far less than the potential operational life of the machine(s).

Additional benefits from an automated system are:

- The cost savings due to the increased production per unit time, reducing the need for overtime with its higher labour costs.
- Increased system capacity such that it can respond to varying demand for the products processed.
- Cost savings because it is no longer necessary to train humans for one function and then have to fire or retrain them when product demand is low.
- A reduction in the number of orders lost because of stockouts due to erratic human production methods. Similarly a saving in the inventory of products and components needed to cover for scrap and erratic production by human operatives.

Deterministic and stochastic factors

Typically when a project is being proposed, the working environment in its *total* form is implicitly assumed as being deterministically known. An examination of some of the tangible and intangible factors involved in the environment may indicate the magnitude of erroneous values that can be applied to a project cost, and simultaneously show that intelligent application of probability values to these values will yield a better and more accurate indication of the expected costs[10].

Tangible factors (mainly technical-economic)	*Intangible factors* (mainly socio-organisational)
Cost of machines	Cost of quality control
Cost of jigs and fixtures	Cost of supervision and manpower control
Cost of power supply	Cost of absenteeism
Cost of labour	Cost relative to job satisfaction
Cost of maintenance	Cost due to grievances and strikes
Cost of layout	Cost of organisational flexibility
Cost of set-ups	Cost of occupational disease

Other factors that have a dramatic effect upon the cost of the system are:

- □ useful life of the system,
- □ number of pieces required per unit time,
- □ time to produce a specific number of pieces by human,
- □ time to produce a specific number of pieces by robot,

320   PROGRAMMABLE ASSEMBLY

☐ number of executed operations per piece,
☐ life cycle of product,
☐ number of products for which the system was designed, and
☐ flexibility parameters of system (geometric, functional, power).

From the above listings, it is possible to visualise the degree of error that could arise from erroneously assuming that all of the values assessed for these factors were absolute. It is therefore necessary to determine which, if any, of these factors are more influential than the others, and which are not deterministically known.

When a fact is not absolutely known in quantitative terms (deterministic) it is often known in qualitative terms, whereby it can be said that such-and-such is liable to happen under a given set of circumstances. An allocation of numerical values to those qualitative terms and conditions is known as stochastic or probabalistic analysis.

When using any form of statistics it is wise to take note of the adage:
'Statistics can be used as a drunk uses a lamp post – for support not illumination'.

Recognition of this danger should lessen the risk of unquestionly using any data so computed.

A stochastic statement that could be formulated is:
'Given that an alternative system is more consistent, there is a 0.80 probability that the quality level will be increased by 20%. It is also assessed that there is a 0.60 probability that the additional products could be sold without long storage or price reduction.'

The resultant expected revenue from the increased quality level would be:

P(increased quality) × 20% of annual production × P(of selling) × sell price
+P(no better quality) × 0% of annual production × P(no selling) × sell price

In the general cases that will be met, the probability statements will usually offer only two alternatives at each stage of analysis. Therefore it will often be of benefit to use 'decision trees' to determine the outcome of the various options.

Example of use of expounded ideas

When comparing alternative production systems it is important to know what you are comparing and against what criteria, i.e. it is essential that one compares operating costs for the same quantity of production or operating efficiencies for the same period.

The determination of an alternative production system can be on the basis of either the proposal for automating an existing process without necessarily wanting to increase the production quantity, or it could be the comparison of two automated systems to produce a given production quantity.

Case 1 – Manual vs automation for same (existing) quantities

Existing system: 4 operatives @ £4,000 p.a. each.
Single 8 hour shift working.
Reject level 5%.
Inspection: 1 @ £6,000 p.a.
Allocated overhead £1,000.
Cycle time 15 seconds.

Flexible automation: System cost £40,000.
Depreciation period 5 years straight-line.
Cycle time 10 seconds.
Hard automation: System cost £100,000.
Depreciation period 5 years straight-line.
Cycle time 1 second.
Product: Annual production 480,000.
Labour content £0.05.
Material content £2.95.
Selling price £5.00.

Item	Manual	Flexible automation	Hard automation
System cost	£500	£40,000	£100,000
Depreciation	N/A	£8,000	£20,000
Fixed cost (F)	£7,000	£15,000*	£27,000*
Variable costs (V)	£3.00	£2.95	£2.95
Sales price (S)	£5.00	£5.00	£5.00
Break even F/(S-V)	3,500	7,317	13,170
For production of 480,000			
Production cost	£1,447,000	£1,431,000	£1,443,000
Cost reduction vs manual	N/A	£16,000	£4,000
Payback period	N/A	2.50 years	25 years
Cost per item	£3.0145	£2.9812	£3.0062

* At this stage it is assumed that the inspection facility will be required in total

From the above table it can be seen that flexible automation offers the lowest per unit cost, but does not comply with the mandatory (?) two year payback period. Therefore let us look at other cost savings that *might* occur.

Reduction in reject rate. It is anticipated that there is a 0.80 probability that a 20% reduction in the reject rate will occur. If this is not so then the reject rate will reduce by only 5%.

Expected savings $= [(0.8 \times 0.2 \times 0.05) + (0.2 \times 0.05 \times 0.05)] \times 480,000 \times £2.95$
$= £13,036$

Revenue from the 'non reject' items. It is anticipated that there is a 0.50 probability that the increased production resulting from the reduction of rejects can be sold without recourse to extended stocking or reduction of prices.

Expected savings $=$ expected reduction in rejects $\times 0.50 \times £2.05$
$= £4,182$

Reduction in cost of inspection facilities. It is anticipated that there is a 0.4 and 0.6 probability that the inspection costs could be reduced by 75% and 50%, respectively.

Expected savings $= [(0.04 \times 0.75) + (0.06 \times 0.50)] \times £6,000$
$= £3,600$

Total expected savings. The anticipated savings from these three non-manpower activities is £20,718, which is 29.5% greater than the savings from manpower alone. The total anticipated savings from this *simplified* analysis of a flexible automation system is £36,718, which results in a payback period of 1.09 years.

Case 2 – Comparing two automated systems

Automation *per se* equates to consistency of output and quality. Therefore, the only comparator between two alternative systems is that of the average cost per assembly produced.

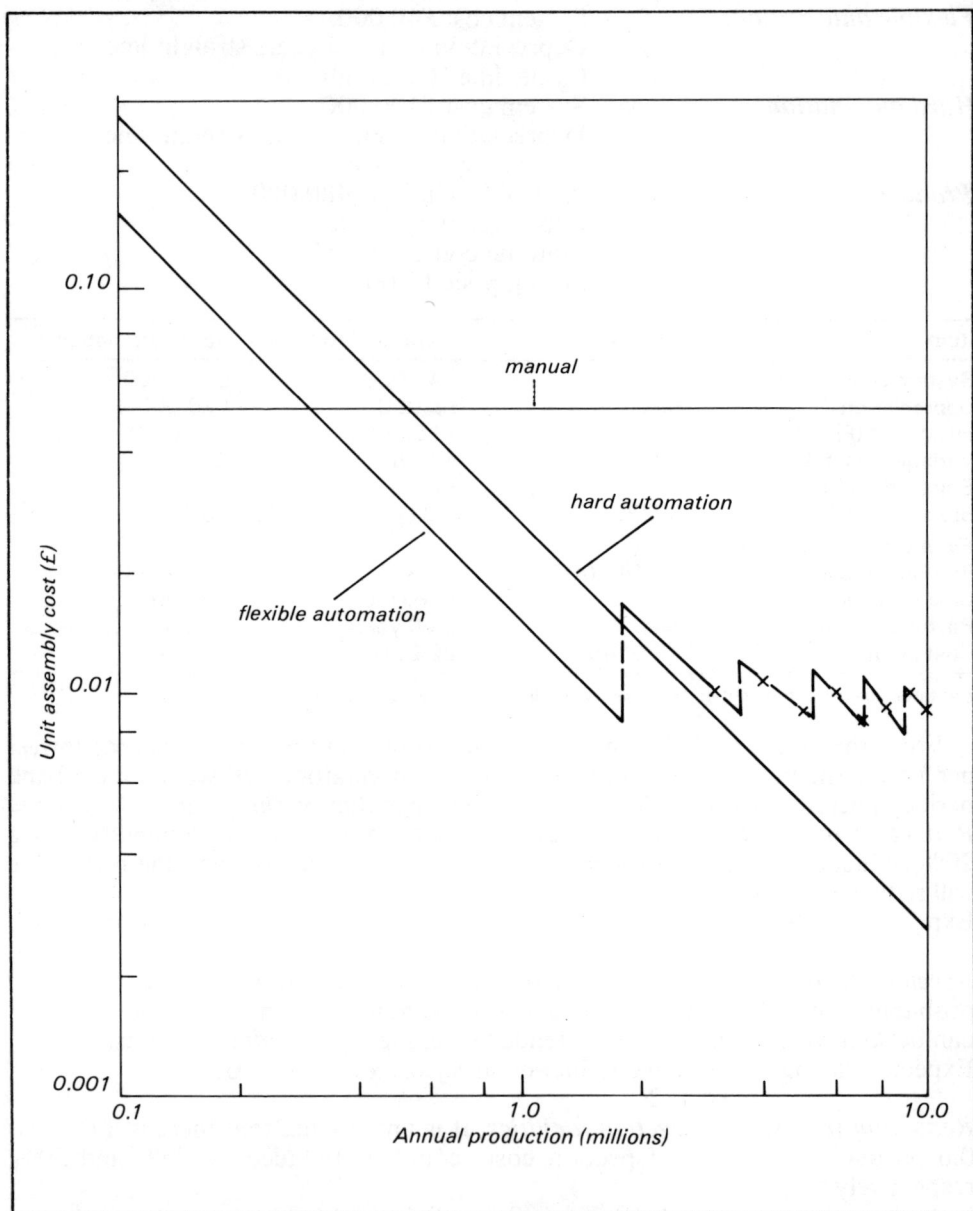

Fig. 1 Econometric graphs

When the annual demand for the product is large, the fixed costs of the system become an insignificant proportion of the assembly's prime cost, and the cost advantage of hard automation shows itself only when the annual demand is in excess of the capability of a single alternative system.

The following tabular data show how the per unit production cost changes for selected quantities in both flexible and hard automated systems. The total relationship is shown in Fig. 1.

Fixed cost	£15,000	£27,000
Cycle time (seconds)	10.00	1.00
Output per 8 hour shift @ 100%	2,880	28,800
Output per 200 day year (single shift)	576,000	5,760,000
Output per 200 day year (three shifts)	1,728,000	17,280,000
Per unit cost per system year	£0.0086	£0.0015
Per unit cost per 0.1 million	£0.1500	£0.2700
Per unit cost per 1 million	£0.0150	£0.0270
Per unit cost per 10 million	£0.0090	£0.0027

Conclusion

This paper has identified those costs that affect the financial efficiency of any automated assembly system. It has shown how important, albeit often ignored, factors can be quantified such that the total benefits of an alternative production system can be derived. Because many of these factors can only be anticipated in terms of their direct effect on the financial efficiency of the system this paper has shown how the application of stochastic quantification enables the expected benefit to be computed.

The future of automated assembly systems is assured by the ever increasing cost of human labour as a production element. The sophistication of the automated solutions is limited only by the technological state of the art. A diagram of some of these social, technical and economic forces that are shaping and influencing automated assembly in general, and flexible automation in particular, is shown in Fig. 2[11].

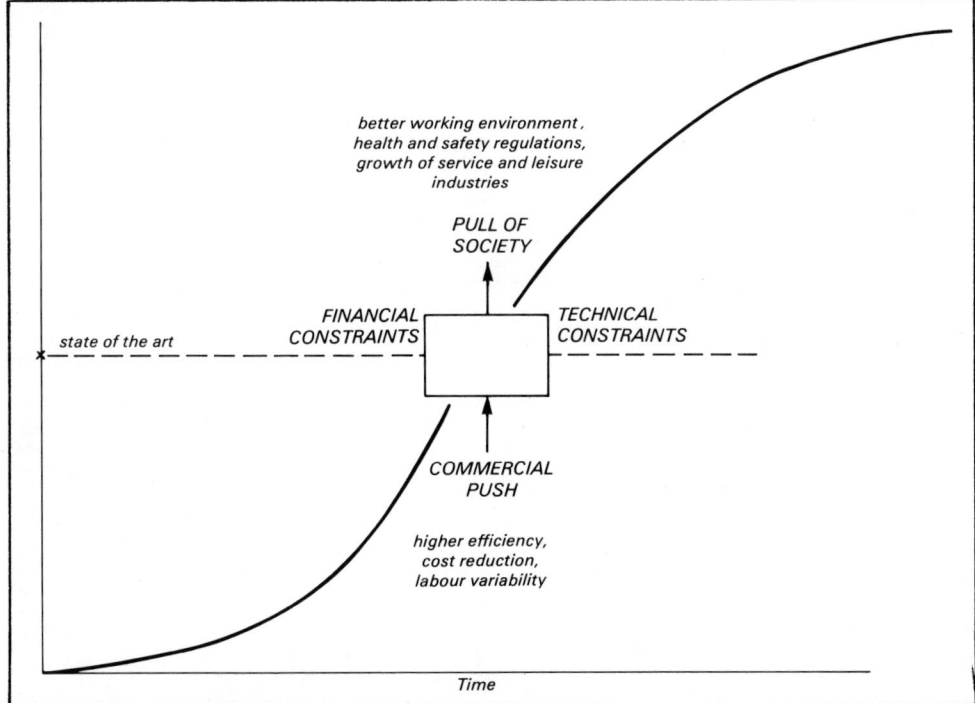

Fig. 2   Social, technical and economical forces affecting automated assembly[11]

References

[1] Owen, A. E. The integrated factory. In, Proc. 2nd International Conference on Assembly Automation, 18–21 May 1981, Brighton, UK, pp. 25–36. IFS (Publications) Ltd. 1981.
[2] Carnegie-Mellon University, Dept. of Engineering and Public Policy. The Impact of Robotics on the Workforce and Workplace (preliminary draft), June 1981.
[3] Bestwick, P. and Owen, A. E. The impact on production management. In, The Management Implications of Microelectronics, Chap. 10. Macmillan, November 1981.
[4] Bache Hasley Stuart Shields Inc., Institutional Research. Robotics Presentation, June 1981.
[5] Lee, M. Boom in industrial robotics. Industrial Robot, Vol. 8, No. 1, March 1981, pp. 66–67.
[6] Sugarman, R. The blue-collar robot. IEEE Spectrum, September 1980, pp. 53–58.
[7] Owen, A. E. Economic criterion for robot justification. Industrial Robot, Vol. 7., No. 3. September 1980, pp. 176–177.
[8] Dallas, D. B. The advent of the automated factory. Manufacturing Engineering, November 1980, pp. 66–76.
[9] Owen, A. E. Management of the Manufacturing Industry in the Microelectronics Age. MBA Dissertation, University of Bradford, UK. 1979.
[10] Ciborra, C. and Romano, P. Economic evaluation of industrial robots – A proposal. In, Proc. 8th IFIR/4th CIRT, Stuttgart, Germany, 1978. pp. 15–21.
[11] Rathmill, K. Industrial Robotics – A Major Challenge for the University and British Industry, Inaugural lecture. Cranfield Institute of Technology, UK, March 1981.

Chapter 7
SOCIAL ASPECTS

Automating a process makes even greater demands on the people involved, simply because there are fewer of them. The three papers in the chapter give just some of the aspects of people involvement in assembly – people performance, future implications for labour and management, and the changes taking place in the work environment.

PEOPLE PERFORMANCE IN AUTOMATED ASSEMBLING

J. M. Wood, Rank Xerox Ltd, UK

First presented at the Institution of Production Engineers Automated Assembling/Robotics Working Party Seminar on Automated Assembling Systems Choice – Dedicated vs Flexible, 2 February 1984, London, UK

Abstract

This paper is not written by an individual who claims any expertise in human behaviour related to automated assembling, but rather to present some views in the hope it will promote some discussion and debate. It is likely that human reaction and performance will vary according to the environment, management direction, and local and company culture.

Four major areas of human interface with automated assembling systems are focused on. First, the people within the system, i.e. a worker who performs an operation which is an integral part of the automation of the assembly; secondly the people who feed the system; thirdly the people who are fed by the system; and lastly those people who service the system.

In these four areas of activity, all of which are vital to the effective use of an automated assembling system, the effect of human behaviour, company culture, resistance to change, selection criteria, trades union attitudes, workplace design, satisfaction of 'the ego', boredom, absenteeism, frustrations, together with attitudes to quality, service and understanding of the total system and its objectives are discussed.

The importance of performance feedback, audit, publicity and the involvement of the people to gain improvement is discussed.

At the best of times predicting human behaviour is difficult, and very difficult if you are entering an area, such as automated assembling, for the first time. Therefore employee involvement backed by contingency planning may form the best basis to give both employee and management the advantages required from automated assembling so that the business will survive.

Introduction

One's first thoughts on the subject of people in automated assembly suggests that if the assembly is automatic then there are no people. However, it is recognised that emerging systems can be populated with people who are interfacing with automatic assembly devices in the overall task of assembly; and there are those who indirectly interface with such systems.

Those who interface are those who feed the system either with parts or previous assembly operations, those who are fed by the system, and those who service it.

I do not claim any special expertise in this area other than that my company has installed assembly lines which create problems of human interface with devices that are controlled by computers, micros and other electromechanical devices. In order to improve efficiency they can be seen as de-humanising work and as a threat to employment. I propose to touch upon a range of topics related to this problem in the hope that it will promote discussion and an exchange of experiences.

Man–machine interfaces

When discussing the performance and problems of humans in the workplace, it is very difficult to predict what will happen. The culture of a company is a significant factor in what is likely to happen when new ways of assembly are adopted, based upon the emerging technology in this field.

Firstly the introduction of new radical technology must look at a number of factors. Without a doubt the change will only exchange one set of problems for another.

There is no reason for secrecy, but a need to have a full disclosure and discussion about what one is going to do. If the production schedule can stand it, introduce the system in the production area and let the workforce see the developments, the problems, and encourage them to participate in finding solutions. This will go some way to heading-off the problem of fear of job losses which secret development and rumour brings.

However, such a way of introducing will increase the risk of disruption and this will lead to an impact on schedule and possibly quality. Extra demands are made from some sections of the workforce who have no direct connection with the new system, and resentment follows especially amongst supervisors and management whose performance record is affected.

Regarding the attitudes of the threat to jobs, one needs to stress the benefits to the business as a whole and cast the communication net as wide as possible. Acceptance of change can go to a factory or shop vote. If the majority see and understand that their job security is improving, then acceptance may be achievable.

The four major areas of human interface with automated assembling systems are:

☐ the people within the system,
☐ the people who feed the system,
☐ the people who are fed by the system, and
☐ the people who service the system.

People within the system

Common sense would tell us that not everybody is suited for the task of attending an automated assembly line. The job can be boring and has the problem of being fed or feeding a demanding mechanised system, which is stressful to some people, yet many people are 'drafted' into such conditions with little thought by management. A company may need to develop selection criteria, not only for the physical requirements but also the mental attitudes required.

This is easier said than done. The use of 'head shrinks' is equally resisted by management and unions, but some discussion and a review of the options is required before large capital sums are committed. Perhaps the best that can be achieved is to review all existing similar jobs in the manual assembly/machining field within a plant and evaluate the performance of existing staff. Those who can handle the routine, repetitive, paced environment, can be safely moved into the robotic/automated assembly interface task.

Many of us have developed, for the reasons of working conditions, very acceptable redeployment agreements with the trade unions. Most trade unions prefer 'last in/first out' (LIFO) as a way of selecting people, especially if they suspect that cut backs are in the offing. Their desire to take any bias out of the selection process does not lead to an ideal marriage of personal skills and temperament and the needs of the job. This suggests that early discussions are required to allow the trade unions to understand what is required, to agree a fair selection process and to agree the need and scope of training, and if possible an escape route for operators who do not respond to training or whose performance on the job falls below the required standards.

The biggest problem for most people is the relentless pacing of the robot/automated system. If the interface between operator and systems is limited then you may find sufficient people who can handle this situation. If the numbers increase to a significant proportion of the type of skills required, one has to review the workplace design and the duties of the operator throughout the day. The way the operator shows his disenchantment is by absence, usually backed by a medical reason.

There are probably two major alternatives to look at when considering the operators interface with the workstation and the duties he has to perform. Based on McGregor's Theory X and Y concepts, we can take the view that we require an operator who has a passive attitude; that is, he or she does not seek responsibility, will be happy to be directed and have his/her routines and objectives set out. Because all this is precisely defined he can expect some form of penalty for mistakes. He or she is the Theory X group and the workstation can be designed to suit that type of individual. It is thought that many industrial engineers design for this type of operator as it appears to give a highly controlled and predictable situation and optimises pay-back, etc. The individual almost becomes part of the machine. The problem seems to be managements difficulty in selling the Theory X type of operator for reasons already discussed.

Many people in our workforce are not happy working in this way; they seek responsibility and need a challenge, and they want to participate in goal setting and require to perform some part of the management of their role in operation. So if these people are to be satisfied and not cause the disruption which can lead from boredom and constraint, designers and industrial engineers need to design to satisfy the expectations of people who contribute best in this type of environment. This is one of the major challenges for us to solve in the introduction of automated assembly systems.

One should look at job rotation, the use of buffers before and after his station, workplace orientation to give social contact, decor and lighting, music, good canteen facilities, and careful consideration to the days work pattern to give variety and relieve boredom. We should try and give the operator some sense of being in control of events, rather than the robot. Manufacturing engineers need to research this aspect of factory design and layout.

Everyone likes to succeed; we like our peers to know we are successful and we like praise for a job well done, it is good for our ego. Some attention needs to be paid to this basic human need for recognition. Interfacing with hopefully highly efficient electromechanical devices can make one feel pretty inadequate. Make sure you lavish the attention on the human link you have interposed. If you do not they will find ways of making you recognise they are there. The attitude of 'get another one' often prevails if there is something wrong with the human being, something you are not prepared to do with the mechanical/electrical functions, at least not until you have done everything possible to correct the problem.

The same professional approach in fixing the human failures should be taken. This is not so easy, but at least this part of the operation can 'talk'.

Job satisfaction means different things to different people. People are satisfied if they view coming to work as a reasonable pleasure for the majority of the time they attend. Because people are different and have various aspirations it is as well to identify what gives them job satisfaction and try to meet those needs. Some needs may be impossible to meet and discussion may ensure that the individual fully understands what he can fully expect. False promises may be a short-term solution but the problems will come later; if people are not satisfied, and all has been done to meet their needs with the funds available, then a change of staff is required. This needs constant review by supervision, after all you may well be very satisfied if you have just been given a new and therefore interesting job. However this may soon wear off and give way to boredom. Absenteeism, lowering performance, and workmanship defects, signal a possible loss of job satisfaction.

Some people seem capable of dealing with repetitive and boring jobs, they find their pleasures and satisfaction elsewhere. If the job can be matched to the 'intellect' of the operator then boredom may not be a problem. If this is not the case then you may wish to consider how the day is organised for the operator.

In-cycle tasks are a way of helping the boredom problem, as is job rotation. Giving the operator the total management of his or her workstation may be a possibility. Here the operator would audit incoming quality, complete performance data, do tool changes and set-ups, do preventative maintenance tasks, and improve the productivity of the station. The attitude to be fostered, is that which we as individuals take if we were self-employed. In many cases this is just not possible, but at least it can be considered and discussed with the people involved.

People who work with 'mechanical things' get very frustrated with the failure of such devices. Therefore, reliability is a paramount need, but this is not a perfect world and we need to consider the opportunity to correct the fault by operators, and the way the operator can continue his output if the system has failed. It is surprising how we forget to ensure that output will continue when failures occur.

People who feed the system

First, it is necessary for this group to look upon the automated system and its operators as a customer — a customer that has to be satisfied, and managements' job is to make this group understand fully that this is its prime role. It has to foster by education and clearly defined objectives, an attitude of reliable and efficient service to the system it supports. This means taking the trouble to find out what its customer requires and then to maintain a dialogue so that people react to change quickly.

Secondly the major thrust in satisfying its 'customer' is to have a full understanding of how the system works and what its objectives are. So these 'servants' of the system are fully integrated into the team whose objective is to ensure that the system's objectives are met.

People who are fed by the system

The next group are those who are fed by this automatic assembly system and again the system has a customer, i.e. this group. This group need to know if any changes will affect them; it's nice for the back legs to know what the front legs are up to! Also it is important that their feedback is sought in the performance of the system and if they have suggestions to improve the interfaces.

This group are usually the first to see the quality of the product from the automated system and fast feedback on quality problems is essential to avoid excessive scrap or rework costs.

People who service the system

With any such system there are a significant number of people who 'service' it: maintenance, programmers, materials movement, storemen, etc. Some of these people can appear remote from the actual operation, they may never see this miracle of efficient assembly, yet their performance is vital. Again it is managements' job to ensure an attitude of service to give customer satisfaction. So all the remarks related to the group who directly interface by feeding the product apply to these groups. Again it is bringing these people to see what they serve, to review and discuss problems, to allow them to make suggestions to improve their service, that is so important and so often neglected.

Concluding remarks

In summary, to make a success of the task of people interfacing with automated systems it's really a matter of telling people what the system requires of them, letting the different groups interface in some way, and encouraging team spirit to improve performance. In facing the boredom problem, either plan to minimise it or choose people who can handle it. Make sure management audit the performance of the system and those who service it. If you audit properly, people will know you are interested. If the audit news is good news then tell the world about it, if it is bad news get involved and put it right; either way people will be glad to see you, for studies show that many of the forms of negative behaviour, fears and antagonisms stem from not knowing what others are doing. Most of us have a fear of being lost from view, forgotten, by-passed or left out; in our management style and machine design and layout we must cater for the human recognition needs to ensure a successful interface between man and machine.

PERSPECTIVES AND SOCIAL IMPLICATIONS OF ASSEMBLY AUTOMATION IN WEST GERMANY

D. Seitz and V. Volkholz,
Gesellschaft für Arbeitsschutz-und Humanisierungsforschung (GfAH),
West Germany

First presented at the 5th International Conference on Assembly Automation, 22–24 May 1984, Paris, France

Abstract

Over the next 10 years industrial series assembly will be the centre of automation. In order to assess the perspectives of automation correctly the status quo must be analysed. Investigations in 355 firms in West Germany and 43 case studies provided representative data on strategies of rationalisation, on automation and human factors of assembly work.

Flexible machinery enlarges the field of assembly automation but does not substitute 'traditional' strategies of rationalisation and the flexibility of manpower; conventional automatons will remain an important factor.

Work organisation, which is often underestimated, must be regarded as a supporting measure.

A large potential for future automation can be prognosticated, but the discrepancy between the future investments in automation and the automatons to be involved, as well as uncertain economical factors, complicate exact assessments.

Automation implicates changes in the structure of qualifications. A decrease in the quota of unskilled labour for the benefit of skilled labour is to be expected. As the shifting of qualification structures is linked with increasing displacements, there will be higher demands for personnel planning on enterprise level and labour policy on the level of industries and national economy.

Introduction

The problems of flexibility in production have increased since flexible marketing strategies have become more relevant. Within this context there have been officially financed projects concerning work organisation and job design in West Germany. The task to make production more flexible often was linked with efforts to improve the quality of working life. To a certain extent flexibility of manpower, has been newly discovered. Meanwhile the traditional possibilities of rationalisation in this sector with its high wage cost seemed to be exhausted. New possibilities of flexible automation seemed to be suitable to improve the efficiency and competitive power of the companies as well as to remove some grievances concerning working conditions. On the other hand, one must consider whether or not the expectations of technical flexibility have been exaggerated and subsequently the flexibility of manpower underestimated. Other problems are the economic and technical risks – especially for smaller companies – and negative effects on the job situation – with backlashes on economy.

Due to the relevance of the questions about tendencies of technical, organisational and social development in assembly, and other questions related to the subject, the Bundesministerium für Forschung und Technologie (BMFT) ordered a study on flexible assembly automation from the Arbeitsgemeinschaft Handhabungssysteme (ARGE HHS). It was conducted by an interdisciplinary team consisting of collaborators from the Institut für Produktionstechnik und Automatisierung (IPA), the Gesselschaft für Arbeitsschutz und Humanisierungsforschung (GfAH), the Institut für Arbeitswirtschaft und Organisation (IAO) and some other institutions. Some results of the study[1], completed in December 1983, are presented.

Status quo of assembly work

Most prognoses on development, dispersion and implementation of new technologies have been either too optimistic or too pessimistic. Among other reasons, this is mainly due to the lack of a realistic assessment of the status quo in the field in which new technologies will be used. Thus we first need a realistic analysis of today's situation in assembly. In this context, technical and economic matters as well as social and personnel aspects are of interest.

First of all assembly in many aspects differs from other sectors of production – a statement which may be commonplace at first sight, but is often neglected. Compared to parts manufacturing, for example, assembly is a more differentiated field. Experience in the advanced sectors such as the motor industry cannot simply be transferred, as is often suggested.

Social and qualification structure

Until now the lack of sufficient reliable data on the number of employees in assembly sectors made it difficult to estimate the potential of the future rationalisation process and other consequences. Projections on the basis of our sample and official statistic data showed that about 665,000 people are employed within the assembly sector: about 286,000 in the electrical industry, 129,000 in the motor industry, and 154,000 in the machine building industry.

Assembly work is predominantly women's work. This is valid for all situations except the machine building and motor industries where the assembly departments and the total companies employ about the same quota of women. On average 24% of all employees in our sample are women, but in assembly departments there are nearly twice as many (44%). The percentage of women is a function of qualification demanded in the respective sector of production. Companies with an above average employment of women are those with extremely restricted work and a low level of qualifications.

The consequences of these facts are that: on the one hand it is often easier to substitute unskilled labour by automatons, but on the other hand, as a result of dequalification, the flexibility of manpower is lost to a great extent. This deficiency is, nowadays, attempted to be eliminated by technical means primarily. But by the same token, a low qualification level in assembly provides an obstacle for innovations.

Qualifications
The specific characteristics of the various industries do not only depend on traditional product-orientated factors but are also overlapped by internal (company's policy of recruiting and qualifying personnel, etc.) and external factors (job situation, regional structural policy, etc.).

In the machine building industry the quota of skilled workers is dominant (50%); this group is larger than the two groups of semiskilled workmen put together (Fig. 1). In future the quota of skilled workers is expected to decrease in those sectors of the branch where there will be possibilities for (partial) automation.

In recent years in the motor and electrical industries an extensive taylorisation and, in some sectors, a one-functional automatisation have already taken place – with hard consequences on the structures of qualification. The quota of skilled workers cannot be diminished further without negative implications on security and quality of production.

Cycle times and paced work
Paced, restricted work is typical and constitutive of most sectors of assembly. The cycle time for 42% of workplaces is less than 1.5 minutes, for 26% it is less than 30 seconds. In the electrical, precision mechanics and optical industries, and in the sheet-metal industry, there are an outstandingly high number of short-cycle jobs (Fig. 2). This is the main factor of stress, but at the same time the distribution of cycle times gives an indication of the future priorities of automation. The different structures of cycle times produce different preconditions for future strategies of rationalisation and for changed working conditions.

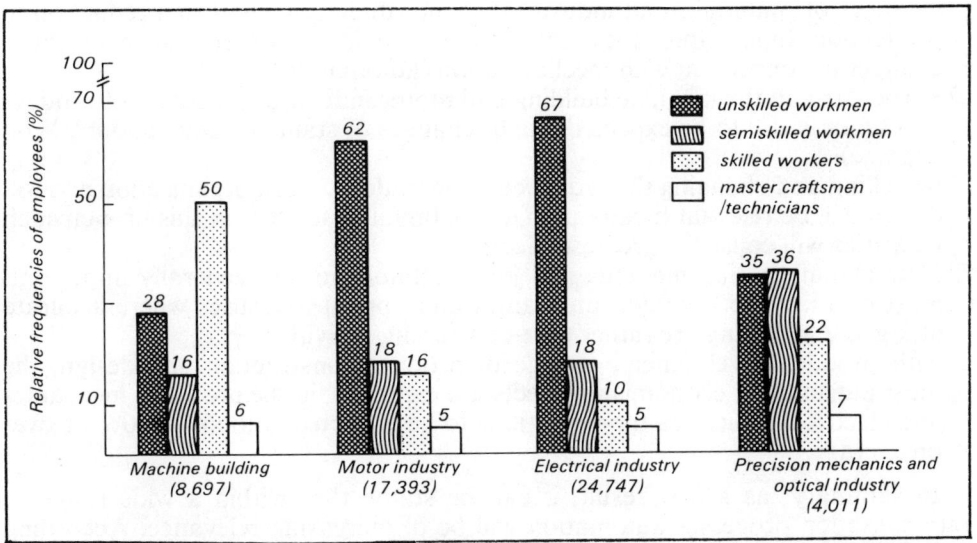

Fig. 1   Workforce qualifications (source: GfAH)

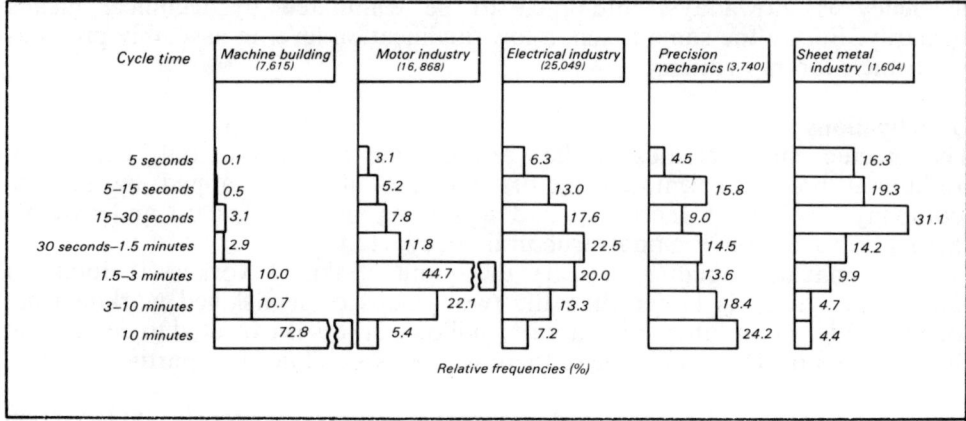

Fig. 2 Distribution of cycle times (source: GfAH)

Strategies of rationalisation

According to the actual opinion of the management in the 355 companies of our sample there will not be 'one best way' of rationalisation in the immediate future. A matter of interest is the relative importance of the various methods of rationalisation (see Fig. 3).

Besides the 'substitution of material' and 'checking/changing of standard times' for more than 75% of companies each other measure is relevant. However, most companies (88%) assess 'product development/product design' as necessary. This result is most important as the majority of companies seem to realise that by improving product construction preconditions for automation are simultaneously improved.

Greater differences between the importance of each measure occur when we analyse the possible reductions of the total standard time expected by the companies. The results shown in Fig. 3 are as follows:

☐ Changes of material in all industries are not directly related to a reduction in production times; this does not exclude the increased relevance of those changes in connection with mechanisation (automation).
☐ Particularly in the machine building and motor industries reduction of production times is still to be expected due to changes of standard time (about 9% on average).
☐ Checking and changing the sequence of operations, work organisation and job design will be relevant mostly in machine building sectors; in this branch such measures will cause the greatest effects.
☐ New manufacturing methods and joint technologies are generally important, most of all in the precision mechanics and optical industries where accurate fitting and adjusting are rather time-consuming activities.
☐ with product development, change of product construction and design, the most outstanding economising effects are expected; in the precision mechanics and electrical industries the reductions to be expected amount to 20% or over on average.

In summary, as a first result, it can be stated that within a wide range of rationalisation processes automation will be of increasing relevance. According to specific demands of the particular industry and of the company, combined

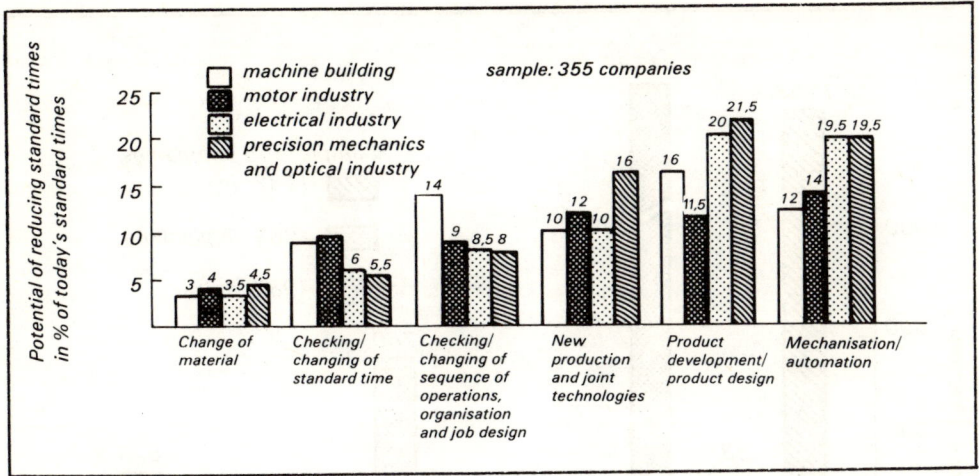

Fig. 3 Potential of reducing total standard times in assembly by different means of rationalisation over the next five years (source: IAO)

strategies will become more important. Complementary measures will multiply effects by reciprocal action. It is within this context that the perspectives of assembly automation are discussed.

Perspectives of assembly automation

Diffusion of automation technologies

The experts are expecting an impetuous increase in assembly robots and flexible, modular-designed assembly automatons. In 1982 about 3% of the companies in our sample used assembly robots; in 1983 13% were already planning their implementation, and in 1987 about 28% will be using robots in their assembly departments. In comparison, the number of companies using special purpose automatons will only increase from about 35% in 1982 to about 39% in 1987.

The quantitative significance of the various types of automatic machinery can be illustrated by comparison with the number of employees. In 1982 for 5,000 assembly workers there were about 125 special purpose automatons, 20 flexible modular-designed automatons, and only one assembly robot in use. Although the increase in special purpose automatons is small (14% per annum) compared to that of assembly robots (about 100% per annum), in 1987 their total number will still be four times as high as the number of industrial robots in assembly (Fig. 4).

It follows, therefore, that conventional means of automation will remain relevant in the near future, but there will be appropriate combinations of different types of automation.

This is substantiated by prognoses on the dispersion of assembly robots (Fig. 5). Most of the 26% of companies which will use assembly robots in 1987 will use more than one: in every fifth company there are expected to be 3–4 and in one third of the companies between 5 and 14. In every twentieth company more than 14 robots are estimated to be in the assembly departments. In many companies there will be interlinked robot systems, making a great difference to the actual situation. For example, work organisational problems or means of flexible interlinkage are not so important in the isolated use of single robots.

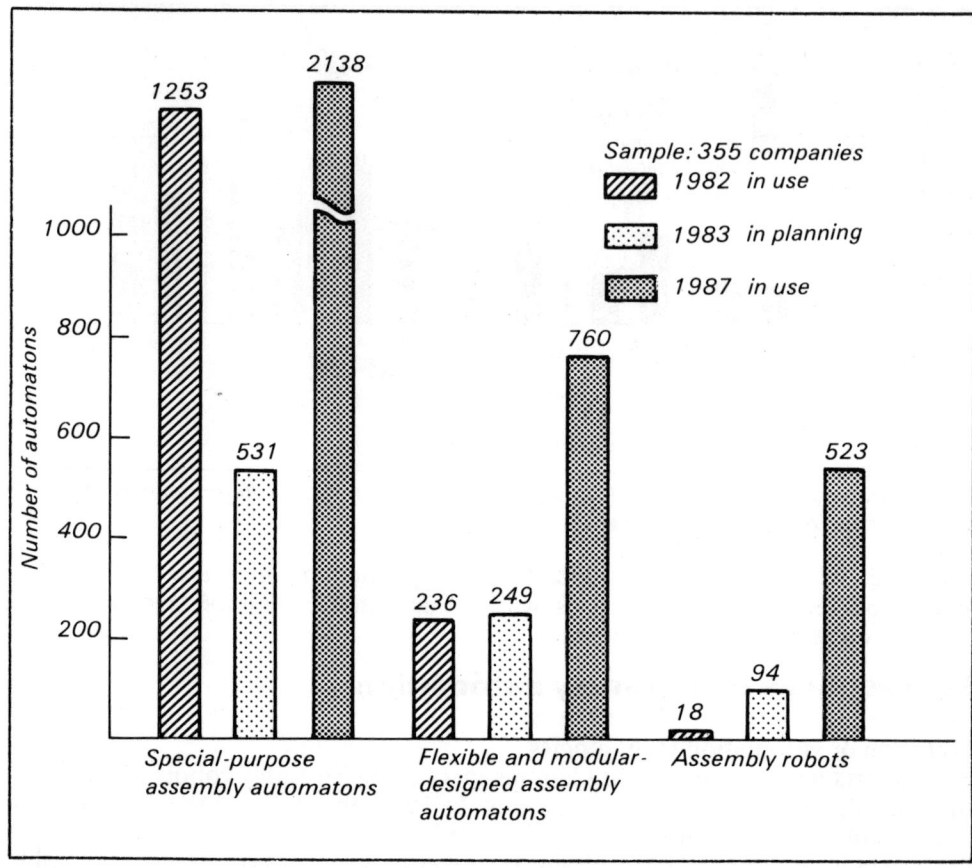

Fig. 4 Number of automatons in use in 1982, planned in 1983, and expected to be in use in 1987 (source: IAO)

The anticipations of the experts however, must carefully be interpreted: our study has revealed a gap between the assessments on automatons to be in use until 1987 and the prognosticated amount of investments within the same period.

The rate of investment in assembly automation has risen from 7% in 1977 to 10% in 1982, and in 1987 will rise to 15%. But in spite of these considerable increases the total amount will be largely insufficient to buy as many automatons and robots as intended. Our case studies showed another problem in this context: the cost has been assessed about 50–100% too low, particularly concerning the cost of planning, testing and implementation. This poses difficulties for companies with less than 1,000 employees most of all, due to small planning capacities. Any assistance offered by producers of automatic devices is often unsuitable for companies' requirements.

Potential displacements

Due to the discrepancy between prognosticated automatons and investments there are different assessments on the future potentials of displacements. In order to get a realistic view of the possibilities of assembly automation we conducted a further calculation based on factors concerning:

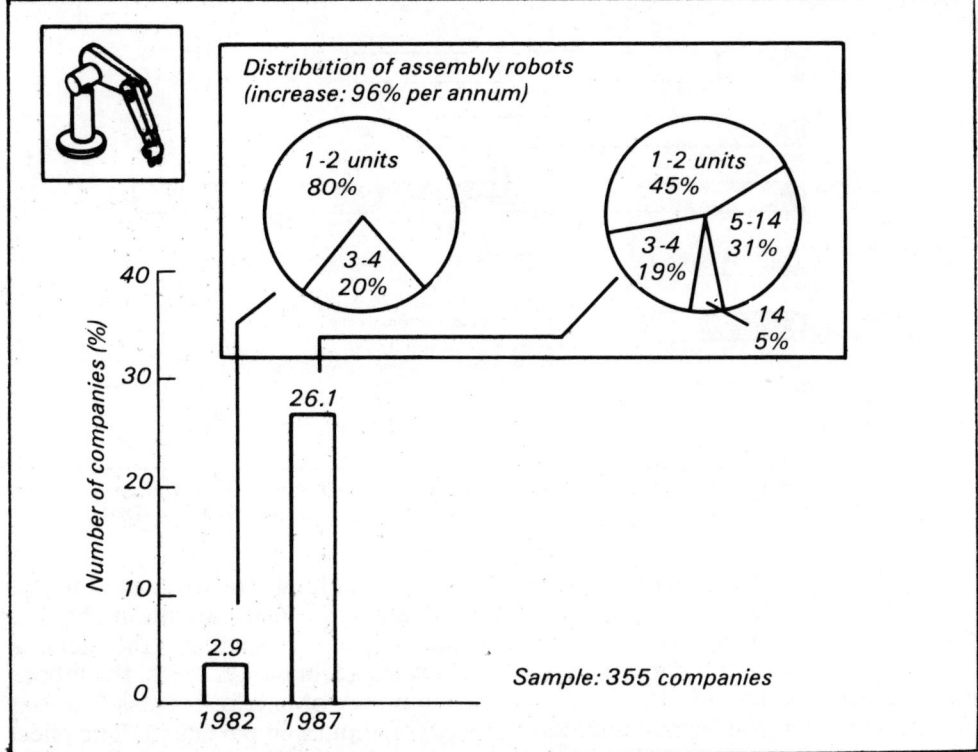

Fig. 5 Distribution of and the number of firms using assembly robots in 1982 and 1987 (source: IPA)

○ the actual cycle times,
○ the number of pieces produced, and
○ the number of components of each product.

This mainly product-orientated approach implicates the thesis that existing technology, at least in the medium-term, is no determining obstacle for assembly automation. Within a first phase of far reaching automation it is changes of constructing products, of joining, materials, etc. up to product conceptions and standardisation of components that will be in the foreground. Measures of this sort can have a relatively autonomous function, and they provide the first essentials to automatise the remaining operations.

Three different calculations on the potential of displacements to be expected are compared in Fig. 6. All of them are based on our representative inquiry, but on different premises. By demonstrating these different approaches – and the respective results – the problems become more transparent since the reputedly exact data are spread out.

For all industrial areas different calculation methods lead to very different assessments of the potentials. In the electrical industry the maximum potential of respective automation displacements caused by automation is 56% over the next 10 years, the minimum is 20%. In the motor industry it is 24% and 6.5%, respectively. These discrepancies clearly point out different developments of productivity in companies and sectors of industry.

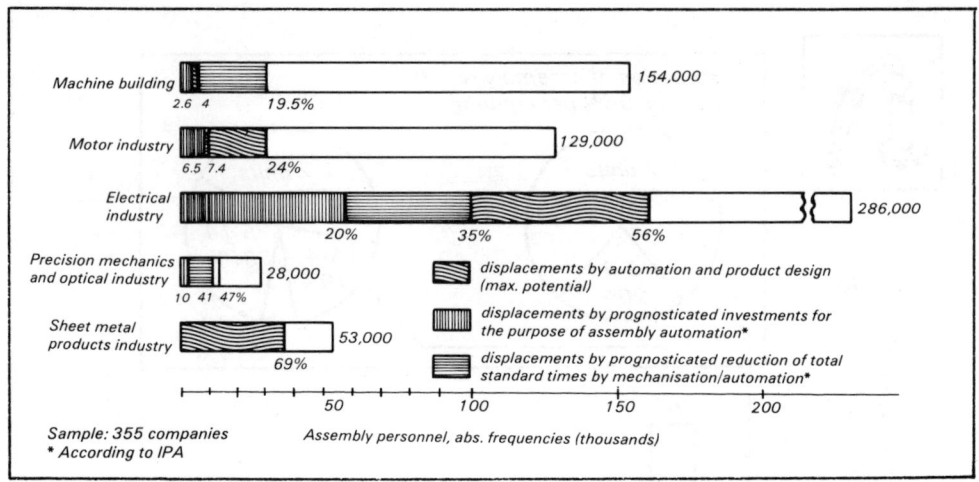

Fig. 6 Potentials of displacement by automation and related processes in industrial series assembly (forecast until 1992) − comparison of different calculations (source: GfAH)

To what extent the potentials of displacements will finally be exhausted by the companies and how displacements will be accomplished operationally and economically, depends on some uncalculated factors, such as the general economic situation, the national policy concerning technology, taxes, the labour market, party policy and last but not least the policy of employers' associations and trade unions concerning reduction of working hours, in particular. The effect of these factors must be neglected here, but some internal factors responsible for diffusion of automatons in assembly must be mentioned.

Personnel and organisational implications of assembly automation
Generally, assembly automation does not imperatively result in a reduction of monotonous, paced work and stress, nor in an improvement of the qualification structures[2]. On the contrary, we have to state an increased need of specific measures to avoid deficiencies caused by automation. Some of the most important subjects are work organisation, qualification and personnel planning.

Work organisation. It has already been mentioned that work organisation and automation can be alternative strategies of making production more flexible. Now we have to stress the interaction of both strategies, which is even more important. For two reasons it seems to be generally a wrong approach not to consider this interaction. First, there is the function of work organisation as a pioneer of flexible automation. For economical reasons, as well as for reasons of working conditions, existing sequences of operation cannot be automated without precise measures and changes of organisation. Secondly, work organisation must be regarded as a supporting measure. Manual and automated sections must be separated. Any separation of manual and automatic units is the precondition to reduce restricted and paced work which may be bound to the rhythm of the automation. Separation of units and integration of buffers are also necessary to reduce malfunctions and interruptions.

Qualification as an obstacle to automation. On average personnel problems due to flexible automation have been underestimated. This is partially due to companies being inexperienced in automation. Companies with experiences in single-purpose automation have had no problems with the qualification of their personnel, but these experiences cannot simply be transferred; flexible automation will need more skilled and less unskilled labour. Obviously those companies which are still using industrial robots in other sectors of production have a more reliable estimate on qualificational requirements.

Fig. 7 distinctly shows that those companies which actually use more robots estimate the quota of skilled workers to be growing. On the other hand the companies which estimate this quota to be constant or reduced in future are actually using very few robots or none at all.

Personnel planning. Largely because of the reduction of unskilled and semiskilled labour to be expected, there will be extensive personnel movements. The demands for higher skilled workers during a first phase of flexible automation cannot be projected linearly. There will be a decrease after finishing the first phase of implementation, and generally it will be more difficult to absorb or equalise personnel movements in future; assembly departments no longer will be buffers, because assembly itself will be a main field of rationalisation. In order to equalise personnel movements and to avoid undesirable social effects this makes personnel planning a more difficult and more important requirement.

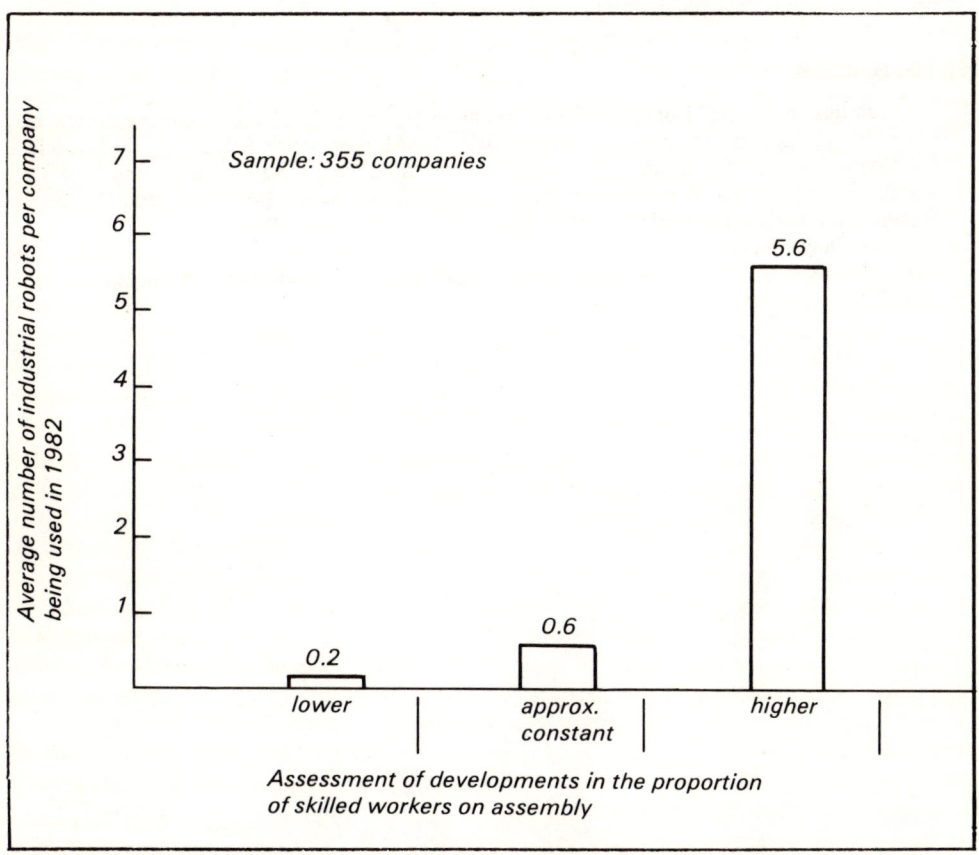

Fig. 7   Estimate of companies' developments in the proportion of skilled workers on assembly, according to the average number of robots installed in 1982

Conclusions

Without underestimating the increasing possibilities of program-controlled machinery, any modernisation of assembly will not be successful by technical means only. In the long run it would be sumptuous for companies, as well as the national economy, to maintain that the problems in assembly could primarily be solved by robots. Every overestimation and overcharge of flexible automation as a partial strategy is detrimental to both economy and working conditions.

Therefore, we need technically and organisationally integrated solutions. Experiences on the *application* of new assembly technologies, including supportive scientific investigations and workers' codetermination, are more important than the promotion of purely technological development. For this purpose pilot projects must include measures of qualification and personnel planning. Actually there are comparable researches in the sector of parts manufacturing; our institution is also active in this field. Those projects first of all should support smaller companies which do not have research departments of their own and which could lose their competitiveness soon.

As far as we know there have not been any studies on international comparisons and documentations concerning this subject. An exchange of experiences would be important in order to reduce the costs of learning and to avoid undesirable developments.

References

[1] Bundesministerium für Forschung und Technologie (ed.). Studie zur Untersuchung der Einsatzmöglichkeiten von flexibel automatisierten Montagesystemen in der industriellen Produktion unter Beachtung der technischen, arbeitsorganisatorischen und sozialen Voraussetzungen einer menschengerechten Gestaltung der Arbeit (Montagestudie Assembly Study). VDI-Verlag, Düsseldorf (due to be published in October 1984).

[2] Seitz, D. Developments in assembly work. In, Proc. 1st International Conference on Human Factors in Manufacturing, 3–5 April 1984, London, pp. 137–148. IFS (Publications) Ltd., 1984.

CHANGES IN ASSEMBLY WORK ENVIRONMENTS

R. G. Davison, University of Salford, UK

Abstract

The roles of modern assembly operatives are much different from their earlier counterparts. The improvement in consistency of parts quality has eliminated the skill required by the former fitter. A new breed of unskilled assembly operatives has been created, through the division of labour, to perform repetitive, mundane tasks but in many instances, these tasks are now done by machines of varying complexity.

The development of modern assembly techniques is discussed, together with future trends in manual and automatic assembly. Particular emphasis is placed upon the changing requirements of the people directly involved in these assembly operations.

Introduction

There is no doubt that in the 'advanced' manufacturing nations there has been a rapid increase in living standards throughout the twentieth century; this can be attributed to the application of technology to manufacturing. The mass production of goods has made available many items at economic prices. Housewives now have a multitude of labour saving devices to reduce the amount of time spent on household chores. This has enabled many women to work in factories which produce these goods. Many workers gain satisfaction from mastering a short assembly task and take pride in performing it, without fault, for long periods. Working with other people on an assembly line often brings a sense of cooperation into a joint effort.

In recent years there has been much criticism of the assembly line technique. Many people argue that repetitive work is boring and tedious. They claim that workers no longer gain satisfaction from doing their job. It is stated that operators never see the finished product and that the repetition of movements create boredom. The increase of industrial unrest in certain companies has been associated with this dissatisfaction of assembly line workers. Many manufacturers now realise that the economic benefits of the division of labour have to be judged alongside the social disadvantages.

The use of automatic assembly equipment for the manufacture of products eliminates worker dissatisfaction which results from repetitive work, since many of the repetitive operations are performed by machines. Manual operators are used to fill magazines and feeders and to maintain the equipment. The reduction in labour requirements often leads to a reduction in the cost of the finished goods. The end product of this process is an increase in leisure time through a reduction in the working week and emphasis must then be placed on how people are to spend their leisure time. This must be the subject of major reform in our education establishments.

Technology

Technology is the systematic knowledge of the industrial arts. Industrial engineers have been applying technology to the workplace for over two centuries in various forms. Manufacturing systems analysed by method and time studies have been improved by the division of labour, automation and robotics. Large productivity improvements have been achieved by applying technology to various manufacturing processes. From the mechanisation of flour production to the robotic assembly of vehicles, the process cost has been reduced. The application of technology to the motor industry can be seen to have resulted in vast increases in productivity.

Method study is concerned with the dissection of a relatively complex operation into its single constituent parts which are then systematically analysed. The method-study engineer synthesises the complete operation using components which optimise certain factors, such as symmetry and rhythm of movement.

The time-study engineer is concerned with measuring the time taken to perform an operation. The operation must be carried out in a systematic manner, and this makes this form of study only suitable for simple and repetitive tasks. Often, time study exposes inefficient operations and these can then be analysed using method study.

It was the use of both method and time studies that led to the widescale use of the division of labour and the concept of the assembly line. Operatives grouped on lines achieve productivity levels many times greater than single operatives making the entire product.

Automation has also produced large productivity increases by replacing men with machines. In highly automated manufacturing plants, the operator controls and supervises the process. According to Burnham[1], the main holders of power in future societies will not be communists, socialists, or capitalists, but people who possess expert technological skills. In this way, power will be passed to the technostructure.

Automation

Automation in the manufacturing industries encompasses a whole range of electrical and mechanical equipment. In the field of automatic assembly, devices are used for automatic feeding and insertion. In addition, work transfer is by conveyor or rotating table. The type of system used for the assembly of a product is dependent upon many factors. The local labour cost affects the economic justification of using automation to replace labour. The frequency of design changes and the number of product styles will decide how flexible the equipment needs to be. The market life of the product has an influence on the amortisation period of the capital investment. Finally, the annual product volume determines the required cycle time.

In addition to the above economic considerations, another reason for employing automatic assembly may be one of necessity. In certain areas, where the right kind of labour is scarce, the use of automatic assembly methods may be imperative. Certain operations are hazardous or must take place in bad working environments. For example, the handling of dangerous chemicals or working in extreme temperature conditions may exclude the use of manual operators. A further reason may be associated with the scheduling of the assembly operations: more flexible control over production can be achieved with the introduction of automation and product quality will be more consistent.

The majority of assembly operations can be broken down into the two basic activities of handling and insertion. When a product is assembled automatically, clearly, thought has to be given to the economics of these activities. The automatic feeding of relatively simple parts is usually carried out using a vibratory bowl feeder. Components in bulk random orientation are placed into the feeder and the parts are presented to the workhead in an ordered manner. More difficult parts may be fed by special feeders, hoppers or by magazines. The insertion process is defined as being the action where one part is assembled to another part or group of parts. High speed operations, where the same parts are inserted for long periods of time are normally effected by standard pick-and-place units. For more difficult operations involving the assembly of a number of different parts and/or a number of different operations, assembly robots may be used. The flexibility of the robot is created by using a number of programs run through the robot's computer. The generally accepted difference between a robot and a pick-and-place unit is that the path of the robot arm is not restricted by mechanical means whereas at best, pick-and-place units rely upon mechanical stops to determine the path they adopt.

Division of labour

The division of labour is the process whereby a complex operation is broken down into a number of simpler tasks. These shorter operations are carried out using one person for a limited number of operations. In this manner a complex task peformed by one operator is replaced by a number of operators working in series. This allows operations to be carried out simultaneously, instead of the single operator having to complete one task before commencing another different one. Unskilled operators can then be used to carry out these simple operations and they soon become efficient at the particular task[2].

Assembly systems

An assembly method can be conveniently classified into one of six types[3], and most systems contain a number of different methods.

The oldest form of assembly is *manual*, and for high volume production, the operatives usually work on an assembly line. Other forms of manual assembly include a single operator assembling a complete product and groups of operatives assembling a portion of the product.

When the range of products is more limited, a *manual assisted* method may be used whereby operatives are assisted by mechanical devices such as parts feeders. The feeders present the parts to the operator in an ordered manner, and the assembly time is reduced by eliminating the time taken to separate the parts from bulk random orientation.

The third form of assembly is by the use of *automatic indexing* assembly machines. This consists of a rotary or in-line system with a number of workstations. Automatic feeders supply components to workheads which assemble the part to the fixture or part-built assembly. The workstations are 'special-purpose' and are dedicated to the assembly of one product only. Production volumes need to be high for the economic justification of these machines. Component quality must also be high to avoid workstation downtime due to jamming, etc.

The efficiency of an *automatic free-flow* assembly machine is less dependent on parts quality. The transfer of workpieces between each workstation is non-synchronous. Small buffer stocks are held between each workstation so that other workstations may operate whilst one is stopped due to a fault, e.g. defective part jammed in the escapement mechanism.

The *automatic programmable* assembly machine consists of a non-synchronous transfer line with a series of programmable (robotic) workstations to assemble the parts. Parts are presented to the workheads by automatic feeders or, in the case of difficult components, part magazines may be used. The workheads may perform one or a number of operations. Flexibility is provided by using different programs for each product to be assembled.

The final type of assembly system is a variation of the above. *Robotic* assembly is used for the assembly of products manufactured in low production volumes. This method may also be used when there is a large amount of product variety. Work transfer is not by conveyor as all the assembly operations are carried out by a single, two-armed robot. The robot itself may transfer the completed product onto the next operation.

The amount of direct labour in assembly is reduced as a move is made from manual assembly to robotic assembly. However the complexity of the equipment increases as operators are replaced by machines, and indirect labour is increased to maintain and control the equipment.

Economic aspects

There is no doubt that the application of technology to manufacturing increases productivity. In the early nineteenth century the output per employee was raised dramatically in the 50 years between 1810 and 1860 in the USA, and the output for the three million industrial workers increased from US $200 million to US $2,000 million[4].

Selection of a system for the economic assembly of a product is based on a number of factors[3]. The final selection must take account of all the following variables:

- *Market life of product* – The product life will affect the decision of the company on investing in capital equipment. Products with short market lives are usually assembled manually.
- *Variations in demand* – Automatic assembly machines are designed to operate on fixed cycle times. A low demand would lead to increasing stock levels or stopping the machine. Both these actions are expensive. Flexibility to assemble different products is required for large variations in demand. This flexibility is provided by manual assembly or programmable machines.
- *Proportion of defective parts* – With an automatic machine the appearance of a defective part at a workhead may cause a station to breakdown. Whilst interstation buffers will reduce the effect on efficiency, manual assembly is required for products having poor quality parts.
- *Number of products* – The number of products to be assembled by a system will determine how flexible it needs to be. Many products manufactured in high volumes can be assembled using programmable workheads. Smaller volumes require manual assembly.

- ☐ *Major design changes* – Products subject to frequent major design changes need flexible assembly systems, in a similar way to systems assembling a variety of products.
- ☐ *Company investment potential* – The selection of an assembly system is heavily influenced by the company's attitude towards investing in automation. If the company requires a payback period of less than one year then it is highly unlikely that any form of automated assembly system can be justified.
- ☐ *Annual production volume* – The annual production volume will determine the cycle time of the system. Automatic systems must run continuously to be justified. If the annual volume is low then the product must be assembled manually.
- ☐ *Number of parts* – The number of parts in the assembly determine whether the product should be assembled in a series of simple operations or in a single complex operation. Automatic indexing machines cannot be used for the assembly of more than ten parts on a single machine. The downtime due to defective parts increases dramatically for every part above this value. Where investment allows, free-flow transfer should be used for products containing a large number of parts.

Social aspects

The application of technology to the assembly environment creates many social effects. It has often been said that the economic advantages of certain assembly systems produce serious social disadvantages. These social effects are not limited to the confines of the factory, as they affect the whole of society.

Assembly line work can provide work for people who have limited abilities. They can soon acquire skill at a particular task and take pride in doing a job that would seem uninteresting to the majority of people. Working with others on an assembly line often brings a worthwhile feeling of cooperation in producing goods required by society. Some people enjoy the fact that they can start a job, and with minimal training soon be earning a bonus on piece-rate lines; a highly specialised assembly task provides this opportunity. The right kind of person can be selected for the assembly line by using aptitude and vocational tests. There is scope for job rotation and many managers circulate operators so that they are not performing the same operation for long periods. Job rotation also provides the manager with a labour force skilled in all operations. This is particularly beneficial when there is a high rate of absenteeism. The operators on an assembly line soon adopt a rhythm of working as they do not have to set aside one tool to pick up another.

Many operators do not want to use mental effort or have any responsibility in their job. They prefer to perform a task which allows them to talk to their workmates or listen to a radio. The operators are also able to take advantage of the reduced selling price of goods assembled by the flow-line method. They can buy goods that would normally be beyond their means were it not for the division of labour. Cheap household appliances like washing machines, vacuum cleaners, etc. reduce the amount of time required to carry out housework. Many housewives find that they are now able to work on an assembly line to earn money and to find companionship.

The social advantages of assembly line work must be considered alongside the sometimes serious social disadvantages. There can be a loss of job satisfaction because the operative is not involved in all of the assembly processes that lead to the finished product. The task is repetitive and some operators do not take much pride in the task itself as they rarely see how important their operation is to the successful completion of the product. Boring work may destroy the creative ability of operators and their time out of work may be spent so passively that they have no objectives in life. The effect of carrying out monotonous work often leads to excessive fatigue. With the decline in individual craftsmanship many unskilled operatives have no opportunities to display their artistic talents or merits at work. Goods built on an assembly line are of a more limited range than those built by craftsmen. This can lead to dull uniformity of products which has an effect on the operators.

Assembly lines are usually installed in factories with a large workforce. Each group within the factory is dependent upon the other for the manufacture of the product. Strike action by one group of workers may affect the production of the whole factory. It is known that the operator output is affected mostly by the time spent performing a task. The cycle time is fixed and so it is the periods of time spent off the job that reduce the production volume.

The social problems mentioned above often manifest themselves in the form of avoidable delays in which the operator endeavours to gain control of the rate of work.

The social effects due to automation are different from those caused by the division of labour. Many of the simple operations carried out by operatives can be carried out by automatic workstations. By replacing operators with automation, these repetitive tasks can be performed by machines. The displaced operators are then available to carry out other, less tedious tasks like supervision and inspection. The automatic assembly machines must be fully utilised to be justified. Dedicated automatic assembly machines are less flexible than manual assembly lines. The products must be assembled in relatively large batch sizes. Overproduction and underconsumption leads to inefficiencies. This particularly applies in times of severe fluctuations in demand or large-scale unemployment.

Many behavioural scientists claim that technology can be applied to assembly without employing automation. These people believe in job enlargement and job enrichment. They argue that the division of labour has been taken too far in producing boring, repetitive jobs.

Job enlargement increases the number of tasks peformed by a single operator and this is intended to provide more interest and variety in the job. The same grade of worker would carry out the more complex operation, but the net effect of job enlargement is to reduce the number of operators on assembly lines to increase the cycle time, to give more flexibility but to increase the cost.

Job enrichment increases the responsibility of an operator by giving more opportunities to make decisions. If the grade of operator is increased it is anticipated that the feeling of job fulfilment will be increased. Job enrichment, by the strictest definition is most suited to employees with more responsible jobs than assembly workers, e.g. technicians and engineers. Nevertheless, forms of job enrichment have been applied to assembly workers in Swedish car factories and Dutch electrical companies, among others, with some success.

There are many critics of the theories of job enrichment and enlargement. The left wing view of these ideas is that the workers are misled into participation and accepting leadership whilst the 'conflict' between men and managers remains unchanged. Others claim that the nature of the work is only one of the many factors contributing to the attitude of workers towards their jobs. It is the nature of work that enrichment theorists believe is the difference between satisfied and dissatisfied workers. By changing the nature of work, social attitude will also be changed. Others claim that the nature of work is not the top priority and that other factors such as pay, working conditions, job security and the attitude of the supervisors must be considered. Assembly operators have individual preferences for the nature of work. Some prefer routine work whilst others enjoy performing complex tasks. Many workers do not choose the job that they would most enjoy, in return for high pay. Others, in periods of high unemployment, would accept any job to support their families.

The future

A short-term effect of the use of automation is the increase in unemployment. There is a transition period between workers being displaced by machines and their subsequent re-employment in the service sector. The less qualified workers are particularly affected. In-house training is required for educating existing personnel on the maintenance of these new machines. The higher technical content of jobs generally means that there will be more opportunities in full-time education because greater expertise and higher qualifications will be required.

There will be a change in the proportion of labour employed in the manufacturing and service industries. More leisure time will demand ways of spending that leisure time. New leisure activities will be developed and the existing service sector will expand.

Automatic assembly reduces the cost of producing goods. The availability of goods at economic prices creates a higher standard of living. The lower labour content in producing goods leads to a shorter working week and more holidays to relax. The greater availability of human resources if used wisely would be available to cater for a better quality of life for all, but in particular for those who because of age or circumstance are vulnerable.

References

[1] Nobbs, J., Hine, B. and Flemming, M. Sociology. Macmillan, 1976.
[2] Smith, A. The Wealth of Nations. Dent and Sons, 1910.
[3] Boothroyd, G., Poli, C. and Murch, L. E. Automatic Assembly. Marcel Dekker, 1982.
[4] March, P. The Robot Age. Abacus, 1982.